Vivarium

Vienna Series in Theoretical Biology
Gerd B. Müller, editor-in-chief

Thomas Pradeu, Katrin Schäfer, associate editors

Vivarium

Experimental, Quantitative, and Theoretical Biology at
Vienna's Biologische Versuchsanstalt

edited by Gerd B. Müller

The MIT Press
Cambridge, Massachusetts
London, England

This book was set in Times New Roman by Toppan Best-set Premedia Limited. Printed and bound in the United States of America.

Library of Congress Cataloging-in-Publication Data

Names: Müller, Gerd (Gerd B.) editor.
Title: Vivarium : experimental, quantitative, and theoretical biology at Vienna's Biologische Versuchsanstalt / edited by Gerd B. Müller.
Description: Cambridge, MA : The MIT Press, [2017] | Series: Vienna series in theoretical biology | Includes bibliographical references and index.
Identifiers: LCCN 2017001169 | ISBN 9780262036702 (hardcover : alk. paper)
Subjects: LCSH: Vivariums. | Biology, Experimental.
Classification: LCC QH68 .V55 2017 | DDC 570.72/4--dc23 LC record available at https://lccn.loc.gov/2017001169

10   9   8   7   6   5   4   3   2   1

Contents

Series Foreword

Biology is a leading science in this century. As in all other sciences, progress in biology depends on the interrelations between empirical research, theory building, modeling, and societal context. But whereas molecular and experimental biology have evolved dramatically in recent years, generating a flood of highly detailed data, the integration of these results into useful theoretical frameworks has lagged behind. Driven largely by pragmatic and technical considerations, research in biology continues to be less guided by theory than seems indicated. By promoting the formulation and discussion of new theoretical concepts in the biosciences, this series intends to help fill important gaps in our understanding of some of the major open questions of biology, such as the origin and organization of organismal form, the relationship between development and evolution, and the biological bases of cognition and mind. Theoretical biology has important roots in the experimental tradition of early-twentieth-century Vienna. Paul Weiss and Ludwig von Bertalanffy were among the first to use the term *theoretical biology* in its modern sense. In their understanding the subject was not limited to mathematical formalization, as is often the case today, but extended to the conceptual foundations of biology. It is this commitment to a comprehensive and cross-disciplinary integration of theoretical concepts that the Vienna Series intends to emphasize. Today, theoretical biology has genetic, developmental, and evolutionary components, the central connective themes in modern biology, but it also includes relevant aspects of computational or systems biology and extends to the naturalistic philosophy of sciences. The Vienna Series grew out of theory-oriented workshops organized by the KLI, an international institute for the advanced study of natural complex systems. The KLI fosters research projects, workshops, book projects, and the journal *Biological Theory*, all devoted to aspects of theoretical biology, with an emphasis on—but not restriction to—integrating the developmental, evolutionary, and cognitive sciences. The series editors welcome suggestions for book projects in these domains.

Gerd B. Müller, Thomas Pradeu, Katrin Schäfer

Foreword

Where do stories start and where do stories end? Sometimes dates can be precise but sometimes not. Of course, it all depends on how the story itself is defined. Certain dates in this enthralling book are indeed precise. The Biologische Versuchsanstalt (BVA)—the Institute for Experimental Biology—located in the Vivarium building in Vienna, came into existence in 1902, and perhaps its soul can be said to have died in 1938, even though the Vivarium itself survived until April 1945, when it was totally destroyed by fire.

On the other hand, why might the dates for the BVA perhaps be imprecise, or even "never ending"? This rich book answers that question unambiguously. In chapter 1, Georg Gaugusch presents the background and genealogy of the three families from which the three founders of the BVA came. He also sets the background for why the three may have chosen, in the context of their heritage and society in Vienna at the time, to devote themselves so completely, in terms of both commitment and money, to the BVA. So perhaps the start date is imprecise because, for two of the families, reference is made to their ancestors from Prague as far back as the sixteenth and seventeenth centuries! It is persuasively argued here that the nature and motives of the three founders, in the context of Viennese and wider Austrian society at the turn of the twentieth century, are best understood on the basis of the attitudes, traditions, and expectations that had developed and evolved over those preceding 300 years. And why "never ending"? The innovativeness of the approaches initiated in the BVA, and then sustained for 35 years, provides the material for the varied and compelling chapters 2 through 7. Each focuses on specific and distinct research activities that were undertaken, illuminating their global importance in initiating and advancing new ways to consider and address topics within the biological and life sciences. So, the story is "never ending" because it is clear from these accounts that foundations were being laid for different ways of thinking—and thinking evolves and never ends! These chapters contain a rich and broad narrative, not only describing the research undertaken in the BVA but also highlighting the importance of the very many individuals who, in addition to the three founders of the BVA, made it all happen, sometimes under difficult personal circumstances. Chapters 8 to 11 focus particularly on four individuals and give depth to the story of the BVA. In combination, the twelve chapters convey the vision that was at the start of it all in 1902, its development up to 1938, and subsequently, in light of the global development of the biological and life sciences, the BVA's enormous, almost forgotten legacy and its global influence on modern biology and the life sciences. What makes this book special is that it brings into the open an important piece of the history of biological and life sciences that has been hidden and never, until now, fully recorded or even fully recognized. It is an important contribution to the history of these sciences as well as to the histories of all the individuals who were involved. Focusing on the BVA also contributes greatly to our understanding of the sociopolitical context in which science develops.

It is remarkable that the story of the BVA and what it achieved has not until very recently been properly shared or published. In fact, one of the contributors to the book, Klaus Taschwer, wrote the following just three years ago:[1]

It is possible that the botanist [Fritz Knoll] had a guilty conscience about the complete destruction of the research institute [BVM] that had been entrusted to him. When he published two volumes on great Austrian natural scientists, engineers, and physicians in 1950 and 1957 on behalf of the Academy [the Austrian Academy of Sciences], he had two opportunities to recall the great history of the research institute and to pay tribute to its most important researchers. But the principle of repression was stronger. Knoll did not write a single word about any of them. And, in essence, the eradication of the Institute for Experimental Biology from the collective memory of the history of science continues to have an effect to this day; even now there is no comprehensive portrayal of and tribute to this once world-famous institute of the Academy.

Clearly Fritz Knoll was not alone in his reluctance to come to terms with the past.

However, the situation has lately changed. In the same year, 2014, the Austrian Academy of Science addressed this matter by organizing—in collaboration with the Konrad Lorenz Institute for Evolution and Cognition Research—a two-day symposium to commemorate the 100th anniversary of the donation, in 1914, of the BVA by its three founders to the Austrian Imperial Academy of Science. The symposium also, for the first time, celebrated those three individuals and shared publicly the innovative research they facilitated there. This book carries on that commemoration. It helps to highlight a part of the history of science that has been blotted from memory, namely, and in the words of Taschwer, a "comprehensive portrayal of and tribute to this once world-famous institute of the Academy." As readers will discover, the insights and analyses in this book bring to light what had been missing from our collective memories: the immense contribution of the BVA to the evolution of biological and life sciences up to the present day.

Why is it important that the story be told and the achievements of the BVA be recognized and celebrated? The social changes within Vienna and Austria that occurred between the two world wars are not changes that a nation might look back on with pride. The pressures on the Jewish community in particular, and the increased isolation of Jews—for instance their exclusion from various academic positions—are elucidated in the accounts from the various contributors to this book. Never to acknowledge this would have perpetuated the tensions that unfortunately existed during those 25 years between the two world wars.

Clearly, this difficult and problematic period for Vienna should not be ignored as if it never occurred. How can history be understood if pieces of the puzzle are hidden away? Of course, Vienna and Austria are not alone in this: many nations have had periods in history that, with hindsight, they might wish they could change, and that may also have impacted the evolution of science. Moreover, as the book's contributors well appreciate, the development of science always occurs within a social, political, and cultural context. Their research on the BVA provides a model for future research on other hidden pieces of historical narrative, and particularly for the history of science.

Especially important to me (and perhaps worthy of study by future historians) is the courage of those within the current generation of academics who want to bring such missing historical pieces to light, especially those who set up the symposium and those who now, through this book, have provided a reflective account of the BVA. In this they are building on their own research interests, which date back much farther than the 2014 symposium. These scholars are individuals who have recognized—bravely—that this period of Austrian history, and the associated history of science, need to be explored and recorded, not ignored or forgotten. Their work has enormous significance as they look at, provide perspectives on, and share their understanding of the dynamics of the society and content of science at that time, which formed the context for the BVA itself. While this book may present some views that are critical, it is to the authors' credit that they are never judgmental. So, as well as being an important contribution to the history and sociology of science, this overdue book is a contribution to the history and sociology of knowledge itself.

Many phrases in the chapters that follow leap out, succinctly elucidating the importance of the BVA, its achievements, and its legacy. I will leave you, the reader, to enjoy finding your own and hope that you will have as much pleasure as I have had in doing so. However, one particular quotation from the book reinforces my feelings about why this period of history must no longer be ignored. At the end of chapter 5, Laubichler writes that "we have the potential to learn from history precisely because history does not repeat itself." For me, that is true only if we do not choose systematically to ignore or forget history. The academics now engaged in researching this period are attempting to ensure that that will not happen. For this especially, I am very grateful to them and also to the contributors to this book.

And now a personal reflection as the grandson of one of the three founders of the BVA, Leopold von Portheim. How is it that I became involved in the events of February 2014 and then was invited to write this foreword? It all started by the coincidence of our younger daughter, Rachael, meeting the then deputy head of mission for the Austrian Embassy based in Delhi, while both were trekking Southern Laos and Cambodia in late December 2013. None of us could have known what would follow when they first met and talked far from their home countries, nor what would transpire from the subsequent "research" on our family by the deputy head of mission back in India! So many things happened, and so quickly. Our family was "rediscovered" by the Academy of Science, and at very short notice we were invited to attend the symposium. The Academy had previously been unaware that there were direct descendants from the Portheim family. For our family, it was like "coming home"; it was so informative and interesting, and provided us with new perspectives on the period from 1914 to 1938 and also on our family's direct involvement in the BVA. It was emotionally rewarding to finally be able to explore Vienna itself with much greater perspective on all its connections to my family, who, fortuitously, had been able to relocate in 1938 and begin a new chapter in their lives in the United Kingdom. And here, I especially wish to single out my mother Suzana, the daughter of Leopold von Portheim. She is a remarkable survivor of that period in European history who celebrated her 100th birthday here in Cambridge in March 2017. She has taken pride in and also solace from the recognition that her father now is receiving by the publication of this book.

In his introduction to the second session of the symposium, Gerd Müller (chairman of the session and editor of this book) mentioned that some had come a long way to be present. At the time, but even more so having read this book, I thought perhaps my family's journey from the United Kingdom was not so long or difficult as the journey he and his colleagues had been and are undertaking in researching, understanding, and more fully coming to terms with this period of Viennese and Austrian history. Their work is extremely important in order for us all to gain perspective on and appreciation of the dynamics of that era. Science is never detached from the wider society, as the book makes clear; they are highly interdependent. Hence, my family and I are indebted to all that they have done and are continuing to do.

Knowledge of history is essential, even if some of the events discovered are not what we would have wanted to learn, but the new information that is being found and shared is both significant and cathartic. Many families will be grateful to all those who are ensuring that this is occurring at last, and that it will continue. Many of us also look forward to what else emerges from this ongoing research into what clearly is a much deeper story. The symposium and this book are outcomes of historical work that should not and will not ever be brought to "closure"—and hence is "never ending." They are part of a much larger process of exploring, interacting with, and reflecting on the past and the evolution of science and knowledge in society. Together, these create an evolving picture, which now we all are privileged to share, and from which we have so much to learn.

I hope that both the research and the courageous researchers in this book are an inspiration to others around the world who are contemplating similarly challenging and important "archaeological" historical projects.

Rob Wallach, grandson of Leopold von Portheim

Note

1. Klaus Taschwer, in *The Academy of Sciences in Vienna 1938 to 1945*, ed. Johannes Feichtinger, Herbert Matis, Stefan Sienell, and Heidemarie Uhl, pp. 110–111 (Vienna: Austrian Academy of Sciences Press, 2014).

Preface

Only rarely in the history of science is an entire research institute given away. Precisely this happened in 1914, when the three founders of the Biologische Versuchsanstalt in Vienna donated a research facility fully equipped for biological experimentation, which they had established using their own funds, to the Imperial Academy of Sciences in Austria. The generous bequest was accompanied by a considerable endowment to ensure the continuation of experimental investigation. Only 24 years after making what was then the largest-ever private financial contribution to science in imperial Austria, these very same persons would be denied access to the institute, their efforts and investments would be ridiculed, and, together with other members of the institute, they would be forced to emigrate or were deported to National Socialist concentration camps. The history of their institute would be effaced even after the end of World War II, and the scientific achievements of this once internationally renowned establishment would be almost forgotten. This volume is the result of a symposium that commemorated the 100th anniversary of the donation of the Biologische Versuchsanstalt to the Imperial Academy of Sciences, co-organized by the Austrian Academy of Sciences and the Konrad Lorenz Institute for Evolution and Cognition Research in February 2014.

Before its incorporation into the Academy, the Biologische Versuchsanstalt—commonly called the *Vivarium*—had already thrived for twelve years as a private and independent research center. Its scientific output was substantial and continued despite the disruptions caused by World War I and the postwar economic recession. The institute comprised several departments, which addressed independent questions in zoology, botany, physicochemistry, and physiology but were united under the overarching research goal of rigorous quantification and theoretization of experimental data. The chapters in this book recount the incredible and troubled history of the Biologische Versuchsanstalt from its foundation in 1902 to its complete destruction in 1945. In addition to the economic, societal, and political contexts, for the first time this volume investigates the actual science that was performed at the Versuchsanstalt. So far, this work has been clouded by unsubstantiated myths and popular stories about sensational experiments. But what really happened in the laboratories of the different departments? What were the scientific questions, the methodologies, the experiments, and their results? Who were the principal actors, and where was their work published? In short, what was it that made this institution so special and internationally acclaimed, and why did it become famous to some but infamous to others?

The present volume attempts to answer these questions, even though certain domains could not be approached with the necessary depth, because comprehensive historical analysis is lacking. It has been particularly rewarding to reassemble, after such a long period of oblivion, the fragments of a unique intellectual enterprise, piece together the theoretical and empirical components of research, and situate the conceptual debates in which the Biologische Versuchsanstalt and its actors were involved within the larger scientific context of the period. Whereas in recent years several publications have concentrated on selected personalities or provided condensed historiographies of the institute, this volume brings together scientists and historians in a concerted effort to cover both the work and the long-lasting consequences of the institute as a whole.

Many individuals and institutions assisted in this enterprise. I would like to thank the Austrian Academy of Sciences, the ÖAW Archive, the archive of the University of Vienna, the Konrad Lorenz Institute for Evolution and Cognition Research, and the doctoral program "The Sciences in Historical, Philosophical, and Cultural Contexts" at the University of Vienna for their support and inspiration. My special thanks go to the contributors of the chapters, who had to adapt to several phases of the volume's restructuring, and to Rob Wallach for his moving foreword. Klaus Taschwer helped with his profound knowledge of the BVA and with locating rare photographs, and Debbie Klosky assisted with the English style of several chapters. The editors at MIT Press must be thanked for their patience, and Katrin and Vincent for the time they allowed me to spend away from our joint investigations of dinosaurs, angry birds, intergalactic humanoids, and the heroes of ancient Greece.

Above all, I thank Sabine Brauckmann, who was one of the coordinators of the commemorative symposium that inspired this volume. She has been a tireless researcher of the archival sources extant at the beginning of this project, and her profound historical knowledge and rich personal contacts have been invaluable. Without her this volume would not have happened.

I   INTRODUCTION

1 Biologische Versuchsanstalt: An Experiment in the Experimental Sciences

Gerd B. Müller

The transformation of natural history into the scientific field of biology was intimately related to its progressive institutionalization and theoretization in the late nineteenth and early twentieth centuries. One aspect of this process was the foundation of dedicated chairs and institutes at universities, a development related in Austria to the major university reform of 1848 (Kniefacz et al. 2015; Fröschl et al. 2015). Other, nearly contemporaneous impulses led to the establishment of scientific facilities outside the universities, especially directed toward the experimental research for which universities were often ill equipped. Well-known examples exist in the case of physics (e.g., Marie Curie's Institut du Radium in Paris or the Institut für Radiumforschung in Vienna) but also in biology, such as the establishment of marine biology institutes in Concarneau, Naples, Trieste, and Roscoff. Often such biology-oriented institutions were created for pragmatic reasons, such as the requirements of long-term animal keeping or the proximity to the habitats of the organisms studied. These developments happened in short succession in several places in Europe and North America and created a particular model of institution at which, besides the permanent staff, visiting scientists would spend a certain period of time, possibly renting a work space or ordering specific kinds of organisms to be made available for study while there, or making use of the special laboratory equipment. In the case of biology, these features of extra-university establishments and the processes of institutionalization have been thoroughly studied and are well described by several treatments (Maienschein 1985; Benson 1988; Groeben 2002; Ash 2008; Sachse 2014).

The second aspect of the formation of scientific biology, theoretization, is rarely linked to the founding of independent scientific institutions. Yet, in addition to the pragmatic reasons mentioned above, this was a primary motivation in the case of the Biologische Versuchsanstalt (Institute for Experimental Biology; BVA) in Vienna. From the very outset its founders had declared quantification, mathematization, and theory formation a central goal of their institute. As typological thinking in biology was giving way to experimentation and quantification (Müller and Nemeschkal 2015), the BVA had targeted this crucial aspect precisely in order to place biology on equally firm theoretical grounds as the "hard" sciences of physics and chemistry. This explicit grounding in theory sets the BVA apart

Figure 1.1
The three founders of the BVA. From left to right: Hans Przibram, Leopold von Portheim, Wilhelm Figdor.

from almost every other biological facility that existed at the period and makes its study
a distinctive topic in the history of science.

This book lays out the remarkable story of how three inspired biologists from affluent
Viennese families—Hans Przibram, Leopold von Portheim, and Wilhelm Figdor (figure
1.1)—bought, adapted, and equipped a run-down former aquarium building in the Vienna
Prater and built it up to become one of the most advanced research institutes of the
time. The Biologische Versuchsanstalt was founded in 1902 and opened on the first of
January, 1903. Its fame and prestige soared, it became a hot spot of experimental, physiological, and theoretical biology in the first decades of the twentieth century, and then it
disappeared in the turmoils of World War II, left mostly in oblivion until recently. Nick-
named "Vivarium," because of the earlier inscription above the main entrance to the
building, the institute thrived for only three decades, yet its productivity was stunning and
its role as a trigger of numerous developments in experimental morphology, plant physiology, biochemistry, and hormone research cannot be overestimated.

The present volume covers the many facets of the founding, organization, research, and
conceptual transformations of the BVA between 1902 and the beginnings of World War
II. Starting, in part II, with treatments of the Viennese sociocultural context at the founding
of the BVA, it continues in part III with examinations of the scientific Zeitgeist that determined the institution's research orientation. These are followed by chapters in part IV on
the institute's four different departments and their empirical research topics and finally, in
part V, by two examples of the BVA's scientific and international ramifications. A summary
timeline indicates the main events during its short history (table 1.1).

The concept of the BVA was significantly shaped by the sociocultural conditions that
characterized turn-of-the-century Vienna. As has often been described (e.g., Janik and

Table 1.1
Timeline of the Biologische Versuchsanstalt, based on a compilation by Sabine Brauckmann

1902—Hans Przibram, Wilhelm Figdor, and Leopold von Portheim buy the former Aquarium building of the 1873 World Exposition—later renamed Vivarium—and start its transformation into a research facility.

1903—Official opening of the Biologische Versuchsanstalt on January 1, and installation of the Zoological Department (headed by Hans Przibram) and the Botanical Department (jointly headed by Wilhelm Figdor and Leopold von Portheim).

1903—Przibram, Portheim, and Kammerer travel to Egypt and Sudan for the collection of live organisms for the BVA (December 1903 to February 1904).

1904—Habilitation of Hans Przibram; graduation from the University of Vienna of Paul Kammerer, who worked for Przibram at the BVA since its foundation, as did Franz Megusar shortly later.

1907—Opening of the Physicochemical Department headed by Wolfgang Pauli Sr.

1909—Wilhelm Figdor is appointed Associate Professor of plant physiology at the University of Vienna. Karl von Frisch performs experiments for his doctoral thesis at the BVA.

1910—Habilitation of Paul Kammerer and graduation of Karl von Frisch from the University of Vienna.

1911—Eduard Uhlenhuth hired as assistant in the Physicochemical Department.

1912—Przibram launches a new program investigating physical effects on the germ cells.

1913—Start of the Physiological Department under the directorship of Eugen Steinach. Hans Przibram receives the title (without appointment) of Associate Professor at the University of Vienna.

1914—Incorporation of the Biologische Versuchsanstalt into the Imperial Academy of Sciences on January 1, following its donation, along with substantial financial support, by the founders and owners. Wilhelm Figdor ceases the codirectorship of the BVA and becomes director of the newly established Department of Plant Physiology. Wolfgang Pauli Sr. leaves the BVA and the Physicochemical Department is closed. Paul Kammerer receives the position of BVA adjunct (having been an assistant of Przibram since the foundation of the BVA).

1915/16—Temporary cessation of work at the BVA due to WWI, and establishment of a convalescence home for wounded soldiers in some of its rooms, privately financed by Przibram and von Portheim.

1919—Kammerer's application for the title of associate professor is denied by the University of Vienna.

1921—Hans Przibram is appointed "full" Associate Professor (*Extraordinarius* with salary) at the University of Vienna.

1923—On his own application, retraction, and reapplication, Kammerer is retired from his adjunct position at the BVA by the Austrian Ministry of Education. Steinach begins cooperation with the pharmaceutical company Schering

1924—Paul Weiss succeeds Paul Kammerer and takes on the duties of adjunct, but requests for a permanent position are repeatedly postponed.

1926—The US herpetologist Gladwyn K. Noble examines a museum specimen of Kammerer's midwife toad experiments and finds it injected with india ink. The ensuing suggestion that Kammerer's earlier experiments had been faked causes an international scandal and leads to Kammerer's suicide. The applications for habilitation by Leonore Brecher und Paul Weiss are denied by the University of Vienna.

1927—Paul Weiss takes a leave of absence on a stipend to work in the United States, where he remains. His vacant adjunct position is filled by Franz Köck.

1930—Przibram's comprehensive work *Experimental-Zoologie* is completed with the appearance of volume 7. Publications of BVA research in Roux's *Archiv* cease.

1932—Financial considerations prompt the reinstallation of the public aquarium exhibit. Hans and Karl Przibram consider starting a new program at the BVA for the study of the biological effects of radium.

1938—Wilhelm Fidgor dies. Following the *Anschluss*, Hans Przibram and Leopold von Portheim are denied access to the BVA building and stripped of control of the institute's finances. Leopold von Portheim emigrates to Great Britain. Franz Köck is designated substitute head of the BVA.

Table 1.1 (continued)

1939—Hans and Elisabeth Przibram flee to the Netherlands.

1940—The scientific unit of the BVA is ordered closed by the National Socialist head of the BVA Curatorium and rector of the University of Vienna, Fritz Knoll; only the public aquarium is allowed to stay open.

1942—Leonore Brecher and Helene Jacobi are killed in the extermination camp of Maly Trostinec near Minsk.

1943—Contract with the German Kaiser Wilhelm Society regarding the use of the BVA building for the Kaiser-Wilhelm-Institut für Kulturpflanzenforschung.

1944—Hans Przibram dies in the concentration camp of Theresienstadt, and Elisabeth Przibram commits suicide. Eugen Steinach dies in Switzerland. Henriette Burchardt is killed in Auschwitz.

1945—German troops occupy the BVA building, which is destroyed by fire during the last days of WWII.

1947—Leopold von Portheim dies in London.

1948—The Austrian Academy of Sciences sells the ruin of the Biologische Versuchsanstalt to a Viennese architect.

Toulmin 1973; and, with particular relevance to the present topic: Coen 2006; De Waal 2010; Logan 2013; Nemeth and Stadler 2015), that environment provided a state of rich ferment for major advances in the arts and sciences. Consider, for instance, the secessionist movement in painting (e.g., Klimt, Schiele), art nouveau architecture (e.g., Loos, Wagner), the developments in music and musicology (e.g., Mahler, Schönberg, G. Adler), the foundations of psychology (e.g., K. Bühler, C. Bühler) and of psychoanalysis (Freud, A. Adler), as well as the achievements in mathematics (e.g., Menger, Gödel), physics (e.g., Mach, Boltzmann, Schrödinger), and economics (e.g., Schumpeter, Hayek), the revolutionary philosophy of the *Wiener Kreis* (e.g., Schlick, Carnap), the birth of the social sciences (e.g., Lazarsfeld, Jahoda), and the multitude of developments at the Vienna Medical School (e.g., Semmelweis, Landsteiner). Extensive societal interactions fostered the cross-fertilization of ideas, for instance through the influential "salons" organized in many homes of the *haute bourgeoisie* of Vienna, such as those of Berta Zuckerkandl, Alma Mahler, and Mathilde Lieben. Thus we find depictions of early embryos in Klimt's paintings (Gilbert and Brauckmann 2011), reflections of evolutionary morphology in Adler's musicology (Breuer 2011), social considerations in Kammerer's views on inheritance, and allusions to contemporary physics in Przibram's biological concepts.

The facilitating environment of the Viennese *haute bourgeoisie* of the early twentieth century was primarily rooted in the Jewish upper class that had developed in the nineteenth century. In the case of the BVA, the liberal Jewish background of Przibram, Figdor, and von Portheim also had a decisive influence, as shown by Georg Gaugusch in chapter 2, which traces the genealogies, economic backgrounds, and societal interconnections of the three founders' families. The overall intellectual influence of fin-de-siècle Vienna is laid out in chapter 4 by Johannes Feichtinger, who also examines the affiliation of the BVA with the University of Vienna and the Imperial Academy of Sciences, its funding sources, and the path of the institute's decline. The historical conditions for establishing the institute in a former building of the Vienna World Exposition of 1873 in the Prater area (figure 1.2) are examined by Klaus Taschwer in chapter 3. In addition to the changing uses of the building, the rise of aquarianism, the amateur science related to it, and the connection of this movement with Paul Kammerer, chapter 3 also describes the contemporaneous state of zoology at the University of Vienna. Overall, part II of the book presents the BVA as a prime example of how developments in the sciences are shaped by the societal and cultural conditions of a given period and location.

Scientific Inspirations

The BVA became a very special place, to which international scientists flocked from as far as the United States and Japan. For instance, the British pioneer of mathematical biology D'Arcy Thompson, although skeptical of certain aspects of Przibram's later views (see chapter 8), sent students such as Dorothy Wrinch and Joseph Woodger to work with Przibram at the BVA. But whence the attraction? Why was the BVA so important in its sphere? The chapters in this volume suggest that the main draw was the pervasive ambition to provide answers to crucial biological questions of the time. In other words, from the outset, work at the BVA was theory driven, an achievement mostly due to the visionary mind of Hans Przibram, as can be seen from the strategic essays (Przibram 1903, 1908, 1913) in which he clearly states this objective. Several strands of the scientific Zeitgeist fed into this particular conception of a research facility, as we shall briefly examine below.

Hans Przibram was familiar with the intense theoretical debates of the time concerning evolution, development, and inheritance through his doctoral advisor at the University of

Figure 1.2
The Biologische Versuchsanstalt in the Vienna Prater. Courtesy of ÖAW Archive.

Vienna, the embryologist Berthold Hatschek (Müller and Nemeschkal 2015). Hatschek, a student and close friend of Ernst Haeckel, was involved in the disputes regarding the role of development in evolution, in particular as it concerned larval organization. He was an early adopter of Mendelian inheritance and, in his search for the biological vehicles of transgenerational transmission, had formulated his own chromosomal "hypothesis of organic inheritance" (Hatschek 1905a). Developed to some extent in contrast to Weismann's germ plasm theory, the concept had a distinctly systemic flavor, postulating a critical influence of both environment and functional activity within the organism on the germ cells. His systemic, organismal, and strictly anti-typological view of evolution, which allowed for nongenetic factors to act in concert with genetic ones, certainly had been a motivation for Przibram to focus on these topics. Hatschek, although a firm evolutionist, was rather critical of certain facets of Darwinism (see also chapters 3 and 8). Furthermore, Hatschek's outspokenly liberal stance, which also led him to publicly denounce the anti-evolutionary, vitalistic, and pan-Germanic ideas of Houston Chamberlain (1905) in the Vienna newspaper *Neue Freie Presse* (Hatschek 1905b), must have made a deep impression.

The second major strand of influence came from the experimental movement in the late nineteenth and early twentieth century. As shown by Heiner Fangerau in chapter 5, different takes on the experimental method had evolved at the time of the founding of the BVA and had strong interconnections with medicine. In biology, Wilhelm Roux's work had been a major engine for the propagation of the experimental method, in particular as applied to what he termed *Entwickelungsmechanik*, or developmental mechanics. Przibram credits a lecture by Roux as the critical experience that made him decide to systematically elaborate this field of investigation. During his time as a professor of anatomy at the University of Innsbruck, Roux was also the founder, in 1895, of the scientific journal *Archiv für Entwickelungsmechanik der Organismen*, which was devoted to publishing results of experimental developmental work and which has continued to appear, under different names, up to the present. Significantly, between 1907 and 1925 the Zoological Department of the BVA had its own section in the journal, entitled "Arbeiten der Zoologischen Abteilung der Biologischen Versuchsanstalt in Wien." One hundred seventy-five articles from the Zoological Department appeared until 1930 (accounting for more than 10 percent of the output of the journal during that period), when the publications by the BVA stopped abruptly, for reasons not entirely clear. Przibram followed through with his programmatic experimental approach and, in addition to the numerous individual articles by members of the BVA and by himself, eventually published a comprehensive seven-volume treatment of *Experimental-Zoologie* (figure 1.3). In these overviews of different subjects of experimentation, Przibram also included all information available to him from work outside of the BVA.

Rigorous testing through experiment was the overarching theme of the entire institute, as reflected in its name. It required an institutional setting rather different from the tradi-

Figure 1.3
Cover of volume 1 of Hans Przibram's *Experimental-Zoologie* (1907). Courtesy of the author.

tional establishments at the universities. This included architectural and technical adaptations of the building, the construction of apparatuses and animal facilities, and the invention
of instruments, methods, and protocols. It meant an extraordinary methodological expansion of the possibilities of organismal research over what was available in the standard
university environment. Christian Reiß, in chapter 7, explores the international context of
experimental research in nineteenth-century zoology and draws comparisons with similar
institutions in Germany, such as Semper's institute at Würzburg. Highlighting specific
features of what he terms techno-natural installations, Reiß demonstrates the new importance of long-term animal husbandry in facilities dedicated to experimentation.

A third strand of influence concerned the quantification and theoretical abstraction that Przibram had conceived as a central obligation of the BVA from the very outset. It reflected a desire for a higher scientific legitimation of biology. Measurement practices were a hallmark of the "exact" sciences and had authoritative connotations, because quantification suggested the possibility of mathematical formalization. Przibram explicitly cited physics and chemistry as "brilliant examples" of the achievements of quantification and mathematical formulation and argued that elaboration of these tools was the only possible way forward in order "to transform biology into an exact science" (Przibram 1931). Here the most obvious influence was that of physics and its formidable successes in Vienna at the turn from the nineteenth to the twentieth century. More specifically, Hans Przibram's brother Karl, a physicist at the University of Vienna who also was vice chair of the Institut für Radiumforschung, another fascinating independent institution in Vienna (Rentetzi 2007), must have been a major inspiration. The two brothers even attempted to join forces at one of the conceptual turning points for the BVA. The goal of rigorous quantification was also achieved, even if no advanced mathematization but rather strategies for the uses of measurement data and mathematical characterizations in biology resulted. Manfred Laubichler, in chapter 6, examines how these developments were related to the wider scientific contexts of generalization and theoretization in biology. He reviews the concurrent debates on experimentation and general biology in Germany and places particular emphasis on how this discourse influenced the establishment of new research institutes, such as the Kaiser Wilhelm Institute of Biology in Berlin. In Austria, the BVA came to embody the quest for theoretical foundations of biology in many ways. A further similarity with the German developments is the interconnection with popular education, another hallmark of the BVA's science.

The Departments

Whereas the overall goals of the BVA were programmatically elaborated, the concrete empirical research topics in its four (briefly five) departments differed. They were influenced by the interests and preferences of the department heads and by the major contemporary debates in the respective scientific fields. Not surprisingly, the Zoological Department, headed by Hans Przibram, adhered most closely to the defined goals, as discussed by Müller in chapter 8. The relationships between heredity, development, and environment in the evolution of animal form were studied in an extensive series of experiments, the most common of which concerned regeneration, transplantation, coloration, growth, homeosis, and the inheritance of environmentally induced variation. The chapter investigates the formative intellectual influences on Hans Przibram, the design of the zoological experiments—especially those concerned with multigenerational breeding—Przibram's ultimate quantitative intent, his pioneering work in mathematical biology, and his attempt to develop a comprehensive theory of organismal form.

The Botanical Department, initially headed jointly by the two cofounders of the BVA, Leopold von Portheim and Wilhelm Figdor eventually split into a Botanical and a Plant Physiology Department. In chapter 9, Kärin Nickelsen introduces the protagonists of the work on plants and examines how the experimental research of these departments was organized and how it related to the overall goals of the BVA. Two examples, the investigation of anisophylly and the study of regeneration processes in plants, reveal how the department's research was guided by some of the most hotly debated contemporary questions in botany and plant physiology, while at the same time the findings also contributed to Przibram's program. The chapter highlights the importance of the technical infrastructure at the BVA for successful experimentation with plants and the central relevance of the work at the Botanical Department to gain a full appreciation of the achievements of the BVA.

The purpose of the Physicochemical Department was to apply physical and chemical methods to biological problems and to collaborate with the zoological and botanical departments. The department's head, Wolfgang Pauli Sr., was mostly concerned with what was then called colloid science, today probably best characterized as a forerunner of the study of macromolecules such as proteins. Chapter 10, by Heiko Stoff, shows how the later-denigrated nineteenth-century idea of colloids had been a topic at the forefront of chemical research in the first decades of the twentieth century. Wolfgang Pauli, one of the main protagonists of colloid science in Europe, even thought that it contained the key to a new theory of life and vitality and envisioned it as a new chemical master science. His department at the BVA generated an impressive amount of new data on colloids and their functions in biological processes. Stoff's chapter describes the fierce battles that ensued between structural and colloidal chemistry and the important collaboration of Pauli with Karl Landsteiner, the later Nobel Prize winner, in the burgeoning field of immunology. Because colloidal chemistry relied heavily on technology-dependent procedures such as filtration methods, electrodialysis, and electrophoresis, one might imagine that the BVA was the ideal institution for Pauli. But although some cross-fertilization with other departments took place, Stoff's analysis indicates that the department did not receive sufficient financial support, which may have been the reason Pauli left the BVA for a position at the University of Vienna after seven years.

The Physiological Department, only established in 1913 after the BVA had become part of the Imperial Academy of Sciences, was headed by the famous and controversial Eugen Steinach, who had been involved with the BVA since its early years. By the time Steinach assumed the department's directorship, he was already an accomplished neurophysiologist. In chapter 11, Cheryl Logan reveals Steinach's continuing transformation into a (neuro) endocrinologist during his period at the BVA and emphasizes how a distinctly comparative and evolutionary approach guided the work at his department. The long-term keeping of animals and the elaboration of ever more complex experimental procedures were essential for Steinach's approach. With its focus on the pituitary gland as a center for mediating neurological input and feedback from the gonads in the control of sex-hormone production, the Physiological Department was at the vanguard of the work of deciphering regulatory hormonal activity and its role in behavioral control. The extensive measurements of behavioral traits in experimental animals, mostly rats, mice, and frogs, emerge as one characteristic aspect of Steinach's approach that, among other issues, led to the "pituitary wars" with the American school of thought and other scientific disagreements. Logan reveals how these opposing intellectual positions were rooted in different laboratory cultures and the economic prospects of basic research.

Many of the BVA's scientists went on to other institutions or to do work in different fields, where they were very often highly influential. Although several BVA biologists were proposed for the Nobel Prize, only one of them, Karl von Frisch, actually received it in his later career, long after his period at the BVA. Tania Munz, in chapter 12, traces Karl von Frisch's early work at the BVA during his doctoral dissertation, under the supervision of Hans Przibram, on the nervous control of pigmentation changes in fish. Munz shows how the experimental attitude, the focus on whole organisms, and the BVA's emphasis on keeping animals under conditions as natural as possible, influenced von Frisch's own observational practices and his later conception and establishment of an institute at the University of Munich. Thus the example of the BVA may have had a lasting conceptual effect on the field of behavioral science.

The BVA scientists were highly connected internationally, entertaining extensive exchanges with other research centers of similar character. An example is the interaction of Hans Przibram with Charles Davenport, director of the Carnegie Institution of Washington's Station for Experimental Evolution in Cold Spring Harbor. In chapter 13, Kate Sohasky explores the long-lasting communication between the two scientists based on their correspondence from 1907 to 1930. Davenport had an interest in the possibility of the inheritance of acquired traits and held the BVA in high esteem. Sohasky also exposes the political undercurrents of this sensitive period, and the rifts that opened up between US science and German/Austrian science following World War I. In the case of the BVA, the estrangement was hardened by the sensationalism of some of its research and by the Kammerer affair (also addressed by Müller in chapter 8). Ultimately, in combination with many other factors—scientific, political, and economic (Logan and Brauckmann 2015)— these events led to the beginning of the demise of the BVA in the 1930s.

Resurrection

The BVA ended in a scientific, institutional, and, above all, human tragedy. The science was discredited, the building was destroyed, and many of the protagonists were forced into emigration or died in Nazi concentration camps (see chapter 4). After World War II, the BVA and the splendid role it had played in the biological sciences of the first four decades of the twentieth century were almost completely forgotten, although occasionally mentioned in passing by some university teachers in Austria. It was not until the end of the past century that a resurrection began (Hirschmüller 1991; Reiter 1999; Hofer 2002; Gliboff 2006; Coen 2006, 2007; Rentetzi 2007; Berz 2009; Taschwer 2013, 2016; Logan 2013; Wald 2013; Feichtinger et al. 2013; Logan and Brauckmann 2015; Müller and Nemeschkal 2015; Taschwer et al. 2016; Walch 2011, 2016). In recognition of the role of Hans Przibram in the foundation of theoretical biology, the Konrad Lorenz Institute for Evolution and Cognition Research (KLI) has offered a Hans Przibram Fellowship since 1998, and the institute organized a symposium at the Austrian Academy of Sciences (ÖAW) in 2002, on the 100th anniversary of the founding of the BVA. Subsequently, the BVA was part of the Academy's analysis of its own history during the period of National Socialism (Feichtinger et al. 2013; Taschwer et al. 2016). A second symposium, co-organized by the ÖAW and the KLI in 2014, commemorated the 100th anniversary of the donation of the BVA to the Imperial Academy of Sciences. On the occasion of that event, a bust of Hans Przibram was reinstalled in the entrance hall of the Academy building, and, under the auspices of the city of Vienna, a memorial plaque was affixed at the location of the former BVA building in the Prater. For the first time since the destruction of the BVA, descendants from the founders' families were able to witness a public recognition in Austria of the generosity and achievements of their ancestors.

The present volume is based on the symposium organized in 2014, at which most of the authors were speakers. The foreword by Rob Wallach, emeritus senior lecturer in the Department of Materials Science and Metallurgy at Kings College, Cambridge, a grandson of Leopold von Portheim, provides a touching connection with the descendants of the BVA's visionaries. Throughout the volume, some repetition of a few of the basic facts about the BVA was unavoidable, not least because the chapters are intended to be intelligible as stand-alone articles as well.

Open Questions

The account delivered by this volume is incomplete. Many open questions remain, and much research still needs to be done. Missing in particular are further explorations of the legacy of the BVA left by students and associates, such as Paul Weiss, Ludwig von Bertalanffy, Eduard Uhlenhut, Walter Hohlweg, and other workers at the BVA who continued scientific careers (see e.g., Brauckmann 2013). More needs to be understood about the international reception (and also the criticism) of the BVA's science, its publication practices (why, for instance, did the contributions to Roux's *Archiv* stop in 1930?), the conceptual consequences of its work for the mathematization of biology, and many other subjects. Another aspect that requires further attention is the role of women at the BVA. Maria Rentetzi (2007) credits Hans Przibram for his active support of participation by female scientists. Leonore Brecher, for instance, one of his students and his later private assistant, was among the most productive scientists at the BVA. But there were many others, such as Irma Pisk-Felber, Rosi Jahoda, Helene Jacobi, and Auguste Jellinek. From 1920 to 1934, out of 109 listed workers, 39 were women (Rentetzi 2007). Given these and other blind spots in the history of the BVA, the present volume should be understood also as a call for further research.

Legacies

Hans Przibram and his codirectors cultivated extensive personal communication and institutional networking with similar centers in Austria and abroad. Through their activities and those of former members and associates, as well as through the large numbers of international visitors (e.g., William Bateson, Joseph Woodger, Theodor Boveri, Erwin Baur, and Philippe de Vilmorin), many of the ideas generated at the BVA were propagated to other institutions and countries. To some extent, the BVA even served as a model for experimental research centers abroad, such as the Carnegie Station at Cold Spring Harbor, the Kaiser Wilhelm Institute for Biology in Berlin, and the Institute for Experimental Biology in Jena. Through its dissemination of laboratory techniques, methodological practices, and conceptual innovations, the BVA influenced many fields of modern biology, most importantly functional morphology, developmental biology, physiology, biophysics, neuroendocrinology, and several more. Notably, it also affected the changes taking place in US biological research of the early twentieth century (Logan and Brauckmann 2015).

To mention a few examples, the regeneration and transplantation studies of the Zoological Department were continued overseas by, among others, Paul Weiss, Eduard Uhlenhut, and Theodor Koppányi, in each case leading to a successful career. The experiments on homeotic transformation, also developed at the Zoological Department, were not directly continued, but their importance was recognized by workers at other institutions and led to the discovery of master regulatory genes in development, an achievement that was eventually recognized with the Nobel Prize in 1995. The work in botany and plant physiology seems not to have had a strong international influence but helped shape the development of plant physiology at the University of Vienna. Colloid chemistry, as practiced in the Physicochemical Department under the leadership of Wolfgang Pauli, was eclipsed by the rise of molecular biology and protein biochemistry, yet it can be seen as an important heuristic step in the development of those fields. In the case of the Physiological Department, Eugen Steinach took over the directorship when he was already widely recognized for his pioneering work in hormone research, but his work at the BVA may be seen as foundational for the field of neuroendocrinology and had its continuation with Walter Hohlweg, for instance. Many others, who spent shorter periods of their scientific careers at the BVA, such as Karl von Frisch, likewise took the inspirations they found in that unique environment into their future endeavors, in turn influencing their students and collaborators.

From the information collected in the present volume it appears that the BVA's lasting legacy lies not so much in the styles of experimentation that were devised there, which nowadays are superseded by other methods, but primarily in the generation of an alternative approach to the study of the physiology–development–inheritance–evolution relationship and in the contributions to the theoretical foundation of biology. In contrast to the then-prevailing concentration of evolutionary biology on assumed genetic variation and selectionist scenarios, the BVA developed tools for precise measurements of the reactive plasticity of developmental and physiological processes to environmental stimuli in the generation of form and function. Yet in an increasingly reductionist climate of evolutionary research, this approach came to be overwhelmed by statistical and, later, molecular studies of genetics. Today, the very same subjects have again come to the fore (see chapter 6) in ongoing theoretical debates on the systemic perspectives introduced by evolutionary developmental biology, epigenetic inheritance, and niche construction, as well as other physical, physiological, behavioral, and cultural components of the evolutionary process. The BVA preceded this systemic view by nearly a full century. The systems perspective can be traced from Hatschek to Przibram, Weiss, and Ludwig von Bertalanffy (who was a friend of both Weiss and Kammerer) and, to some extent, still resonates in Rupert Riedl's (1978) systems account of evolution and in recent suggestions for a revision of evolutionary theory (Laland et al. 2015; Noble et al. 2014).

The theoretical biology cultivated at the time of the foundation of the BVA was still far removed from the way it is understood today and corresponded more closely to our modern conception of philosophy of biology (Nicholson and Gawne 2015). This is because in the early decades of the twentieth century no clear distinction existed between "theoretical" and "philosophical" examinations of biology, and the gathering of biological knowledge was simultaneously a theoretical and a philosophical enterprise. This usage began to be transformed by the new approach to theory developed at the BVA. The rigorous establishment of a quantitative methodology in biological studies that could be used in the formalization of form-generating processes helped to lay the foundation for a modern theoretical biology that supports, for instance, the mathematical modeling of biological processes. This was an achievement certainly not of the BVA alone; rather, it emerged in concert with similar movements in other countries, most notably Great Britain and the work of D'Arcy Wentworth Thompson, with whom Przibram exchanged correspondence and students. Marjorie Senechal (2012) aptly described the differences and commonalities between the British and the Austrian approach: "Where D'Arcy Thompson saw analogies, the Przibram brothers did experiments. Where he found elegance and simplicity, they found chaos and complexity. Where they glimpsed biological laws, he suspected leaps of imagination. But D'Arcy and the Przibrams were allies in the nascent international campaign to infuse biology with physics and chemistry." Joseph Woodger's visit to the BVA resulted in his later collaboration with Ludwig von Bertalanffy on the translation and expansion of Bertalanffy's seminal 1932 book *Theoretische Biologie* (Bertalanffy 1962), thus affecting also the British tradition of theoretical biology (Nicholson and Gawne 2014).

The continuing individuation of the field of theoretical biology had effects on the ways philosophy of biology would be done henceforth. In this sense, the organicist style of thought developed at the BVA had a lasting influence on the transformations of the scientific and philosophical perspectives of biology that took place during the first half of the twentieth century. Because of this and the other innovative contributions documented in this volume, the BVA can rightly be called a cradle of experimental, quantitative, and theoretical biology.

References

Ash, M., ed. 2008. *Mensch, Tier und Zoo. Der Tiergarten Schönbrunn im internationalen Vergleich vom 18. Jahrhundert bis zur Gegenwart*. Vienna: Böhlau-Verlag.

Benson, K. R. 1988. The Naples Stazione Zoologica and its impact on the emergence of American marine biology. *Journal of the History of Biology* 21:331–341.

Bertalanffy, L. v. 1962. *Modern Theories of Development: An Introduction to Theoretical Biology*. New York: Harper.

Berz, P. 2009. The eyes of the olms. *History and Philosophy of the Life Sciences* 31:215–239.

Brauckmann, S. 2013. Weiss, Paul Alfred. In *eLS Encyclopedia of Life Sciences*. Chichester: John Wiley & Sons, Ltd. doi: 10.1002/9780470015902.

Breuer, B. 2011. *The Birth of Musicology from the Spirit of Evolution*. Ann Arbor: UMI Dissertation Publishing.

Chamberlain, H. S. 1905. *Immanuel Kant*. Munich: F. Bruckmann.

Coen, D. R. 2006. Living precisely in fin-de-siècle Vienna. *Journal of the History of Biology* 39:493–523.

Coen, D. R. 2007. *Vienna in the Age of Uncertainty: Science, Liberalism, Private Life*. Chicago, London: University of Chicago Press.

De Waal, E. 2010. *The Hare with Amber Eyes: A Hidden Inheritance*. London: Random House.

Feichtinger, J., H. Matis, S. Sienell, and H. Uhl, eds. 2013. *Die Akademie der Wissenschaften 1938–1945*. Vienna: Österreichische Akademie der Wissenschaften.

Fröschl, K. A., G. B. Müller, T. Olechowski, and B. Schmidt-Lauber, eds. 2015. *Reflexive Innensichten aus der Universität*. Göttingen: V&R Unipress / Vienna University Press.

Gilbert, S., and S. Brauckmann. 2011. Fertilization narratives in the art of Gustav Klimt, Diego Rivera, and Frida Kahlo: Repression, domination, and eros among cells. *Leonardo* 44:221–227.

Gliboff, S. 2006. The case of Paul Kammerer: Evolution and experimentation in the early twentieth century. *Journal of the History of Biology* 39 (3): 525–563.

Groeben, C. 2002. The Stazione Zoologica: A clearinghouse for marine organisms. In *Oceanographic History. The Pacific and Beyond*, ed. K. R. Benson and P. F. Rehbock. Seattle, London: University of Washington Press.

Hatschek, B. 1905a. *Hypothese der organischen Vererbung.* Leipzig: Wilhelm Engelmann.

Hatschek, B. 1905b. Herr Houston Stewart Chamberlain und die Evolutionslehre. *Neue Freie Presse* 7.

Hirschmüller, A. 1991. Paul Kammerer und die Vererbung erworbener Eigenschaften. *Medizinhistorisches Journal* 26:26–77.

Hofer, V. 2002. Rudolf Goldscheid, Paul Kammerer und die Biologen des Prater-Vivariums in der liberalen Volksbildung der Wiener Moderne. In *Wissenschaft, Politik und Öentlichkeit*, ed. M. G. Ash and C. H. Stifter, 149–184. Vienna: WUV.

Janik, A., and S. Toulmin. 1973. *Wittgenstein's Vienna.* New York: Simon & Schuster.

Kniefacz, K., E. Nemeth, H. Posch, and F. Stadler, eds. 2015. *Universität–Forschung–Lehre: Themen und Perspektiven im langen 20. Jahrhundert.* Göttingen: V&R Unipress / Vienna University Press.

Laland, K. N., T. Uller, M. W. Feldman, K. Sterelny, G. B. Müller, A. Moczek, E. Jablonka, and J. Odling-Smee. 2015. The Extended Evolutionary Synthesis: Its Structure, Assumptions and Predictions. *Proceedings of the Royal Society B. Biological Sciences* 282:20151019.

Logan, C. 2013. *Hormones, Heredity, and Race: Spectacular Failure in Interwar Vienna.* New Brunswick, NJ: Rutgers University Press.

Logan, C. A., and S. Brauckmann. 2015. Controlling and Culturing Diversity: Experimental Zoology before World War II and Vienna's Biologische Versuchsanstalt. *Journal of Experimental Zoology* 323:211–226.

Maienschein, J. 1985. History of biology. *Osiris, 2nd Series: Historical Writing on American Science* 1:147–162.

Müller, G. B., and H. L. Nemeschkal. 2015. Zoologie im Hauch der Moderne: Vom Typus zum offenen System. In *Reflexive Innensichten aus der Universität*, ed. K. A. Fröschl, G. B. Müller, T. Olechowski, and B. Schmidt-Lauber, pp 355-369. Vienna: Vienna University Press.

Nemeth, E., and F. Stadler. 2015. Die Universität Wien im "langen 20. Jahrhundert" und das unvollendete Projekt gesellschaftlich verankerter Vernunft—Zum "Streit der Fakultäten" von Kant bis Bourdieu. In *Universität – Forschung – Lehre. Themen und Perspektiven im langen 20. Jahrhundert*, ed. K. Kniefacz, E. Nemeth, H. Posch, and F. Stadler, pp 33-76. Göttingen: V&R Unipress / Vienna University Press.

Nicholson, D. J., and R. Gawne. 2014. Rethinking Woodger's legacy in the philosophy of biology. *Journal of the History of Biology* 47:243–292.

Nicholson, D. J., and R. Gawne. 2015. Neither logical empiricism nor vitalism, but organicism: What the philosophy of biology was. *History and Philosophy of the Life Sciences* 37:345–381.

Noble, D., E. Jablonka, M. J. Joyner, G. B. Müller, and S. W. Omholt. 2014. Evolution evolves: Physiology returns to centre stage. *Journal of Physiology* 592:2237–2244.

Przibram, H. 1903. *Die neue Anstalt für experimentelle Biologie in Wien.* Verhandlungen der Gesellschaft deutscher Naturforscher und Ärzte 74:152–155.

Przibram, H. 1908/09. Die biologische Versuchsanstalt in Wien: Zweck, Einrichtung und Tätigkeit während der ersten fünf Jahre ihres Bestehens (1902–1907). *Zeitschrift für biologische Technik und Methodik* 1: 234–264, 329–362, 409–433; *Ergänzungsheft*: 1–34.

Przibram, H. 1913. Die Biologische Versuchsanstalt in Wien. Ausgestaltung und Tätigkeit während des zweiten Quinquenniums (1908–1912). Bericht der zoologischen, botanischen und physikalisch-chemischen Abteilung. *Zeitschrift für biologische Technik und Methodik* 3:163–245.

Przibram, H. 1931. *Connecting Laws in Animal Morphology: Four Lectures Held at the University of London.* London: University of London Press.

Reiter, W. L. 1999. Zerstört und Vergessen. Die Biologische Versuchsanstalt und ihre Wissenschaftler/innen. *Österreichische Zeitschrift für Geschichtswissenschaften* 10 (4): 585–614.

Rentetzi, M. 2007. *Trafficking Materials and Gendered Experimental Practices*. New York: Columbia University Press.

Riedl, R. 1978. *Order in Living Organisms: A Systems Analysis of Evolution*. New York: Wiley.

Sachse, C. 2014. Grundlagenforschung. Zur Historisierung eines wissenschaftspolitischen Ordnungsprinzips am Beispiel der Max-Planck-Gesellschaft (1945–1970). In *Dimensionen einer Geschichte der Kaiser-Wilhelm / Max-Planck-Gesellschaft*, ed. D. Hoffmann, B. Kolboske, and J. Renn, 215–235. Berlin: epubli GmbH.

Senechal, M. 2012. *I Died for Beauty: Dorothy Wrinch and the Cultures of Science*. Oxford: Oxford University Press.

Taschwer, K. 2013. Vertrieben, verbrannt, verkauft, vergessen und verdrängt. Über die nachhaltige Vernichtung der Biologischen Versuchsanstalt und ihres wissenschaftlichen Personals. In *Die Akademie der Wissenschaften*, ed. J. Feichtinger, H. Matis, S. Sienell and H. Uhl, 105–116. Vienna: Katalog zur Ausstellung.

Taschwer, K. 2016. *Der Fall Paul Kammerer. Das abenteuerliche Leben des umstrittensten Biologen seiner Zeit.* München: Carl Hanser Verlag.

Taschwer, K., J. Feichtinger, S. Sienell, and H. Uhl, eds. 2016. *Experimentalbiologie im Prater. Zur Geschichte der biologischen Versuchsanstalt 1902 bis 1945*. Vienna: Österreichische Akademie der Wissenschaften.

Walch, S. 2011. Sexualhormone in der Laborpraxis: Eugen Steinachs Experimente und seine Kooperation mit Schering (1910–1938). Dissertation, Universität Wien, Vienna.

Walch, S. 2016. *Triebe, Reize und Signale. Eugen Steinachs Physiologie der Sexualhormone. Vom Biologischen Konzept zum Pharmapräparat, 1894–1938*. Wien: Böhlau Verlag.

Wald, C. 2013. Austrian Academy of Sciences faces its Nazi history. *Science* 339:1368.

II VIENNESE CONTEXT

2

The Founders of the Biologische Versuchsanstalt: A Families' Topography

Georg Gaugusch

In the years that preceded the founding of the Biologische Versuchsanstalt in Vienna in 1902, three scientists from affluent Viennese families jointly developed the idea of establishing and financing an independent research institute devoted to the experimental study of biology. These were the zoologist Hans Przibram, the botanist Leopold von Portheim, and the plant physiologist Wilhelm Figdor, all associated in various ways with the University of Vienna. All three protagonists belonged to what may be called the Jewish *haute bourgeoisie* of Vienna (Beller 1989). The composition and influential societal roles of these highly interconnected groups of families have often been described, but they must be understood in detail for an analysis of the factors that contributed to the founding of the Biologische Versuchsanstalt (Institute for Experimental Biology, or BVA). This chapter will describe the genealogical and social backgrounds of the founders of the BVA.

The Jewish Haute Bourgeoisie

The history of the Jewish population in the Austrian Empire—the dominant country in Central Europe between 1718 (Treaty of Passarowitz) and its demise in 1918—and especially in its capital, Vienna, has been written about in a wide range of surveys, the most comprehensive among them Häusler (1985), McCagg (1989), and Brugger et al. (2006) for Austria in general, and Rozenblit (1984), Beller (1989), and Wistrich (1994) for Vienna in particular.

As comprehensive as those studies are, they do not often focus on the Jewish haute bourgeoisie, which came into being in the early nineteenth century, had its golden age between 1870 and World War I, and was eradicated by National Socialism during the 1930s and World War II. Another blank spot in the historiography of the Jews of the Austrian Empire is a conclusive answer to the question of who was considered a part of the Jewish haute bourgeoisie and, even more important, by what criteria one belonged to this affluent and influential social stratum. In the following I will briefly describe the societal processes that formed this new social group, emphasizing the intellectual and economic background of the class. The main objective is to introduce the biographies of the Viennese families of Hans Przibram[1] (1874–1944), Leopold von Portheim[2] (1869–1947), and Wilhelm Figdor[3] (1866–1938), all members of the Jewish haute bourgeoisie.

Within a few decades of the formal declaration of the Austrian Empire at the beginning of the nineteenth century, a new social class developed there that had very little in common with the age-old Jewish communities and ghettos and was distinct from the majority Roman Catholic population. The individuals of this group held influential positions in the Central European economy, in banking as well as in all kinds of industrial production, in trade, and in the newly formed press.[4] It is notable that within a very short time members of a suppressed minority came to make up one of the wealthiest social sets in Austrian society, and that nearly all the cultural and intellectual achievements that became part of the highly admired "Wiener Moderne" culture of Vienna were originated by people from that social group. Nevertheless, it is not easy to define precisely who did and did not belong to the Jewish haute bourgeoisie; even within the group the defining characteristics of inclusion are contradictory.[5]

If we look more closely at two major criteria, namely religion and wealth, we can recognize the problems. An easy attempt at defining the group could posit that everyone wealthy and belonging to the Jewish faith was a member of the Jewish upper class. For several reasons this definition does not work: First, especially after 1900, many Jews left their faith and either converted to Catholicism or Lutheranism or remained without any confession or faith. However, one's paternal religion may change without resulting in a change in one's social environment. Friends or business partners may remain the same, and many Jews stayed within the same social circles even after changing their religion or altering their name to one that did not allude to a Jewish background. For instance, someone who had changed his name from "Moses Basseches" to "Moritz Basse" would still have been known in his social surroundings under the former, Jewish-sounding name, and the new name would have represented a continuous source of covert amusement. These practices also led to another phenomenon: Many Jews, especially the wealthy ones who had left the Jewish religion, still married within their old social group. There were, unsurprisingly, examples of couples, who, although both Jewish-born, were married in a Catholic or Protestant Church.[6]

For the financial criteria, the case is similarly dubious. In that period quite a number of people in Central Europe, especially in countries dominated by Roman Catholicism, believed that wealth in any form was simply the result of luck. They were also convinced that if you are rich, you will stay rich. Both mistaken prejudices led to a strong, open or hidden anti-Semitism, often nurtured by envy, toward the Jewish haute bourgeoisie, which dominated the politics of the Austrian Empire from the 1890s onward and led directly to the National Socialism of the 1930s.[7] As the family biographies will show, the wealth of the families Figdor, Portheim, and Przibram was, rather than the result of luck, the outcome of hard work, innovation, knowledge, and an open-minded attitude toward new developments and international markets.[8] In fact, many wealthy Jewish families who lost their fortunes in the big crash of the Vienna Stock Exchange during the Viennese World Exhibition of 1873 (e.g., the Hofmannsthals or the Scheys), still played major roles in Viennese society, whereas others, who gained great wealth later, never belonged to this network.

Instead of using the criterion of Jewish religion, it is better to speak of members of the Jewish haute bourgeoisie as being born Jewish or generally affiliated with the Jewish community. Those who did not practice Judaism could still belong to that social group, as long as they and their children chose to maintain their place in the social set. Determining specifically whether an individual was a part of the group would require analysis of the genealogical background.[9] Likewise, instead of just "wealth," perhaps the terms "potential" or "cultural capital" should also be used, which could include the idea of having international connections, education, business ties (and loyalty), or artistic or musical skill, all equally important as money in being considered part of the social class.[10]

An example will illustrate this point: The Hofmannsthal family and its founder Isak Löw Hofmann (1759/1761–1849), who came from the small Bohemian town of Prostibor to Vienna in the late eighteenth century, were one of the most important and interesting Jewish protagonists in Vienna before the revolution of 1848. Isak Löw Hofmann came from a rather poor family but was gifted, intelligent, and cosmopolitan. As a young man he went to Prague and learned business there. He married into the Schefteles family, in an arranged marriage that brought benefits to both sides: the Schefteles family, which was old and well known but far from rich, could gain new wealth through its clever son-in-law, and Isak Löw Hofmann could use the connections of the much older family for his own business. He managed to get the *Toleranz*, the permit for staying in the city of Vienna and doing business there—a permit that was not easy to acquire. He established a wholesale business, a so-called *Grosshandlung*—in those days a mixture of banking house, wholesaler, and transport or export business. Because of his achievements in the silk industry he was ennobled in 1835. Isak Löw Hofmann von Hofmannsthal, as he was called from that time onward, was a member of the board of the Jewish Community in Vienna and one of its most prominent members. Except for one, all his children kept the Jewish faith; only the youngest son, August (1815–1881), converted to Christianity and married an Italian widow, the mother of his son born out of wedlock. That son, Hugo von Hofmannsthal Sr. (1841–1915), was brought up as a Christian, married a Roman Catholic girl, and lost most of his money in the crash of 1873. He did not belong to the Jewish haute bourgeoisie at all, although he had a Jewish father. However, his son, Hugo von Hofmannsthal Jr. (1874–1929), the well-known writer, returned, so to speak, into that set thanks to his social contacts, and married into one of the wealthiest and oldest Viennese Jewish families, the Schlesingers. Gerty Schlesinger (1880–1959) had to leave her Jewish faith; she became Roman Catholic and married Hugo von Hofmannsthal in the Viennese Schottenkirche in 1901, in a union that provided benefits to both sides. Hugo von Hofmannsthal had no money but held a noble title and was a promising writer who needed his wife's money for carefree living.[11] The shifting from one religion to another and social movement among a diverse group of faiths created a nonreligious climate. With a few exceptions (e.g., the Gutmann or the Kuffner family), most members of the Jewish haute bourgeoisie were not religious, and in general, religion was no matter to talk about. Indeed, any kind of piety or religious observances would have been looked upon with wonder.

The Origins

To understand the basic social mechanisms in the Jewish haute bourgeoisie it is of some importance to elaborate how this group developed and which factors precipitated its formation. We therefore have to take a closer look at the situation of Jews in the lands of the former Bohemian Crown, mainly Bohemia and Moravia, and its capital. The city of Prague had one of the biggest and most important Jewish communities in Europe from medieval times up to the Second World War, but nevertheless the situation for Jews there and in the Bohemian and Moravian countryside was constrained.[12] In 1726 the *Familiantengesetze* were decreed; their major target was to regulate the numbers of Jewish families in the countries of the Bohemian Crown (Bohemia, Moravia, and Silesia). From then on the number of Jewish families in Bohemia was limited to 8,451, in Moravia to 5,106, and in the Austrian part of Silesia to 119. Only the eldest son could inherit his father's *Familienstelle* (family number), only he could get married, and only he was a protected member of the *Herrschaft* (dominion) he was living in. His younger brothers could stay in his household (as unmarried clerks, for instance), or could try to find and occupy a free place (where someone had died without leaving male descendants), or they had to emigrate. Most of the Jews who were forced to emigrate went to Hungary, where they were warmly welcomed by the aristocratic estate owners. Hungary was still suffering from the consequences of the Turkish occupation; trade and industry had to be brought up from practically nil, and the Jewish immigrants were able to fill this gap. This legal system, which was abolished in 1848/49, caused huge difficulties in the Jewish communities. However, some gain came in the form of an early economic internationalization via large family networks that spread out through Bohemia, Moravia, Hungary, and beyond the Austrian borders. Needless to say, these networks became very important for doing business and were of crucial advantage when the industrial revolution reached the continent in the early nineteenth century.

Position in Society

Although the Revolution of 1848 had failed, things had changed in that year. With the new constitution, proclaimed in March 1849, most of the discriminating laws fell. Now that "the pleasure of civil rights was independent from the religious faith," a new situation evolved—in a country dominated by the Roman Catholic church for centuries. Even the partial reestablishment of neo-absolute rule in 1853 could not turn the wheel completely backward.[13] In the following 20 years, supported by liberal politics and a tolerant atmosphere, Austrian industry developed rapidly. Among the entrepreneurs, following the revolution, were a high percentage of Jews, some successful, others not. There was no historical reference for this group of industrialists in Viennese society: the new class was mostly not Roman Catholic and was obviously wealthy, mainly not from Vienna, open to new ideas, and more independent from the almighty imperial administration; so the old Viennese citizens, mostly Roman Catholic, conservative, and admiring of the grandeur of the military, aristocracy, and administration, were clueless about the new crowd.[14] The easiest approach for them was to attack the new group as social climbers or nouveau riche, but that neglected the fact that some of these supposedly new industrialists had been in business longer than most of the Christian businessmen. For instance, to call Jonas Königswarter (1807–1871) a parvenu made no sense because the Königswarters by around 1800 had already been wealthy businessmen for more than 60 years, or two generations.[15] The Przibrams, Portheims, and Figdors also established their businesses in the eighteenth century, and none of them would have accepted being called nouveau riche in the 1860s. On the other hand, the aristocracy, or the "first society," as it was called, didn't accept anyone nonaristocratic as equal; social communication between them and the wealthier members of the "second society," Catholic or not, was strictly regulated and limited.[16] Since the position of this new Jewish upper class was difficult, and they didn't integrate into any of the old groups, they formed a new one. As a result both older groups turned their backs on them—the petite bourgeoisie often turned to anti-Semitism, the aristocracy toward snobbery.

Literacy, Learning, and Education

The influential societal momentum in Vienna of the late nineteenth and early twentieth century is characterized by an enormous intellectual activity and creative output.[17] The even more intriguing fact is that in most of these activities the Jewish minority played a major if not leading role. Various attempts have been made to explain this phenomenon, and it can be said that several factors caused the situation to develop.[18] The economic background of this development has not been outlined precisely enough. As described above, the Jewish minority was forced into modernism; they had to find new ways of doing business because they were mostly not allowed to use the old ones. Moving ahead and being innovative had to be valued, and in a group with such a forward-looking mentality it is more likely that new ideas will come up, compared to a group where traditional values and ways of thinking are more highly valued. A second reason can be found in the differences between Catholicism and Judaism, although these should not be overestimated. The Roman Catholic religion does not require its followers to scrutinize its content; to believe what religious authorities proclaim is the priority, and it was never a goal of those authorities that wider circles deal with basic religious texts. That is a huge difference from the Jewish religion, where every man is required to read, understand, and interpret religious writings. In order to do so the knowledge of a Biblical language is ineluctable, and to understand the religion on a higher level a certain intellectuality is necessary. Most of the Jewish industrialists of the first generation were not very religious or devotional, but many of them engaged in the Jewish community and practiced charity. There was a common sense within that group that knowledge, intellectuality, and learnedness are of the highest value. Most of the Viennese Jewish upper class left the Jewish faith between 1890 and 1920, but this basic mentality stayed, detached from its religious origin. The third reason, probably the most simple and obvious one, is the wealth that was built up during the nineteenth century. The money had to be spent, and compared to the Christian majority the Jews were more likely to use it not only for consumption, representation, or luxury goods, but also for culture, education, or welfare. This too can be called a heritage of the Jewish religion, but that is still not the full story. It also reflects the position the Jewish haute bourgeoisie was in, because neither the aristocracy nor the Roman Catholic majority was very active in those fields. To present themselves as supporters of the state was probably the best way to try to escape the anti-Semitic hostilities, which became stronger and more aggravating in the last quarter of the nineteenth century.

The Przibram Family

The Przibram family was one of the oldest in the Prague Jewish community, dating back to the seventeenth century (see figure 2.1).[19] Having Judah Loew ben Bezalel (ca. 1520–1609), one of the most famous rabbis in sixteenth-century Prague, among its ancestors, the Przibrams belonged to a rabbinical elite and had family connections to more or less all the important Jewish families in the Holy Roman Empire. The first known ancestor of the branch that later became relevant for the BVA was Salomon Przibram, who died in Prague in 1802 at the age of 66. He was married to Fradel Gomperz (ca. 1744–1822), who came from a widespread Jewish family in Germany. The Gomperz family played a major role in the network of the so-called *Hoffaktoren*, managers of major portions of the finance business of the smaller and larger German courts in the eighteenth century.[20] In the Jewish census of 1793 Salomon Przibram is mentioned as a homeowner and *Leinwandhändler* and was already a big player in the textile business.[21] In 1798 Salomon Przibram established a textile printing mill in Karolinenthal, a small town in the outskirts of Prague, which became one of the largest textile businesses in Austria, especially during the Continental Blockade (1806–1814) and the time of the highly economically restrictive and

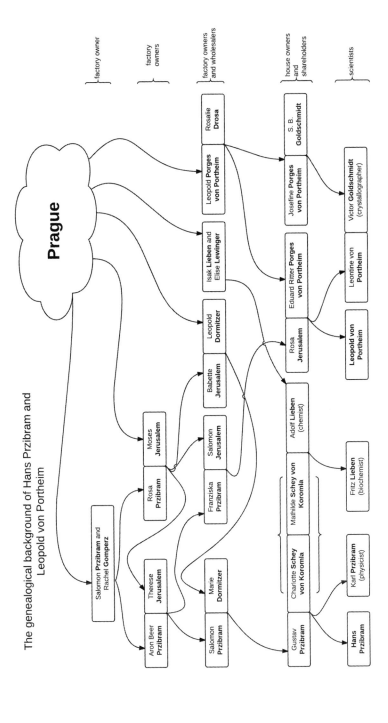

Figure 2.1
Genealogical relations of the Przibram and von Portheim families (© Georg Gaugusch).

prohibitive Austrian government between 1815 and 1848. Przibram's only son was Aron Beer (ca. 1781–1851), who continued his father's textile businesses in Prague. In fact, he and his two brothers-in-law, Moses Jerusalem[22] (ca. 1762–1824) and Ephraim Epstein[23] (ca. 1767–1832), formed the core of what was known from the 1840s onward as the "Prague Jewish textile network," an intriguing web of connections between the leading industrial Jewish families and their companies in Prague. Aron Beer Przibram married his sister's daughter Therese Jerusalem (ca. 1784–1866); the couple had two surviving children: Franziska (ca. 1806–1837), who married her uncle Salomon Moses Jerusalem (1806–1864), but died young, leaving only one daughter, Rosa (1836–1906), Leopold von Portheim's mother; and Salomon (1808–1865), who married his cousin Marie Dormitzer[24] (1818–1886). The Przibrams, Jerusalems, Dormitzers, and Epsteins from Prague, who all ran printing mills or related businesses, played a major role in this closed Austrian textile market, which guaranteed little competition, low labor costs, high prices, and, for these reasons, high profits. By 1818 the factories of Przibram and Jerusalem in Karolinenthal and Smichow, another suburb of Prague, employed nearly 500 workers, in those days a very significant number.[25] By the 1820s Aron Beer Przibram already had close ties to Vienna, where Jews were not allowed to settle prior to 1848. In order to serve the Central European market better he and Moses Jerusalem established a branch of their company in the capital of the Austrian Empire in 1827, basing it right behind the old town hall in the so-called Stoss im Himmel.[26]

After 1848 and the death of Aron Beer Przibram in 1852, part of the family's already huge fortune was invested in real estate, and the magnificent baroque building that housed the Vienna branch was bought as soon as Jews were allowed to purchase real estate in the city of Vienna. Salomon Przibram also continued the inherited textile industries, but the character of the family's entrepreneurship shifted slightly toward a more modern way of investment. After Salomon's death in 1865 the Przibrams left Prague for Vienna, where his widow, Marie, bought two building plots on the newly forming Ringstraße: in 1869 one in the so-called Textile Quarter around the Rudolfsplatz (I., Esslinggasse 13) and later a prestigious house in a more fashionable area on the Parkring. From this time onward Marie Przibram, her children, and their cousins formed an important part of the city's Jewish haute bourgeoisie. In 1871 Gustav Przibram (1844–1904), Salomon's eldest son, married Charlotte Baroness von Schey (1851–1939); her father, Friedrich (1815–1881), was the first Jew not to be christened who was ennobled in Hungary, receiving the title "de Koromla." Up to 1873 when he lost most of his fortune in the market crash, he was one of the big players in the Viennese banking scene and owned a grand palace on the Opernring. Charlotte's sister Mathilde (1861–1940) was the wife of the chemist Adolf Lieben (1836–1914). This family background might have been the reason Gustav's sons Hans (1874–1944) and Karl (1878–1973) had such a strong interest in science. In fact the Przibrams were among the few rich families of the nineteenth century who managed to preserve large parts of their fortune through the First World War and the following hyperinflation. Austrian state bonds became worthless during that period, but their extensive real estate helped to maintain their social level. In 1927 the Przibrams still owned all five houses that had been acquired by their ancestors in the Viennese first district: Parkring 18 and 20, Stoss im Himmel 3, Esslinggasse 13, and Universitätsring 6. Real estate was not very profitable in those days, but still the income was big enough to make it possible for Hans and Karl to do scientific work without troubling much about their income. After the *Anschluss* in 1938, Hans and his brother were forced to sell their real estate but stayed in Vienna until December 1939. Hans Przibram and his wife left Vienna for the Netherlands, where they were caught by the National Socialists in 1943 and sent to the Theresienstadt concentration camp. Hans died there in 1944; his wife committed suicide one day after his death.

The Porges von Portheim Family

Like the Przibrams, the Porges family was one of the most ancient families of the Prague Jewish community, and the name itself was very common in the city (figure 2.1). The Prague *Familiantenbücher*, in which all Jewish citizens had to be registered, mention around 35 different households of that name between 1811 and 1848.[27] The first known member of the branch, which later became the Porges von Portheim family, was Rabbi Wolf Spiro, who died in Prague in 1630.[28] His descendant was Gabriel Porges Spiro (1738–1824), who was married to Esther Kassowitz, the daughter of Löw Kassowitz, another eminent rabbinical authority of the Prague Jewish community. Gabriel Porges had six children, including two sons, Moses (1781–1870) and Leopold Juda (1785–1869), who both became outstanding entrepreneurs. An interesting chapter in the life of Moses Porges was his trip to the court of the so-called Jewish messiah Jakob Frank near Offenbach; his memoirs of this adventure became a very valuable source about the Frankist Christian-Jewish sect, whose rise and fall caused disturbances in the Central European Jewish communities around 1790.[29] When Moses came back to Prague he and his brother Leopold established a tiny textile-printing workshop in the city's center. In the beginning they had only a single printing machine and almost no capital, but they worked hard and were successful, mainly for the same reasons already described for the Przibram family, although neither brother married into the Prague Jewish textile network mentioned above. When production in the center of Prague became too expensive they established a second factory in the Prague suburb of Smichow in 1830, which later became the leading production site.

Brüder Porges, as the company was called, by then already employed around 600 workers, 200 in Prague and 400 in Smichow. In 1841 the Austrian Emperor Ferdinand visited Smichow and the Porges factory, which in those days was one of the biggest industrial facilities in the Austrian Empire. In the same year both brothers were ennobled and acquired the title "von Portheim," which was used by later generations as a single surname. In fact, they were ennobled because of their industrial engagement but also for their extensive spending for welfare and education. Moses Porges von Portheim, for instance, founded and sustained the Josephstädter Kleinkinderbewahranstalt, a kindergarten for the Prague district of Josefstadt (Josefov), the former ghetto.[30] His brother Leopold, the BVA cofounder's grandfather, widened the economic horizon and established, in 1840, a porcelain factory near the Western Bohemian town of Chodau. Leopold's son Eduard (1826–1907) was probably the most important industrialist of the second generation. He married Rosa, the only daughter of Salomon Moses Jerusalem and Franziska Przibram, thus marrying directly into the "Prague Jewish Textile Network." Eduard von Portheim was in equal measure industrialist and politician. He started his career in the 1850s in the Bohemian Chamber of Commerce and was later elected into the Austrian Parliament, where he was a member of the German liberal wing—not unusual, considering his background. When the old firm Brüder Porges was transformed into a stock corporation, the Smichower Kattunmanufaktur AG, he became president of that company although the production in Smichow ended soon after. Eduard von Portheim played a major role in the industrial development of Austria, and especially Bohemia, between 1860 and 1880 and received for his numerous merits the Austrian knighthood in 1879.

Eduard Ritter von Portheim, as he was called from then on, was also famous for his keen interest in the arts. Even at a later age he traveled around Europe obtaining new items for his collection.[31] His seven children—five sons and two daughters—followed their father, but in different ways. The two eldest brothers, Emil and Friedrich, owned the company Kinzlberger & Comp., which Eduard had bought when the factory in Smichow was closed. Their company specialized in the production of aniline dyes and chemicals for tanneries, and had its production plant near Prague.[32] Neither ever married, and both were murdered in Auschwitz in 1942. Their younger sister Franziska (1860–1949) was married to the lawyer Felix Maass (1852–1920) in Berlin, but both spent most of the year in Bad Ischl (Upper Austria), where they owned a large house in Kaltenbach. Their only daughter, Emmy (1884–1953), was married to the physicist and Lieben prize laureate Stefan Meyer (1872–1949).[33] Eduard von Portheim's second daughter, Leontine (1863–1942), or Lola as she was called, was married in 1888 to her cousin Victor Mordechai Goldschmidt (1853–1933), who became professor for crystallography at the University of Heidelberg in 1892. In 1919, as they were without children, he and his wife endowed most of their property to the Josephine and Eduard von Portheim foundation, named after Victor's mother and Lola's father, which was dedicated to supporting the natural sciences and humanities.[34]

Victor Goldschmidt, who had to flee from Heidelberg when the National Socialists came to power in 1933, died in Salzburg in the same year. Between 1933 and the Second World War, the foundation and its fund faced destruction and sank into oblivion. Lola Goldschmidt went back to Heidelberg, where after enduring years of humiliation and abasement, she committed suicide on the day of her deportation in 1942.[35] The three younger sons of Eduard were Leopold, cofounder of the BVA, Viktor (1871–1939), who was a businessman in Vienna but lived and died in Bad Ischl, and Heinrich (1872–1919), who became a lawyer in Prague but died young. The only son of Eduard who married was Leopold. His wife Elise Ungar (1886–1969) came from a German-Jewish intellectual family as well. Her father, Emil Ungar (1849–1934), was an acclaimed professor of forensic medicine and a very busy pediatrician in the small town of Bonn near Cologne. Her mother, Anna Noether, was a descendant of the famous Goldschmidt family from Frankfurt and a relative of the mathematician Max Noether (1844–1921).[36]

Leopold von Portheim had two children. His son, Eduard, born in 1910 and named as usual after his grandfather, finished his studies of law at Vienna University in 1936. In 1940 he was captured by the Gestapo in Salzburg, where he had hidden, and was imprisoned in the Dachau concentration camp, from where he was deported in 1942 and killed in Hartheim. Leopold von Portheim, his wife Elise, and their daughter, Susanne Lily (born 1917), managed to escape to London, where Leopold died shortly after the war, in 1947.

The Figdor Family

In contrast to the Przibram and Porges families, the Figdors did not have any ties to Prague. The first known ancestor was Avigdor, who lived in the first half of the eighteenth century in the small village of Kittsee, in what was then Hungary and is now the Austrian province of Burgenland.[37] His son Jakob (ca. 1740–1808) took his father's name as a surname and called himself Figdor, while other descendants of Avigdor called themselves Kittsee or Kittseer (see figure 2.2). Avigdor must have been a gifted businessman and obviously had strong business relations with Vienna and Pressburg, but it was Jakob Figdor who managed to get the permission to settle in Vienna in 1791. He was married to Regina Sinzheimer (ca. 1745–after 1814), who came from a family tightly connected to most of the Jewish *Hoffaktoren* in the Holy Roman Empire.

Business connections and knowledge about foreign markets were major assets, so it is not a surprise to find Regina Figdor's eight siblings spread all over Western Europe: Johanna Lehmann, who was married to one of the other grand families of the eighteenth century, and two unmarried sisters lived in Vienna; Joseph Sinzheimer was a businessman in Frankfurt; Moses a merchant in London; Magdalena and Marianne, who both had wed into the well-known Schiff family, were also living in Frankfurt and London; and finally Eleonore Gomperz, who was a merchant's wife in Amsterdam, had joined the same family Salomon Przibram wed into. Jakob Figdor established a widespread wool business and mainly sold Hungarian wool to England but did some other wholesale trade as well. Jakob's oldest son, Isak Figdor (ca. 1770–1850), who was married to Anna Schlesinger,[38]

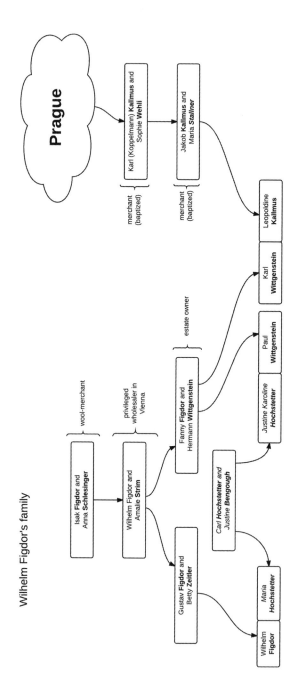

Figure 2.2
Genealogical tree of the Figdor family (© Georg Gaugusch).

continued and expanded his father's business in Vienna. Already in the 1840s he was one of the richest businessmen in Vienna, a fortune his ten children increased. Isak Figdor's most famous grandchild was Joseph Joachim (1831–1907), the son of his daughter Fanny, who had wed the wool merchant Julius Joachim and lived in Frauenkirchen and Pest. Joseph Joachim was, besides Heinrich Wilhelm Ernst (1814–1865), the most significant Austrian violinist of the nineteenth century.

Being such an important artist and having grown up with his Viennese cousins, Joseph Joachim, like all of his family, came in contact with the leading musicians of the time, especially Johannes Brahms, who was a close friend to the Wittgenstein and Figdor families. Fanny Joachim's younger brother was Wilhelm Figdor (ca. 1793–1873), who together with his younger brothers was a partner in Jakob's wholesale business. His two children both converted to Christianity: Fanny Figdor (1814–1890) became the wife of Hermann Christian Wittgenstein (1802–1878) and was the grandmother of the philosopher Paul Wittgenstein.[39] She and her husband left the Jewish faith shortly after their wedding. Wilhelm Figdor's only son, Gustav (1816–1879), from the late 1840s onward, had a liaison with the Bavarian-born Betty Zeitler (1827–1903), but apparently they could not easily get married. After she gave birth to three illegitimate children, Gustav converted to the Roman Catholic faith in 1853; he married her in the Paulanerkirche in the same year. Their youngest son was the BVA cofounder Wilhelm Figdor, who was already brought up as a Roman Catholic. He died on January 27, 1938, in Vienna, just in time not to see the destruction of the BVA.

Summary

The foregoing descriptions of the three founders' genealogical and economic backgrounds make clear why intellectual or cultural potential was far more likely to arise in their milieu than among the Catholic, tradition-oriented majority. All three families gained and held their social and economic status because they also invested in science, not only in shares or real estate. Founding and maintaining the BVA was a major burden, but it seems to have been accepted by most of the family members. Even after World War I, when their great wealth was nearly gone and in a time when for most people fundamental research was an unnecessary luxury, Figdor, Portheim, and Przibram continued their project. They maintained their visionary approach even in hard times—an attitude they might have inherited from their forefathers. In addition, the BVA was a private institution, and after World War I private initiatives lost ground against public institutions, a development that was reinforced by fascistic concepts. In 1938, when the last institutional remnants of the nineteenth century were swept away, the BVA's founders, too, had to surrender their creation. Their world was gone.

Notes

1. *Österreichisch Biographisches Lexikon* 8 (1983), 314–315.

2. *Neue Deutsche Biographie* 20 (2001), 634–636.

3. *Österreichisch Biographisches Lexikon* 1 (1956), 313.

4. Sandgruber (2013), 151ff for further details.

5. Gaugusch (2011), xiv–xv.

6. Gaugusch (2011), xv and 94–96.

7. In fact, most of the Jews in the Austrian Empire were poor, in particular in Galicia, where they formed one of the largest ethnic groups.

8. Sandgruber (2013) p.24ff.

9. Gaugusch (2011) for a broad genealogical survey of around 250 Jewish families in Vienna.

10. See for instance the will of Dr. Felix Joël in Gaugusch (2011), 1283–1284 and 1288. Dr. Felix Joël was in contact with nearly all Jewish industrialists and bankers of his time and was the only member of his family who had not converted to Christianity.

11. See Gaugusch (2011), 1212–1222, for a complete history of the Hofmannsthal family.

12. Kestenberg-Gladstein (1969).

13. Jews were allowed to buy real estate in Vienna in 1849, but it was again forbidden in 1853; see Moriz von Stubenrauch, *Zur kaiserlichen Verordnung vom 2. October 1853, Nr. 190 des R. G. Bl., die Besitzfähigkeit der Israeliten betreffend*, in *Allg. österreichische Gerichts-Zeitung* 4 (1853) 132: 549–550, 133: 553–554, 134: 557–558, 135: 561–562.

14. "Aus der Geschichte der Wiener Gesellschaft im Vormärz" (1867), 4–5; *Ein Duell in Vöslau* (1867), 5–6; *Statistik der Wiener hochadeligen Salons im Vormärz* (1867), 4–5; *Ein Hausball bei einem Wiener Bankier* (1867), 4–5; *Österreichische Adelsverhältnisse im Allgemeinen* (1868), 4–5.

15. For information about the Königswarter family see Gaugusch (2011), 1505–1519.

16. Vasili (1885), 54–71 and 357–399. About the relationship between the aristocracy and "normal" society see *Aus der Geschichte der Wiener Gesellschaft im Vormärz* (1867).

17. Schorske (1980): Kandel (2012), 3–19.

18. Beller (1989).

19. Hock (1892), 291–293.

20. Schnee (1965); Kaufmann and Freudenthal (1907).

21. Ebelová (2006), 159.

22. Gaugusch (2011) pp. 1273–1282.

23. Gaugusch (2011) pp. 576–585.

24. Gaugusch (2011) pp. 411–422.

25. Slokar (1914), 295.

26. Wachstein (1936), 88 (No. 128).

27. Národní archiv (Prague), České gubernium—Knihy židovských familiantů, HBF 160–188 (City of Prague), accessed September 9, 2014, badatelna.eu/fond/2098.

28. Fischel (1906).

29. Seligmann, *Eine Wallfahrt nach Offenbach—Zu Geschichte der Jakob Frankschen Bewegung*, in *Frankfurter israelitisches Gemeindeblatt* 10, 6 (February 1932): 121–123, and 10, 7 (March 1932): 150–152.

30. Wurzbach (1872), 23, 123.

31. See the obituary of Eduard Ritter von Portheim in the *Prager Tagblatt* 16 (February 1907), 4.

32. Hanel (1912).

33. Soukup (2004), 174–182.

34. Engehausen (2008).

35. Marzolff (2007).

36. Gaugusch (2011), 927 (Goldschmidt family from Frankfurt).

37. For the Figdor family see Gaugusch (2011), 610–630.

38. She came from the same old and well-established Viennese family Hugo von Hofmannsthal married into.

39. See Gaugusch (2002), 120–145, for a history of the families Wittgenstein and Salzer.

References

Aus der Geschichte der Wiener Gesellschaft im Vormärz. 1867. Anonymous article (signed, –*r.*) published in the newspaper *Neuen Fremdenblatt* (Vienna), November 1, Morgenblatt, 4–5, I.

Beller, Steven. 1989. *Vienna and the Jews 1867–1938*. Cambridge: Cambridge University Press.

Brugger, Eveline, et al. 2006. *Geschichte der Juden in Österreich*. Vienna: Verlag Carl Ueberreuter.

Ebelová, Ivana. 2006. *Soupis židovských rodin v Čechách z roku 1793*. Volume 6/1 (Prague 1792 and 1794). Prague: Národní archiv

Ein Duell in Vöslau and II. *Die Wiener erste Gesellschaft in den Dreißiger Jahren*. 10. November 1867 (Morgenblatt) pp. 5–6: III.

Ein Hausball bei einem Wiener Bankier. 8. Dezember 1867 (Morgenblatt) pp. 4–5: V. *Österreichische Adelsverhältnisse im Allgemeinen. – Die zweite Gesellschaft Wiens und ihr Verhältniß zur ersten*. 29. Dezember 1867 (Morgenblatt) pp. 4–5: VI.

Engehausen, Frank. 2008. *Die Josefine und Eduard von Portheim-Stiftung für Wissenschaft und Kunst 1919–1955*. Heidelberg: Verlag regionalkultur.

Fischel, Alexander. 1906. *Stammbaum der Familien Porges und Porges von Portheim*. Kassel: privately published.

Gaugusch, Georg. 2011. *Wer einmal war—Das jüdische Großbürgertum Wiens A-K*. Vienna: Amalthea.

Gaugusch, Georg. 2002. Die Familien Wittgenstein und Salzer und ihr genealogisches Umfeld. In Adler. *Zeitschrift für Genealogie und Heraldik* 21 (2001–2002): 120–145.

Hanel, Rudolf, ed. 1912. *Jahrbuch der österreichischen Industrie I: 1913.* Wien: Compassverlag.

Häusler, Wolfgang. 1985. Das österreichische Judentum zwischen Beharrung und Fortschritt. In *Die Habsburgermonarchie 1848–1918*, vol. 4, *Die Konfessionen.* Vienna: Verlag der Österreichischen Akademie der Wissenschaften.

Hock, Simon. 1892. *Die Familien Prags nach den Epitaphien des alten jüdischen Friedhofs in Prag.* Pressburg: Adolf Alkalay.

Kandel, Eric R. 2012. *The Age of Insight.* New York: Random House.

Kaufmann, David, and Max Freudenthal. 1907. *Die Familie Gomperz, J. Kauffmann.* Frankfurt: Main.

Kestenberg-Gladstein, Ruth. 1969. *Neuere Geschichte der Juden in den böhmischen Ländern I, J. C. B. Mohr.* Tübingen: Paul Siebeck.

Marzolff, Renate. 2007. *Leontine und Victor Goldschmidt.* Heidelberg: Mattes Verlag.

McCagg, William O. 1989. *A History of Habsburg Jews 1670–1918.* Bloomington: Indiana University Press.

Neue Deutsche Biographie 20. 2001. Berlin: Duncker & Humblot.

Österreichisch Biographisches Lexikon 1. 1956. Wien: Verlag der Österreichischen Akademie der Wissenschaften.

Österreichisch Biographisches Lexikon 8. 1983. Wien: Verlag der Österreichischen Akademie der Wissenschaften.

Österreichische Adelsverhältnisse im Allgemeinen. – Die zweite Gesellschaft Wiens und ihr Verhältniß zur ersten. 15. Jänner 1868 (Morgenblatt) pp. 4–5: VII.

Rozenblit, Marsha L. 1984. *The Jews of Vienna, 1867–1914: Assimilation and Identity.* New York: State University of New York Press.

Sandgruber, Roman. 2013. *Traumzeit für Millionäre.* Graz: Styria-Verlag.

Schnee, Heinrich. 1965. *Die Hoffinanz und der moderne Staat 5 – Quellen zur Geschichte der Hoffaktoren in Deutschland.* Berlin: Duncker & Humblot.

Schorske, Carl E. 1980. *Fin-de-Siècle Vienna—Politics and Culture.* New York: Alfred A. Knopf.

Slokar, Johann. 1914. *Geschichte der österreichischen Industrie und ihrer Förderung unter Kaiser Franz I.* Wien: F. Tempsky.

Soukup, Rudolf Werner. 2004. *Die wissenschaftliche Welt von gestern.* Wien: Böhlau Verlag.

Statistik der Wiener hochadeligen Salons im Vormärz. 17. November 1867 (Morgenblatt) pp. 4–5: IV.

Vasili, Paul Graf. 1885. *Die Wiener Gesellschaft.* 2nd ed. Leipzig: Verlag von H. Le Soudier.

Wachstein, Bernhard. 1936. Die Wiener Juden in Handel und Industrie. In *Quellen und Forschungen zur Geschichte der Juden in Österreich 11.* Wien: Historische Kommission.

Wistrich, Robert S. 1994. *The Jews of Vienna in the Age of Franz Joseph.* Oxford: Oxford Centre for Hebrew and Jewish Studies.

Wurzbach, Constant v. 1872. *Biographisches Lexikon des Kaiserthums Oesterreich 23.* Wien: k. k. Hof- und Staatsdruckerei.

3 From the Aquarium to the Zoo to the Lab: Preludes to the Biologische Versuchsanstalt in the Viennese Wurstelprater

Klaus Taschwer

In 1908 Hans Przibram, the founder and funder of the Institute for Experimental Biology (Biologische Versuchsanstalt, or BVA) in Vienna, published a comprehensive report on the first five years of his new institution. On more than 100 pages and in six chapters the zoologist gave an overview of the achievements of his new biological research institute. The account, which was published in the German scientific journal *Zeitschrift für biologische Technik und Methodik*, began with a description of the BVA's broad biological research program followed by a detailed account of its new infrastructure. The third and largest part of the report focused on the "living material and its keeping": On more than 30 pages Przibram meticulously described how the institute kept its live stock (Przibram 1908/09, 329–362).

Paul Kammerer, the institute's zoological assistant, provided the data for this chapter—a list of species that had been raised in the institution. Kammerer's tabulation, 24 pages long, included a total of 738 species from all classes of animals: for example, 75 species of protozoa, 255 species of arthropods, 101 species of molluscs and tunicates (such as sea squirts), 73 fish species, 26 snake species, 24 species of turtles, 69 amphibian species, 47 reptile species (such as various lizards and geckos), 7 species of birds, and 23 species of mammals. The zoologists and the technical staff were able to breed 297 of these species (Przibram 1908/09, 410–433). Such numbers would be quite normal for a zoo, but are certainly unique for an institute of experimental biology even a century ago. All other institutes comparable to the BVA were much smaller and therefore limited to fewer species. Why did the BVA start off with a zoo-like variety of species instead of just experimenting with a few model organisms? What was the rationale for keeping and breeding such an immense variety of species in artificial environments?

Przibram's first public announcements after the official opening of the BVA in December 1902 (Przibram 1903) and his report five years later name a few experiments and institutes that inspired the design of the new research facility in Vienna. But the infrastructure Przibram, his two codirectors, Wilhelm Figdor and Leopold Portheim, and their coworkers created was unrivalled in comparison to the facilities existing in Europe and in the United States at the time. The innovations of the new Institute for Experimental

Biology were manifold: it was the first biological research station that was not situated close to the sea or another natural habitat, but created its own environments in a highly artificial environment. The BVA pioneered the interdisciplinary biological research of botany, zoology, physiology, and related fields such as chemistry under one roof together with equipment and instruments that were avant-garde. These specificities together with the impressive unique variety of species were requirements for the ambitious plans to tackle virtually all the major questions in biology. The institute also hosted a museum of *Entwicklungsmechanik* (developmental mechanics), designed not only as a scientific collection but also for a broader public. This exhibition displayed all kinds of zoological abnormities and monstrous animals, which were created by breeding or regeneration. And the BVA was also a local innovation: it was Austria's first privately funded research institute dedicated to basic research.

To fully acknowledge these defining characteristics of the BVA, it is not enough to reconstruct the pre-1900 landscape of experimental biology institutes. We also have to understand the local particularities of the Viennese context: that is, the location of the BVA in Vienna's amusement park.

"Prater No. 1": The Aquarium of the 1873 World Exposition

The starting point for this short prelude to the BVA is the year 1873—just one year before its founder, guiding spirit, and sponsor Hans Przibram was born in Vienna. In that year the World Exposition took place in Vienna's Prater, a large public park in the city's second district that included a huge amusement park, the so-called Wurstelprater (figure 3.1). In this huge economic and cultural spectacle, 50,000 exhibitors from 35 countries, in 194 pavilions, showcased each nation's latest achievements (Pemsel 1989; Kos and Gleis 2014). One of the most expensive investments for the exposition was a pompous building, constructed just off the pavilions in the southern part of the amusement park, in the direction of the Prater Hauptallee, the main avenue crossing the Prater. The popular German zoologist Alfred Brehm (1829–1884), better known as the author and namesake of his multivolume magnum opus *Brehms Tierleben* (Brehm's Life of Animals), was hired as a planner for the building, whose address was Prater No. 1. Brehm had already been founder (and director) of the Berlin Aquarium Unter den Linden, which he headed from 1869 to 1873. The Viennese "fish palace," as newspapers dubbed the building, was constructed in a neo-Renaissance style and opened on June 29, 1873, less than two months after the start of the World Exposition (figure 3.2).

The construction—with its four characteristic towers and spacious hall in the basement—was much more expensive than expected, and the shareholders (mainly local industrialists) wanted to see revenues as soon as possible. This plan initially seemed to work: "Europe's largest aquarium" (figure 3.3) was stormed by visitors right after its

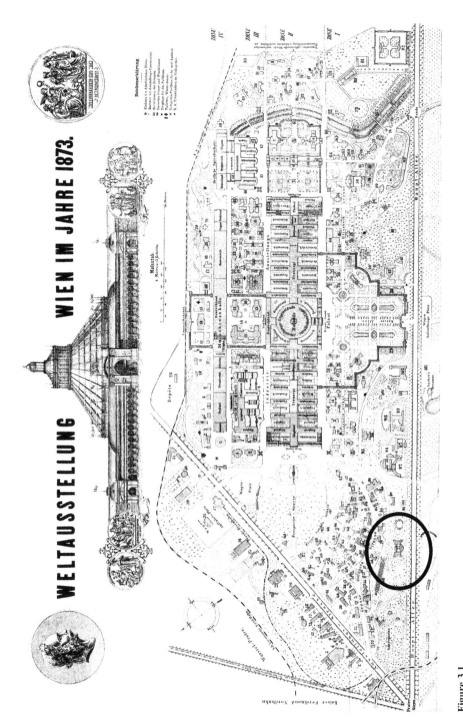

Figure 3.1
Layout of the 1873 Vienna World Exposition area and the Wurstelprater, with the later BVA building indicated (circle). Reproduced with permission from ÖAW Brochure.

Figure 3.2
The Aquarium building in 1873. From *Allgemeine Illustrierte Weltausstellungs-Zeitung*, July 27, 1873 (No. 7), 10. Reproduced with permission from ÖAW Brochure.

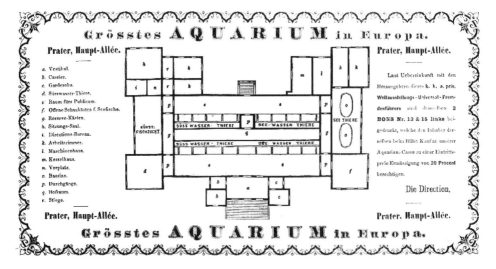

Figure 3.3
The "Largest Aquarium in Europe" and its layout in 1873. Reproduced with permission from ÖAW Brochure.

opening (Knauer 1907, 5). A tour led visitors from the magnificent vestibule to the fresh-water area, and on to the equally large marine department. The sixteen large tanks (3 meters long, 1.7 meters wide, and 1 meter deep), behind glass panes that were about 10 millimeters thick, hosted familiar freshwater fish such as eels, catfish, carp, and tench. The tanks also displayed lesser-known sea creatures including polyps, starfish, crabs, spider crabs, mussels, and snails in many shapes and colors, freshwater and saltwater plants, even sharks and a torpedo. In addition, olms swam in their cave-modeled basins. Large vats contained sea turtles and crocodiles.

The first scientific director of the aquarium was Eduard Graeffe, a Swiss-born zoologist, who later assisted in the establishment of the Zoological Research Station in Trieste. The Polish zoologist Szymon Syrski (1829–1882), who was then director of the natural history museum in Trieste and influential in Sigmund Freud's early studies on the sexual organs of eels (Gicklhorn 1955), was responsible for the procurement of the animals, which came from the Adriatic Sea. Syrski also wrote a comprehensive guide for the exhibition and in his introduction stressed the importance of zoos and aquariums for public education and even for the economy. He felt the benefits of aquariums were particularly significant (Syrski 1874, 2–3). And in fact, for Syrski the commitment to popular scientific education paid off. He was knighted for his efforts to populate the aquarium with fish and people.

But the World Exposition in Vienna and the aquarium seemed to be under a dark cloud: Only a few days after the opening, the Black Friday at the Vienna Stock Exchange put a major damper on the rapid economic growth as well as on the euphoria of liberalism. To make matters worse, a cholera epidemic hit Vienna in 1873. In November of that year, the exhibition ended in financial fiasco. Instead of the hoped-for 20 million visitors, only 7.2 million saw the Vienna World Exhibition, generating a deficit of almost 15 million guilders—a sum that would amount to more than 200 million US dollars today (Kos and Gleis 2014, 572).

The Vivarium: Between Education and Spectacle

After the World Exposition closed in November 1873, the future of the aquarium, which had been trying to find a compromise between enlightened science communication and the attractions of an amusement park, was at stake. After several years under changing ownership and the constant threat of bankruptcy, the aquarium was finally bought by Adolf Bachofen in 1888 (Pemmer and Lackner 1935, 103). Bachofen, the mayor of the village of Nussdorf near Vienna and the president of the Ornithological Society, hired Friedrich Knauer—a prolific zoologist, explorer, and communicator of science from Graz—as the new scientific director. One of Knauer's first initiatives was to rename the building. Having been known as the "Aquarium" for fifteen years, the building at Prater No. 1 was now the

"Vivarium," the name written in golden letters above the entrance. The new name signified that this building would house, and display, much more than just fish.

After a short time the former aquarium was, in keeping with its new name, transformed into a kind of indoor zoo, where a number of different mammals were shown—three orangutans and a chimpanzee—as well as different species of birds, including domestic vultures and eagles. Ungulates, rodents, and bears of all kinds were displayed in the former giant aquarium, and even large cats such as lions, leopards, cougars, and a panther. Parts of the aquarium's fish-tank corridors became the "dark rooms" of the Vivarium (figure 3.4). "The enormously thick glass walls keep both the animals and their smell away from the beholder," noted the 24-year-old zoologist Franz Werner (1892, 23).

However, even with its approximately 1,000 specimens this animal show was not enough for Knauer, the ambitious science communicator. In 1891, he harbored high-flown plans to build a giant zoo in the Prater; this would be a modern challenge to the imperial menagerie at Schönbrunn on the western outskirts of the city of Vienna (Knauer 1891). The Wiener Tiergartengesellschaft (Vienna zoo society) was constituted in 1893; it acquired the Vivarium and opened a zoo in August 1894. But the competition for visitors to the Prater grew rapidly in those years: In 1895 one of the first theme parks in the world opened nearby, the show "Venice in Vienna," in which Venice was reconstructed on a small scale.

Fig. 1.   Erſter Dunkelſaal des Wiener Vivariums.

Figure 3.4
The "dark room" of the adapted Aquarium building (from Knauer 1907). Reproduced with permission from ÖAW Brochure.

Between 1896 and 1897 the 65-meter-high Ferris wheel was erected in the Prater and soon became one of the city's landmarks.

The zoo society took countermeasures to increase revenue by offering more spectacles, especially by displaying the so-called "human zoos" at the site of the Tiergarten. In the summer of 1896, the infamous show "African Gold Coast and its Inhabitants," by Carl Hagenbeck, a German animal merchant and controversial pioneer in displaying humans next to animals, drew sustained attention and was a ubiquitous subject of conversation. Following its enormous success, a similar exhibition was shown the following year (Schwarz 2001; Hagenbeck 1948, 70–71). In 1897 the Vivarium was advertised as housing the "largest collection of reptiles in the world" with animals also provided by Hagenbeck. But when the public snake feeding was banned, the Vivarium lost its main attraction. In 1898 and 1899, new spectacles were offered at the former aquarium, featuring a lion tamer named Ella Frank and a certain Luigi Moglio with his four trained monkeys (Pemmer and Lackner 1935, 103).

Despite many efforts in those years to revive the place with exotic freshwater fish in saltwater aquariums, boas by Hagenbeck, different types of alligators, geckos, iguanas, turtles, and several other reptiles and amphibians, the Vivarium again faced numbered days in the fall of 1900. On October 1 it was closed and the heating was turned off, which spelled death for most of the tropical animals. The zoologist Franz Werner, a careful observer of the Vivarium, feared that in the future even more spectacle could be necessary to secure its survival (Werner 1901, 4–5): "Even the fiercest optimists must have realized that in Vienna, as in Munich, a zoo relying on a paying audience can not exist without 'fun.' That is why one has to ask for the Vivarium: Where is the huge 'fun' for 1901?"

But Werner, now a lecturer in zoology at Vienna University, had come up with another alternative. The heads of the biological institutes of Vienna University were to try to take advantage of the Vivarium and gain influence over its management: "first, in that they encourage, advocate, and mediate the purchase of rare and biologically interesting species, and would then be in a position to breed such animals in the large, available rooms equipped with heating." The animals should be made available for physiological studies, including the dissection of cadavers, in adapted rooms (Werner 1901, 4).

During the year 1901, the Vienna Zoo Society finally went bankrupt. The imposing Vivarium, with its four characteristic corner towers, stood empty. Given its proximity to the Wurstelprater, any serious zoological project carried out there had practically become a caricature of itself—and then still failed. Each of the entrepreneurs was compelled "to flatter the sensationalism of the crowd, because he was not able to exist by their hunger for education," as the journalist Friedrich Lorenz put it. "Even the noblest entrepreneur was demoted to an owner of a show booth, the most learned researcher degraded to a kind of carnival barker" (Lorenz 1952, 126). But in 1902 fortunes were to change. Nevertheless, this prelude, and its unique location, left distinctive marks on the institute of experimental research: in its efforts to popularize biological knowledge, in a certain sensationalism that survived in some of the research, and in very many of the early zoological experiments, which were based on the successful keeping and breeding of a multitude of different species in artificial environments.

Zoology in Vienna before 1900

When the young zoologist Hans Przibram bought the Vivarium in January 1902, he had very ambitious plans. Together with his cofounders, Wilhelm Figdor and Adolf Portheim, both botanists and both—like Przibram himself—from wealthy Jewish families, he soon would convert the nearly 30-year-old institution into a pioneering research facility of experimental biology. The founder and funder of the venture gave his own version of his motivations for establishing this unique institution and creating its singular space for biological research, namely, to tackle all biological questions by experiments with living organisms, as proposed by Wilhelm Roux (Przibram 1903; Przibram 1908/09; Przibram 1935). But to fully understand the scientific as well as many other prerequisites of this endeavor, we also have to go beyond these self-proclaimed intentions and examine both the general and the specific context of this unique institute.

One overarching precondition was the impressive boom in both science and technology taking place at the end of the nineteenth century in both Europe and the United States, fueled by the rise of the experimental method in many disciplines. In 1902, when the Institute for Experimental Biology was founded, Eduard Suess, an internationally renowned geologist and president of the Imperial Academy of Sciences in Vienna (Kaiserliche Akademie der Wissenschaften in Wien), looked back on the biggest changes during his scientific career and came to a simple conclusion: "During these 44 years, much has happened on earth, but nothing has been so pervasive and so crucial for the whole culture of the human race as the progress of the natural sciences in this period. They have penetrated every area of human life and work; they influence and change our social relations, our philosophical views, economic policy, the power of the states, everything" (Suess 1916, iv).

The institutional foundations for scientific progress in Austria during the second half of the nineteenth century had been laid around the time of the 1848 revolution. In 1847 the Academy of Sciences was founded, and two years later a new provisional law provided the universities with their first degrees of freedom from the throne and the church. What is more, the reform of 1849 provided a revaluation of the faculty of philosophy (including both the natural sciences and the humanities), which was equated to the three formerly superior faculties of theology, law, and medicine. As part of this reform, the first chair of zoology was established at the University of Vienna in 1849. Two years later the k. k. Zoologisch-Botanische Gesellschaft (zoological-botanical society) was founded in Vienna.

In the height of the liberal era, from 1861 to 1903, there were even three professorships for zoology, and from 1873 to 1883 three separate zoological departments (Salvini-Plawen and Mizarro 1999; Salvini-Plawen 2009). In this phase another important milestone was realized in the history of the life sciences in late Habsburg Monarchy: the foundation of the k. k. Zoologische Station Triest, a zoological marine station, in 1875. Modeled after the zoological station of Naples, it was co-led by the Viennese professor for zoology Carl Claus, an early and strong supporter of Charles Darwin's and Alfred Russell Wallace's new theory of evolution. One of the earliest tasks of the German-born Claus was to supervise the first scientific research by a medical student named Sigmund Freud: the nineteen-year-old spent the spring of 1876 in Trieste rather unsuccessfully dissecting 400 eels in search of their testicles, a study strongly influenced by Szymon Syrski, the director of the natural history museum in Trieste (Gicklhorn 1955).

Claus's assistants Karl Grobben and Berthold Hatschek took over the two chairs for zoology at Vienna University in 1893 and 1896 and continued the tradition of their mentor with evolutionary and comparative anatomical studies of animals. They reconstructed pedigrees of extant animal species and constructed phylogenetic relationships between extinct species to clarify Darwin's theories. But they worked exclusively with dead material, for several reasons: A trivial one was the fact that it was forbidden to keep living animals, and thus pursue experimental zoology, in the new main university building, which was opened in 1884 on the Ringstraße, the newly built boulevard around the inner city (Salvini-Plawen and Mizarro 1999, 29).

Two more Viennese particularities have to be taken into account when considering why the new experimental approach, already practiced in a few zoological institutes in Germany, could not gain a foothold at Vienna University before 1900. On the one hand, the two professors of zoology, Hatschek and Grobben, were rather old-fashioned scholars and not particularly keen to take up new methodological developments. Hatschek was more familiar with the new experimental method than Grobben; this was reflected in his participation in the marine biological research station in Trieste. In addition, he supported his former student Przibram with the BVA at least as well as possible.

On the other hand, the faculty of philosophy suffered from massive underfunding around 1900. One of the consequences was the abolition of the third chair of zoology in 1903; another was a memorandum by scientists of the University of Vienna in 1902. The main criticism stated in this *Denkschrift* concerned the inferior equipment of their laboratories, which made it impossible to appoint renowned chemists for a vacant chair of chemistry: the candidates, mainly from Germany, did not want to come to Vienna because of poor working conditions.

According to the memorandum, the infrastructure of the physics laboratories at the University of Vienna was the worst of all the physical laboratories of the monarchy, and "anyone representing an institute of the University of Vienna, who has participated at the Naturforschertag 1894 in Vienna, will not forget the humiliation that came over him during the visits of the institutes by the foreign guests" (Universität Wien 1902, 18–19). In other words, if a researcher wanted to perform biological experiments on living animals in Vienna after 1900, he or she had to set up a private laboratory, for financial and many other reasons—as Hans Przibram did in 1902.

Experimental Biology and the Crisis of Darwinism

Twenty-year-old Hans Przibram was one of the participants in the Naturforschertag, the 1894 scientific meeting in Vienna. Presumably, the young student had not yet recognized the sorry state of laboratories and experimental infrastructure at his own university. But one of the lectures at the meeting was to be decisive for his future life, as he remembered 40 years later. Przibram listened to Wilhelm Roux's presentation on developmental mechanics and immediately decided to follow this new direction of research (Przibram 1935, 1). The German zoologist and embryologist Wilhelm Roux, who taught from 1889 to 1895 at the University of Innsbruck, was one of the protagonists of experimental zoology and embryology. In the same year of the meeting in Vienna he founded the journal *Archiv für Entwicklungsmechanik der Organismen*, which became the main forum for the nascent field of experimental biology. A large proportion of the investigations that were carried out at the BVA were published in Roux's journal, which was edited by its founder for 30 years.

Roux's own investigations were performed mainly on frogs' eggs to study the earliest structures in amphibian development. One of his goals was to show Darwinian processes at work on the cellular level. His mechanistic program and his reductionist conclusions were very controversial but at the same time highly influential. Together with his colleagues and competitors such as Hans Driesch, Oskar Hertwig, and Jacques Loeb, Roux was able to establish the new experimental methodology. The experiments helped to provide new insights into the early development of organisms, but they also served another important purpose around 1900: giving more detailed insight into the mechanisms of inheritance, thus leading to confirmation or refutation of the theory of evolution put forth by Charles Darwin and Alfred Russel Wallace.

By several years before the rediscovery of the works of Gregor Mendel in 1900, biologists had begun to challenge one of the central tenets of Darwin's theory of natural selection. Among other concerns, critics questioned the claim that speciation occurs only through random processes, which seemed to contradict the pace that led to the evolution of new species in a relatively short time. Authors such as George John Romanes, Ludwig Plate, and Vernon Kellogg published influential books asserting that alternative mechanisms would have to be taken seriously (Bowler 1983, 4). In particular, environmental adaptations were discussed, as already proposed by the French naturalist and philosopher Jean-Baptiste Lamarck 50 years before the publication of Darwin's epoch-making masterpiece. The debates in the years between the early 1890s and the late 1930s were later dubbed the "eclipse of Darwinism" (Bowler 1983, quoting the term from Julian Huxley), as if Darwin's evolutionary theories had obscured by the speculations of anti-Darwinists. But this misrepresents the fact that many influential and pro-Darwin biologists took part in these lively discussions (Largent 2009).

Vienna around 1900 was one of the centers where this "crisis of Darwinism" was discussed. In the academic year 1901/02, a series of lectures by prominent researchers took place in the Philosophical Society of Vienna University. The series was opened by a controversial address by Max Kassowitz, professor of pediatrics at Vienna University, who distanced himself "definitely and without reservation" from Darwin's theory of natural selection (Kassowitz 1902, 11), praising and embracing Lamarck instead. Kassowitz had published a whole book on that topic already in 1899, defending Darwin's later Lamarckian views against August Weismann's strict neo-Darwinism (Kassowitz 1899). The second speaker was the highly respected botanist Richard von Wettstein, who pointed out that for the time being, more botanists would be inclined toward Lamarckism than toward Darwinism. Wettstein himself saw the theories of Darwin and Lamarck "not as mutually exclusive opposites, but as complementary theories" (Wettstein 1902a, 24, 32).

But how did evolution and evolutionary change really work? And what were the mechanisms that led to speciation? Wettstein in particular hoped that experimental biology could answer these big questions. Consequently, he was one of the most prominent public supporters of Hans Przibram's plans to found an institute for experimental biology (Wettstein 1902b). The early research program of the BVA clearly reflected the theoretical debates on Darwinism that were ongoing in Vienna and many other places (see chapter 8 in this volume). Experiments could be expected to prove whether the inheritance of acquired characteristics, as proposed by Lamarck, was a viable proposition. And for that kind of research, the keeping and well-being of animals over many generations and years would be a critical prerequisite.

Aquariums and Amateur Science

In addition to the situation of Vienna University around 1900, the rise of experimental biology, and the demanding theoretical questions, another factor that aided in the founding of the BVA must be taken into account. It is a phenomenon that brings us back to the location of institute: the popular explosion of aquariums and animal fancying before 1900, especially in Germany (Nyhart 2009; Reiß 2012). By the middle of the nineteenth century, maritime organisms were on exhibit in the earliest public aquariums—the first of these at the World Exhibition in London in 1851. This was the beginning of the "discovery of the sea," both in popular culture and in science. Writers such as the historian Jules Michelet (*The Sea*, 1861), Jules Verne (*Twenty Thousand Leagues under the Sea*, 1869/70), and

Victor Hugo (*Toilers of the Sea*, 1866) explored the unknown depths of the sea. A few years later the first zoological research stations were established in many European coastal towns. The most famous one was the Zoological Station of Naples, which was opened in 1872. Thanks to its founder, the German zoologist Anton Dohrn (1840–1909), it soon became a stronghold of evolutionary and embryological studies. And it is certainly no coincidence that Hans Przibram founded his own institute right after a long research stay in Naples.

But there was another way to approach the world of marine animals. From the 1880s onward, amateurs in Germany founded clubs for aquariums and terrariums, starting with a club simply named Aquarium, which was established in 1882 in the Thuringian town of Gotha. By 1900, the number had grown to 31, including the club Triton in Berlin, founded in 1888, which soon became important for the rest of Germany because its members edited the widespread organ *Blätter für Aquarien- und Terrarienkunde*. In the following years similar clubs opened in Switzerland, in New York, and finally in Vienna in 1895. The name of the Viennese association was Lotus; its members were amateurs, animal lovers, but also professionals such as the zoologist and herpetologist Franz Werner from the University of Vienna. One of the youngest members was Paul Kammerer, who started his studies of zoology at the University of Vienna in the fall of 1899 and was also a member of the club Triton in Berlin (Reiß 2012).

When in 1901 the fiftieth anniversary of the zoological and botanical society was celebrated, all the notable figures from academia were present. Even Wilhelm von Hartel, the minister of education, and Eduard Suess, the president of the Academy of Sciences, addressed the audience. The last speech documented was given by Paul Kammerer, representing the club Lotus. The young student stressed the "intimate relationship" between amateurs and scientists: on the one hand, the amateurs looked up to the experts with reverence and tried to get close to them; but on the other hand, aquarium and terrarium keepers in particular would be capable of providing significant assistance to the scientists, for instance by keeping the living material alive as long as possible. Sometimes the aquarium and terrarium amateurs would even "go ahead in research," as had happened when the interaction between the animal and plant kingdoms was discovered in an aquarium (Kammerer 1901).

Only a few weeks after the celebration, on May 5, a long article appeared in the newspaper *Neues Wiener Tagblatt*, describing Kammerer's menagerie in his parents' apartment in Karlsgasse as a unique collection, including more than 200 different species of animals from all over the world. Probably by way of this impressive report, Hans Przibram became aware of Kammerer's talents and offered the young amphibian and reptile expert a job at his new institute. Kammerer took the opportunity, and he contributed to the success of the institute in its first years with his vast experience as an animal keeper and breeder; one might suspect that the abundance of species kept at the BVA was Kammerer's idea.

It was no coincidence that the members of the association Lotus were the first who learned of the dissolution of the "old" vivarium and the establishment of a new research institute in the same place. The club had already discussed the new foundation by February 1902, and there were even plans to integrate Lotus's activities in the research institute. In 1904 Przibram wanted the member of the club to exhibit their aquariums in an open space within the institute. Although the project failed, there were many other initiatives at the BVA—such as the museum of developmental mechanics and the guided tours for amateurs, but also the spectacular findings themselves—that paid tribute not only to the genius loci at the Wurstelprater but also to the close interrelationship of amateur science and experimental biology around 1900.

The main mediator between those two fields was again Paul Kammerer, as both a passionate animal lover and a gifted experimental biologist with a public mission. On various occasions, Kammerer stressed the mutual relevance of the two practices and tried to reconcile professional biologists and amateurs. In a talk given at the club Triton in Berlin in 1905, he highlighted the affinities between animal fancying and experimental biology. The successful keeping and breeding of animals over a long time would be a precondition for the study of the plasticity of phenotypes or behavior but also of mechanisms of inheritance (Kammerer 1905; Wessely 2013; Klausner 2015). Kammerer repeated this argument in numerous publications and even in his long experimental research reports (e.g., Kammerer 1913, 153).

Przibram's young assistant at the BVA tried on various levels to bridge the gap between professional research and the hobbies of amateurs. In July 1908 he became editor of the (nonscientific) magazine *Blätter für Aquarien- und Terrarienkunde*. To give it a more scientific appeal, Kammerer tried to persuade notable scientists such as Richard von Wettstein and his son to contribute to the journal. And in a letter to Wettstein (from July 8, 1908)[1] Kammerer also sketched his plans to found a new section for the study of terrariums and aquariums within the k. k. Zoologisch-Botanische Gesellschaft.

But it was too late to reconcile two fields that already had begun drifting apart: most of these plans failed. One of the consequences of this separation of the amateur keeping and breeding of animals from experimental biology was that the "zoo" of the BVA remained unique in the history of biology. No other institute for experimental biology would keep and breed hundreds of different species instead of restricting itself to just a few model organisms. Another consequence was that most of Kammerer's experiments, which took years of diligent care of animals, could not be and probably never will be repeated.

Post Scriptum: From the Lab Back to the Aquarium

In a tragic irony, the place ended as it had begun. In June of 1932, a public aquarium was reopened at the institute, mainly because of its deep financial crisis. The visitors' admission fees were meant to contribute to the shrinking budget of the BVA and help the researchers to continue their work. Long after Hans Przibram, his codirectors, and most of their collaborators had been expelled by the National Socialists and the research institute had been abandoned, the place was still used as a public aquarium for entertainment. In January 1945 the press reported on the aquariums in the Vivarium for the last time—more than 70 years after the building had opened as the "fish palace" and just three months before bombs irreparably destroyed the unique birthplace of modern biology in Austria.

Note

1. Archiv der Universität Wien, Nachlass Richard von Wettstein (131.32).

References

Bowler, Peter J. 1983. *The Eclipse of Darwinism. Anti-Darwinian Theories in the Decades around 1900.* Baltimore, London: Johns Hopkins University Press.

Gicklhorn, Josef. 1955. Wissenschaftsgeschichtliche Notizen zu den Studien von S. Syrski (1874) und S. Freud (1877) über männliche Flußaale. *Sitzungsberichte der mathematisch-naturwissenschaftlichen. Klasse, Abteilung I* 164 (1–2): 1–24.

Hagenbeck, Carl. 1948. *Von Tieren und Menschen. Erlebnisse und Erfahrungen.* Leipzig, Munich: Paul List Verlag.

Kammerer, Paul. 1901. Rede in der Festversammlung zur Feier des 50-jährigen Bestandes der K. k. zoologisch-botanischen Gesellschaft. *Verhandlungen der K. k. zoologisch-botanischen Gesellschaft zu Wien* 51: 260.

Kammerer, Paul. 1905. Die Aquarien- und Terrarienkunde in ihrem Verhältnis zur modernen Biologie. *Blätter für Aquarien- und Terrarienkunde* 15 (9, 10): 83–86; 94–96.

Kammerer, Paul. 1913. Vererbung erzwungener Farbveränderungen. Vierte Mitteilung: Das Farbkleid des Feuersalamanders (Salamandra maculosa Laur.) in seiner Abhängigkeit von der Umwelt. *Archiv für Entwicklungsmechanik* 36 (1–2): 4–193.

Kassowitz, Max. 1899. *Vererbung und Entwicklung.* Volume 2 of *Allgemeine Biologie.* Vienna: Verlag von Moritz Perles.

Kassowitz, Max. 1902. Die Krisis des Darwinismus. *Wissenschaftliche Beilage zum 15. Jahresbericht der Philosophischen Gesellschaft an der Universität Wien*, 5–18. Leipzig: Barth

Klausner, Ursina. 2015. "Kein Gelehrter mit Scheuklappen." Paul Kammerer und die Popularisierung der Biologie. *Spurensuche* 23/24:99–112.

Knauer, Friedrich Carl. 1891. *Zur Gründung eines großen zoologischen Gartens in Wien.* Vienna: 1. Wiener Vereinsbuchdruckerei.

Knauer, Friedrich Carl. 1907. *Das Süßwasser-Aquarium. Seine Herstellung, Einrichtung, Besetzung und Instandhaltung. Regensburg: Verlagsanstalt vorm.* München, Regensburg: G. J. Manz, Buch- und Kunstdruckerei A.-G.

Kos, Wolfgang, and Ralph Gleis, eds. 2014. *Experiment Metropole. 1873: Wien und die Weltausstellung.* Vienna: Czernin Verlag.

Largent, Mark A. 2009. The So-Called Eclipse of Darwinism. In *Descended from Darwi: Insights into the History of Evolutionary Studies, 1900–1970*, ed. Joe Cain and Michael Ruse, 3–21. Philadelphia: American Philosophical Society.

Lorenz, Friedrich. 1952. *Sieg der Verfemten. Forscherschicksale im Schatten des Riesenrades*. Vienna: Globus.

Nyhart, Lynn K. 2009. *Modern Nature: The Rise of the Biological Perspective in Germany*. Chicago: University of Chicago Press.

Pemmer, Hans, and Ninni Lackner. 1935. *Der Wiener Prater einst und jetzt. Nobel- und Wurstelprater*. Leipzig, Vienna: Deutscher Verlag für Jugend und Volk.

Pemsel, Jutta. 1989. *Die Wiener Weltausstellung von 1873. Das gründerzeitliche Wien am Wendepunkt*. Vienna, Cologne: Böhlau.

Przibram, Hans. 1903. "Biologische Versuchsanstalt" in Wien. *Österreichische Botanische Zeitschrift* 53:83–84; *Zoologischer Anzeiger* 26, 697 (April 14): 373–375.

Przibram, Hans. 1908/1909. Die biologische Versuchsanstalt in Wien: Zweck, Einrichtung und Tätigkeit während der ersten fünf Jahre ihres Bestehens (1902–1907). *Zeitschrift für biologische Technik und Methodik* 1:234–264, 329–362, 409–433; Supplement: 1–34.

Przibram, Hans. 1935. Unpublished manuscript, May 9 (AÖAW, Biologische Versuchsanstalt [Vivarium], box 4, file 4).

Reiß, Christian. 2012. Gateway, instrument, environment: The aquarium as a hybrid space between animal fancying and experimental zoology. *NTM Zeitschrift für Geschichte der Wissenschaften. Technik und Medizin* 20 (4): 309–336.

Salvini-Plawen, Luitfried. 2009. Die Zoologie in der Habsburger-Monarchie. *Mitteilungen der Österreichischen Gesellschaft für Wissenschaftsgeschichte* 27:63–82.

Salvini-Plawen, Luitfried, and Maria Mizarro. 1999. 150 Jahre Zoologie an der Universität Wien. *Verhandlungen der Zoologisch-Botanischen Gesellschaft in Österreich* 136:1–76.

Schwarz, Werner Michael. 2001. *Anthropologische Spektakel. Zur Schaustellung "exotischer" Menschen in Wien 1870–1910*. Vienna: Verlag Turia + Kant.

Suess, Eduard. 1916. *Erinnerungen*. Leipzig: S. Hirzel.

Syrski, Simon. 1874. *Wiener Aquarium*. Vienna: Selbstverlag des Aquariums.

Universität Wien. Philosophische Fakultät. 1902. Denkschrift über die *gegenwärtige Lage der Philosophischen Fakultät der Universität Wien*. Vienna: Adolf Holzhausen.

Werner, Franz. 1892. Das Vivarium in Wien. *Der Zoologische Garten* 33 (2): 22–26.

Werner, Franz. 1901. Noch einmal das Vivarium in Wien. *Der Zoologische Garten* 42 (1): 1–5.

Wessely, Christina. 2013. Wässrige Milieus. Ökologische Perspektiven in Meeresbiologie und Aquarienkunde um 1900. *Berichte zur Wissenschaftsgeschichte* 36:128–147.

von Wettstein, Richard. 1902a. Die Stellung der modernen Botanik zum Darwinismus. *Wissenschaftliche Beilage zum 15. Jahresbericht (1902) der Philosophischen Gesellschaft an der Universität Wien*, 19–32. Leipzig: Barth.

Wettstein, Richard von. 1902b. Österreichische Biologische Stationen. *Neue Freie Presse*, August 24: 14–15.

4 The Biologische Versuchsanstalt in Historical Context

Johannes Feichtinger

By the nineteenth century, Vienna had become a hot spot for comparative biological research based at the Botanical Garden, the Schönbrunn Menagerie, the Natural History Museum, and the institutes of biology at the University of Vienna. The experimental research carried out by the zoologists Karl Claus (1835–1899), Berthold Hatschek (1854–1941), and Karl Grobben (1854–1945) and by the botanists Julius Wiesner (1838–1916) and Richard Wettstein (1863–1931) produced results that improved the systematic classification of species by describing differences and similarities. Organic development was explored using comparative and historical methods. Since the University of Vienna in particular lacked sufficient laboratory facilities, scientists' research programs did not include efforts to gain a better understanding of organic development directly, through interventional experimentation (i.e., controlled changes in environmental conditions) (Müller and Nemeschkal 2015; Kühnelt 1985). In 1935, Hans Przibram recalled that "back then there was no institute dealing with living animals. At these institutes, comparative studies were carried out with dead objects, genealogy studies of the now-living based on the dead [of earlier epochs]."[1]

These deficits were overcome on the initiative of three young biologists trained at the University of Vienna. On January 1, 1903, the 28-year-old zoologist Hans Przibram (1875–1944), together with Wilhelm Figdor and Leopold von Portheim, officially opened the Institute for Experimental Biology (Biologische Versuchsanstalt, or BVA; Logan and Brauckmann 2015; Taschwer 2014a; Przibram 1959). At the suggestion of his mentor Berthold Hatschek, Przibram in 1902 purchased the former Vivarium building directly opposite the Giant Ferris Wheel in the Vienna Prater, the city's most popular recreation and amusement area, and transformed it into one of the world's leading biological research centers (see Taschwer et al. 2016).

The Vivarium had been built on the occasion of the Vienna World Exhibition of 1873. Afterward, it was mainly used as an aquarium and for exhibiting exotic animals (see chapter 3). The founders of the BVA provided the building with technical facilities (e.g., self-constructed instruments for controlling temperature, humidity, pressure, and light levels) for keeping live animals and plants under stable conditions that could be modified for experimentation (see Coen 2006 and chapter 7 in this volume). Przibram explained the institute's purpose in an address given in September 1902 at the 74th Assembly of the Gesellschaft deutscher Naturforscher und Ärzte (Society of German Naturalists and Physicians): "This institute, founded and directed by the botanists Wilhelm Figdor and Leopold Ritter von Portheim and myself, should serve for studying the newest branches of biology, experimental morphology, evolutionary physiology, and the bordering area of biochemistry and biophysics" (Przibram 1903, 153).

Their style of research provided the groundwork for the shift from a descriptive, historical, and comparative approach to a testing method through which—it was hoped—organic development would be decoded. By carrying out biological experimentation, the BVA sought to respond to the "big questions in biology," for example, the mechanisms of formation (*Formbildung*), or the impact of external influences on evolution. The laboratory in the Prater broke new ground in funding, organizing, and practicing science. It thus became a model for the advancement of experimental biology in the United States, Germany, France, the United Kingdom, and Russia around 1900. For example, the Institute for Experimental Zoology of the Moscow Zoo used the BVA as a template for its research program (Stock 1945, 26). Research was privately financed to a large extent, while additional funding was provided by the Austrian federal government and after 1914 by the Imperial Academy of Sciences in Vienna.

In the first decade, the research program was carried out independently to a considerable extent. State funding provided for the founding of a board consisting of the four chairs of zoology and botany of the University of Vienna. The board was tasked with playing an intermediary role as well as maintaining a supervisory function; after 1914, the latter duty was taken over by the Academy of Sciences. The Academy board did not intervene in the BVA's day-to-day work. The institute's groundbreaking research program was highlighted in an article published in *Nature* in 1932: "The future historian of biology will doubtless count among the more important tendencies of the first quarter of this century the extended application of the experimental method to all kinds of biological problems. More especially the phenomena of growth and form, which previous generations had, for the most part, been content to observe unfolding in *Nature*, are now studied in the modifications induced under artificially controlled conditions. In this department of research, Prof. Hans Przibram has been one of the pioneers" (W.T.C. 1932, 298–299). Earlier, Przibram's pioneering achievements had been both held in high esteem and criticized, as documented by the affair surrounding Paul Kammerer (Taschwer 2016; Koestler 2010/1971; Gliboff 2006; Hirschmüller 1991; Przibram 1926). However, already in 1910 Jacques Loeb (1859–1924) paid tribute to the outstanding achievements of Przibram's institute: "You certainly have succeeded in making it the leading site of biological research."[2]

After the *Anschluss* by Nazi Germany in 1938, the directors of the BVA and its scientific staff were forced to leave the country. Przibram and at least six other former researchers at the BVA (Leonore Brecher, Henriette Burchardt, Martha Geiringer, Helene Jacobi,

Heinrich Kun, and Elisabeth Przibram) died in Nazi concentration camps (Gedenkbuch 2015; Taschwer 2014b, 109; Taschwer 2014c). In the last days of World War II, the building was destroyed by fire. In 1945, Alexander Stock, a former BVA staff member who had found refuge in the United Kingdom, noted that he hoped postwar Austria would recreate the institute of which Przibram had stated: "We are proud that this institute was established in Austria and in Vienna as the first one in the world" (Stock 1945, 28).

Historical research on fin-de-siècle Vienna has shown how modernism came into being (Schorske 1980; Csáky et al. 1996–2008; Coen 2007; Feichtinger 2010). At the turn of the century, Vienna was home to luminaries not only in the fields of the fine arts, music, and philosophy (Gustav Klimt, Egon Schiele, Gustav Mahler, Arnold Schönberg, and Ludwig Wittgenstein) but also in the field of science, including such names as Ernst Mach, Ludwig Boltzmann, Franz Serafin Exner, Erwin Schrödinger, and Sigmund Freud. Here, I will draw special attention to "scientific Vienna," centering on the Institute for Experimental Biology, which was involved in the emergence of a modern culture of experimental research in biology. This chapter traces the history of the BVA, its rise to become an outstanding center of research, its incorporation into the Imperial Academy of Sciences, and its destruction by the National Socialists.

Fin-de-siècle Vienna Science

In fin-de-siècle Vienna, the state system of scientific research was characterized by its differentiated structure (Feichtinger 2015). Besides the University of Vienna, there were state colleges for technology, economy, veterinary medicine, arts, and music, whose leading professors were elected members of the Imperial Academy of Sciences. Additionally, a number of independent institutes had been founded in the aftermath of the Austrian revolutions of 1848. They included, for example, the Imperial Geological Institute (Geologische Reichsanstalt) founded in 1849, the Central Institute of Meteorology and Earth Magnetism (Zentralanstalt für Meteorologie und Erdmagnetismus, 1851), the Austrian Institute of Historical Research (Institut für Österreichische Geschichtsforschung, 1854) and the Zoological Station in Trieste (1875). Alas, the political will to support scientific endeavors through the public sector died away after these institutes were founded. The research system consistently suffered from insufficient funding by the state government.

Nevertheless, in 1905 both the university and the Imperial Academy of Sciences in Vienna had more than a dozen private endowments at their disposal, worth at least one million crowns (Philippovich 1906, 25; Höflechner 1990).[3] The state government's structural underfunding of the University of Vienna troubled the professors of the humanities and of the natural sciences to such an extent that in 1902 they initiated a petition for sustainable funding. They published a memorandum entitled "Über die gegenwärtige Lage der philosophischen Fakultät der Universität" (On the present state of the university's faculty of philosophy) in which they complained strenuously about the almost shabby treatment of the university system and its perceived insignificance both among the Austrian public and by the international scientific community: "Back then [more than 50 years ago] Austria was 20 years ahead of its neighboring states; today it lags far behind them." The "impertinent" professors blamed the Austrian federal governments of the last decades for the decline of the national science system. From their point of view, politics had "not been willing or able to adequately acknowledge scholarship in any decisive areas; yet where this acknowledgement is lacking, the means lack too" (Universität Wien 1902, 19–24). In their memorandum the Viennese professors, in a reproachful and aggressive tone, concluded that the political sector had evaded its responsibility, leaving the government's obligation to promote science to private investors. The botanist and science organizer Richard Wettstein, who was as important for Austria as Adolf von Harnack (1851–1930) was for Germany, took a different argument. In an article in the Vienna daily *Neue Freie Presse*, he underlined the initiatives "that individuals performed for scientific endeavors also here" (Wettstein 1902a, 14), admonishing those who regarded the funding of scientific research as a task solely belonging to the state government. Meanwhile, in the United States private science funding was in its infancy: in 1902 Andrew Carnegie endowed 10 million dollars for scientific purposes, while the Rockefeller Foundation was established in 1913. Wettstein could remind the public of the huge amounts of money that had already been invested by private sponsors for the promotion of science in Austria. He referred, for example, to Joseph Treitl, Johann Graf Wilczek, and Ignaz Lieben (see chapter 2). Furthermore, Wettstein urged wealthy entrepreneurs, "our natural science institutes could develop their performance through private funding to a larger extent when in wealthy circles the conviction was more widespread that the full exploitation of scientific capabilities is of paramount importance in the big rat race between peoples and nations" (Wettstein 1902a, 14).

Wettstein's appeal for a demonstration of loyalty to the Austrian state through the private promotion of science struck a chord. In 1906, the industrial magnate Karl Kupelwieser (1841–1925) founded the Biological Research Station at Lunz am See, which he personally equipped throughout his life. In 1908, Kupelwieser donated 500,000 crowns for the establishment of the Institute for Radium Research (opened in 1910). Wettstein might have had these endowments in mind when in 1912 he published his programmatic declaration on the necessity of establishing independent research institutes, which should "provide the opportunity for in-depth research and its expansion to areas that cannot sufficiently be attended to by universities, and in this sense add to the scientific work of universities but not replace them" (Wettstein 1912, 24).

On January 1, 1914, the Biologische Versuchsanstalt was handed over to the Academy of Sciences in Vienna. Thanks to this endowment, the BVA became an institute belonging to Austria's most renowned state-run, non-university research body. The donation was preceded by intensive, rather lengthy negotiations between Hans Przibram, the other cofounders, and the Presiding Committee of the Academy of Sciences in Vienna. Three years earlier, in 1911, Przibram, von Portheim, and Figdor had come forward with the proposal to donate their research facility to the Academy. What was the strategic background to this generous offer?

The sources tell us that Hans Przibram wanted to ensure the BVA's ongoing existence. Working toward that goal, he reminded the Academy's Presiding Committee of the advantages to the Academy of running research institutes itself, as already envisaged in the neighboring German Empire. In their offer of January 5, 1911, the founders strategically noted that "the new Kaiser Wilhelm Society for the Advancement of Science in Berlin has been established to found, run, and promote research institutions for the various experimental sciences."[4] On the occasion of the centenary of the University of Berlin's founding in October 1910, Emperor Wilhelm II (1859–1941) announced the plan to "establish and maintain, besides the Academy of Sciences and university, independent research institutions as an integral part of the scientific system." As he stated: "We are in need of institutions which go beyond the framework of universities and which, unimpaired by teaching purposes but in close connection with the Academy and university, solely serve research purposes" (Jahrhundertfeier 1911, 37). The German science organizer Adolf von Harnack had recommended the establishment of further research institutions to prevent the German Reich from losing its role as the leading nation in science (vom Brocke and Laitko 1996).

Still, with respect to the establishment of non-university research institutes, the Habsburg Empire was one step ahead of the Wilhelmine Empire. In 1887, the Imperial Physical-Technical Institute was founded in Berlin, but in Vienna several extra-university research institutions had been established from the late 1840s. In the first decade of the twentieth century, three more institutes were opened: the BVA (1903), the Biological Research Station at Lunz am See (Lower Austria, 1906), and the Institute for Radium Research in Vienna (1910). The BVA was funded generously by Hans and Karl Przibram; the Biological Research Station in Lunz and the Radium Research Institute were founded and funded by the affluent Austrian businessman and lawyer Karl Kupelwieser. The hugely wealthy entrepreneurs of fin-de-siècle Vienna, such as Karl Wittgenstein (1847–1913), father of the philosopher Ludwig (1889–1951), more often promoted the arts and founded new cultural institutions (e.g., the Secession). There were some business families, however (such as the Kupelwiesers and Figdors—both families acquainted with the Wittgensteins—and the Przibrams and the Portheims), who invested large parts of their fortunes to further scientific endeavors (Sandgruber 2013). As a result of these funding initiatives in Austria, experimental research institutions were already operating "before the plans of Kaiser Wilhelm to implement independent research institutions had been realized in Germany," as Przibram, von Portheim, and Figdor argued in their generous offer to the Academy in 1911.[5] However, after 1910, the German federal government invested money in an extensive research program. Austria abstained from initiating a parallel effort: a Kaiser-Franz-Joseph-Society was never established, and the funding of research largely remained a matter of private patronage (Reiter 2014; Broda 1979).

Richard Wettstein, the Intermediator

The process of handing over the Biologische Versuchsanstalt to the Academy of Sciences took three full years. Negotiations were led by Richard Wettstein, who had been elected a full member of the Imperial Academy of Sciences in 1900. On February 8, 1912, the professor of botany at the University of Vienna published an article in the *Neue Freie Presse* promoting the establishment of "independent research institutions." The article covered a lecture given at the fourth convention of German university teachers in Dresden. Wettstein argued that experiments could not be conducted at universities on a large scale because of the "unavailability of sufficient resources, ... the large amount of time required for conducting experiments, [and] the thus needed concentration of work [, each] being incompatible with teaching at a, to some degree, larger university" (Wettstein 1912, 22).

In scientific matters, Wettstein had involved himself deeply in the debate on divergent views on evolution. Although he had placed himself in the Darwinian framework, he showed sympathy for Lamarck's theory of the inheritance of acquired characteristics. In a lecture given at the ceremonial session of the Imperial Academy of Sciences on May 28, 1902, Wettstein concluded:

I am thinking of what major importance the theory of selection associated with Darwin has gained with respect to every field of human thinking, yet it does not seem futile to me to draw attention to the fact that not only the breeding principle of selection theory makes a reconfiguration in the world of organisms possible, but that along with it there is a second principle that slowly but steadily influences the world of creatures in a particular way: the direct adaptation and the inheritance of characteristics acquired by such processes of adaptation. (Wettstein 1903; Wettstein 1902b, 333; Wettstein 1902c, 19–32)

Wettstein assumed the position of an intermediary between the adherents of Darwin and the neo-Lamarckists; he regarded the Darwinian selection theory and the Lamarckian adaptation theory not as mutually exclusive opposites but as complementary schemes of evolution. He was not the only Austrian biologist of his time who pleaded both at home and abroad (for example, at the 74th Assembly of the Gesellschaft deutscher Naturforscher und Ärzte in Karlsbad, 1902) for a reevaluation of Lamarck's legacy. Berthold Hatschek, who was both a professor of zoology at the university and a member of the Academy of Sciences in Vienna, also attributed particular importance to "the principle of direct modification of the species by the hereditary effect of functional adaptation." (Hatschek 1902, 35) Wettstein and Hatschek would each play a decisive role in the development of the BVA, whose research program was in fact partly dedicated to devising experimental proof that evolution was determined by both selection and acquired adaptation (see chapter 8 in this volume).

In his 1902 article in the *Neue Freie Presse*, Wettstein explicitly addressed the "special task" the BVA was taking up:

The biological disciplines are today governed by the theory of descent. The epochal achievements of Lamarck, both Darwins, and their numerous allies made the understanding of an evolutionary interrelatedness of all creatures a general one. However, much remains left to research when we establish not just the descendent-theoretical problem in its concrete manifestations but also wish to clarify its nature. Questions as to the reasons of formal changes, which are a prerequisite of any new constitution of a form, as to the ways that organisms are influenced by the external world, and many other questions of fundamental importance can only be answered subsequently after a precise clarification of the general factors on the basis of a comprehensive experimental program. (Wettstein 1902a, 14)

Double Affiliation: The BVA, the University, and the Imperial Academy

When Hans Przibram announced the forthcoming opening of the BVA in Karlsbad in 1902, he seemed to echo the vision that Wettstein, who himself was to benefit from the freshwater laboratory established at the BVA, had published in the *Neue Freie Presse* a month earlier: "Since the random modification of external factors, such as the surrounding medium or temperature or light, lies in our hands, we can similarly use these in order to modify forms and try to determine the influence of the individual factors on the development of forms" (Przibram 1903, 153).

As already mentioned, the understanding of the term "experiments" at the universities, museums, and biological stations established in the second half of the nineteenth century had been limited to descriptive, historical, and comparative studies. The BVA provided what was needed for a new approach to experimental research, characterized by the concept of testing the impact of external factors on evolutionary processes. In order to organize long-term test series efficiently, a particular focus was put on the structuring of the laboratory. The BVA was organized into different departments: one each for zoology, botany, and the physiology of plants; from 1907 an additional one for physical chemistry; and from 1913 another for animal physiology. On the basis of an agreement, the state government supported the BVA by maintaining four laboratory spaces: two each for botanical and zoological research. Use was limited to one year and was assigned by the university board that acted as the BVA's scientific management body. Independent researchers working under the supervision of the heads of department paid an annual fee of 1,000 crowns. From 1907 onward the government funded a further laboratory space for the Physical Chemistry Department, which was assigned to the medical faculty of the University of Vienna. This department was directed by Wolfgang Pauli Sr. (1869–1955), father of the Nobel Prize winner Wolfgang Pauli (1900–1958; see chapter 10). The Physiology

Department, directed by Eugen Steinach (1861–1944), would build a global reputation after the Great War based on its sensational sexual physiology experiments and rejuvenation surgery (see chapter 11).

Although the BVA was established as an independent research institute, it was Przibram's utmost concern from the beginning "to connect the station with the university."[6] Affiliation with the chairs of botany and zoology at the University of Vienna was ensured on the one hand by the scientific board consisting of the heads of the four biological institutes—the zoologists Berthold Hatschek and Karl Grobben, and the botanists Julius von Wiesner and Richard von Wettstein—and on the other hand by the qualification to teach courses at university level (*Habilitation*) held by Hans Przibram (since 1903), Wilhelm Figdor (since 1899), and Paul Kammerer (since 1910). They regularly taught courses at both the University of Vienna and the BVA building in the Prater, for example holding tutorials on plant physiology, experimental morphology and developmental physiology, reproduction and fertility, and so on. As discussed below, after becoming a research institute of the Academy the BVA would no longer serve as a teaching institution.

This affiliation guaranteed a government subsidy (6,000 crowns per year; 7,000 from 1907 on). Government funding was linked to annual reporting on performance and to the obligation to provide the University of Vienna "with working and teaching material from the experimental station within the limits of state subsidization."[7]

In the period between 1903 and 1913, the BVA made available to the university all the living material for the lab work undertaken by the university students of zoology and botany (Przibram 1908/09, 1913). In the first annual report on the work of the BVA, Berthold Hatschek emphasized that "the laboratory provided an opportunity for comparative physiological research, which has been lacking at the zoological institutes of the University of Vienna so far."[8]

In his report on the academic year 1905/06, Eugen Philippovich (1858–1917), pro-vice-chancellor of the University of Vienna, presented the BVA as a model case for the prolific "cooperation between the government, the university, and noble-minded individual private men." He described the institute as a "remarkable creation of three young biologists, Doctor Hans Przibram, Doctor Wilhelm Figdor, and Leopold Ritter von Portheim," stressing that they had purchased the former Vivarium building in the Prater entirely with private assets to transform it into a modern scientific institute (Philippovich 1906, 28–29). He also emphasized the substantial sum Przibram paid for the maintenance of the BVA. Hans Przibram personally invested an amount of 198,000 crowns in infrastructure during the years between 1903 and 1910; he also paid the scientific staff (among them, Paul Kammerer). An additional 12,671 crowns had been covered by the operators of the BVA. Within the same period, the public sector funded the scientific enterprise with 51,000 crowns.[9]

In 1911, the US zoologist Charles Lincoln Edwards reported from the Viennese hotbed of biological research in *Popular Science Monthly*, pointing to its "remarkable scientific productivity." Edwards had also reported from other biological stations in Europe. Comparing them, he concluded that "such institutions as this of Vienna will do much to solve the great problems of biology" (Edwards 1911, 584; see also Kofoid 1910). In the same year in which Edwards's enthusiastic report was published in the American journal, the founders of the BVA took a decisive step to assure the permanent existence of the institute. On the occasion of the announcement that new state-sponsored scientific institutes were to be established in Germany, they submitted their offer to donate the BVA to the Imperial Academy of Sciences. At that time, the Academy was the only organization in Austria running scientific research institutes outside the universities. The donation included the Vivarium building and an endowment of 300,000 crowns, the idea being that the Academy would cover the costs of both the operation of the BVA and the necessary restoration work from the return on the endowment, including income from fixed-interest securities.

The Academy of Sciences expressed its gratitude for the offer and installed a commission to investigate "whether a research institute of the given orientation meets a demand and whether there is the possibility to safeguard the institute financially in a way that any permanent financial burden on the Imperial Academy seems eliminated."[10] The scientific committee consisted of the president of the Academy of Sciences, Eduard Suess, the secretary-general Hofrat von Lang, the deep-sea researcher Franz Steindachner, the plant physiologist Julius Wiesner, the physiologist Siegmund Exner, the histologist Viktor von Ebner-Rofenstein, the botanist Hans Molisch, the zoologists Karl Grobben and Berthold Hatschek, and the botanist and science organizer Richard von Wettstein.

In the report by the scientific committee of the BVA (Vivariumskommission) drafted by Richard Wettstein, the necessity for such an institute was—hardly surprisingly—affirmed. The wording was as follows:

It is without doubt that biological experiments that require the long-term rearing of plants and animals or cultivation under conditions that are difficult to produce can hardly be carried out in the existing physiological institutes. ... In particular, the modern theories of genetics and adaptation, the problem of speciation, and many more, demand experimental treatment. Extensive rearing endeavors are prone to fail in the existing university institutes due to their lack of equipment. In addition, they easily conflict with the demands of teaching. This is why the establishment, or rather the maintenance of an institute that solely serves experimental biological research must be considered worth pursuing and promises success.[11]

Since the scientific committee had confirmed the need for such an institution, the Academy of Sciences agreed to take over the BVA, given that the Ministry of Education was willing to cover a share of the costs for permanent pensionable scientific staff. After lengthy negotiations with various government offices and ministries, the presidents of the Academy of Sciences, Eduard Suess (1898–1911) and Eugen von Böhm-Bawerk (1911–1914), were able to achieve the best possible deal with the Ministry of Education. After consultation with the Ministry of Finance, the Ministry of Education agreed to employ one academic official (adjunct) and three non-academic assistants, and to grant a surprisingly high annual state subsidy.

Plainly stated, the agreement resulted in a research institute hosted by the Academy and fully funded by government subsidies and private donations. The Academy itself refrained from providing subsidies from its own assets. In March 1914, for instance, the Section for Mathematics and the Natural Sciences rejected an application by the new BVA board for a one-time subsidy of 5,000 crowns for construction measures.[12]

On January 1, 1914, the contract between the founders, the Academy of Sciences, and the Austrian Ministry of Education came into effect. As recorded in the takeover document, the BVA became an institute of the Academy. For the Academy and for Hans Przibram, the handing-over resulted in a win-win situation: the Academy gained a world-renowned research institute, while Przibram's long-held wish was fulfilled when the BVA officially became a research institute of Austria's leading research body. In 1945, Alexander Stock, a former staff member, recalled that Przibram had become familiar with the scientific institutes run by the Royal Society on an undated visit to the United Kingdom. Stock reported that the British approach to the organization of scientific research had made a deep impression on Przibram. Since then, Przibram had had a similar solution in mind with respect to the BVA and the Academy of Sciences in Vienna (Stock 1945, 27).

Thanks to its acquisition of both the Institute for Radium Research (1910) and the BVA (1914), the position of the Academy of Sciences in Vienna shifted. It became an organization that hosted not only the scholarly commissions of its members but also research institutes. Since the founding of the Academy of Sciences in 1847, only the Section for Humanities and the Social Sciences had employed scientific staff to do research in the framework of the scholarly commissions. After the integration of the research institutes, the Section for Mathematics and the Natural Sciences was decisively strengthened in comparison to the humanities section, which had dominated research for more than half a century. Thus, the integration of the BVA into the Academy also triggered an important change in the character of the Academy of Sciences.

According to the new statute drafted by Richard Wettstein, Hans Przibram and Leopold von Portheim were appointed directors of the BVA for life (Statut 1914, 232). Przibram remained head of the Zoology Department, and Portheim and Figdor became the joint heads of the Department of Botany and Plant Physiology. Figdor was entitled to use two rooms of the building. According to the new management agreement (Leitungsordnung 1914), each—Wilhelm Figdor, Eugen Steinach, and Wolfgang Pauli—were to pay a rent of 2,000 crowns a year. Steinach had become head of the Physiology Department in 1913, carrying out research supported by his income as a professor at the German University of Prague. Pauli Sr. had become head of the Physical Chemistry Department in 1907. He left the BVA in June 1914 and moved to the University of Vienna after having been appointed associate professor at the Faculty of Medicine in January 1913. After Pauli's departure, the Physical Chemistry Department was closed, and in 1915 Hans Horst Meyer (1853–1939), professor of pharmacology at the University of Vienna, retired from the new scientific board (Kuratorium der biologischen Versuchsanstalt).[13]

After the BVA became an institute of the Academy of Sciences, the University of Vienna no longer had a supervisory function. The university board was dissolved, and according to the new statute the BVA became "a scientific research institute." It was not designated to serve as an "institute for teaching" (Statut 1914, 231). Since it was dedicated to "experimental research on organic life in the broadest sense, including experimental morphology and developmental physiology" (Statut 1914, 231), the BVA was officially no longer allowed to offer courses at the Vivarium building. Some teaching was moved to the university, but in the interwar period practical basic courses were still taught at the laboratory in the Prater. On December 11, 1913, the Academy of Sciences installed a new scientific board, which took over the task of direction from the former board. The Kuratorium der biologischen Versuchsanstalt (the Scientific Board of the BVA) was nominated by the Section for Mathematics and the Natural Sciences, and it acted as the BVA's senior scientific and administrative body. The board included the following members of the Academy of Sciences: Friedrich Becke, professor of mineralogy and until 1929 the Academy's secretary-general; Sigmund Exner, professor of physiology; Rudolf Wegscheider, professor of chemistry; Hans Molisch, professor of plant physiology; and Hans Horst Meyer. Apart from Grobben, the members of the BVA's former university board, Wettstein and Hatschek, retained their positions, now as members of the Academy of Sciences. Julius von Wiesner, who had been elected the first chairman, retired for health reasons in June 1914.

The new board immediately demonstrated its power by rejecting the directors' application for the establishment of a new section for radiobiological research. Even though the nominated young radiobiologist, Dr. Walter Hausmann, was willing to pay the charge of 2,000 crowns a year, his offer was rejected on the grounds of his insufficient scientific reputation.[14] The board turned down the application for several reasons: because they disliked the plutocratic character of the deal; because Hausmann lacked the necessary scientific reputation to be appointed head of department of an institute of the Academy, Austria's foremost scientific body; and for financial reasons. However, the board agreed to Figdor's application to divide the Department of Botany and establish a separate Department of Plant Physiology.

To obtain a more precise idea of the size of the subsidies the BVA received from the government as an institute of the Academy of Sciences, it is useful to compare the BVA's funding with that of the Institute for Radium Research.[15] The Institute for Radium Research was granted 2,500 crowns a year of direct government funding. In addition, four staff members were on the payroll of the Ministry of Education. The BVA was supported with 22,000 crowns a year from public authorities; additionally, the Ministry of Education approved the employment of four tenured staff members, including an academic adjunct position, and a maintenance staff of three. However, the amount dedicated to the BVA included salaries and wages within the total rather than from a separate payroll (see Über-gabsdokument 1914). That funding model proved to be an enormous disadvantage for the BVA after World War I, since funding was not increased to the extent needed to cover salaries as money lost its value in Austria's galloping inflation.

Steps toward Decline: The BVA between Imperial and Nazi Vienna

In the aftermath of the collapse of the Habsburg Empire in 1918, the existence of the BVA was seriously compromised. Prices for gas, electricity, and lab material, as well as the costs for rent, salaries, and wages, had considerably increased. At the same time, the institute's board was unable to persuade the Ministry of Education to directly pay sala-ries and wages, which had increased steadily since 1914.[16] Thus, the annual budget limits were constantly exceeded, while deficits could not be balanced by the deposited reserves because the Böhmisch-Mährische Hypothekenanstalt (Bohemian-Moravian Mortgage Bank) had suspended the payment of interest on Przibram's endowment. In this dire situation, the department heads made an urgent plea to the presiding committee of the Academy: "May the Academy as appointed custodian of scientific research not forget its institute in this critical time!"[17] In the interwar period, such appeals were documented again and again in the annual reports of the BVA. The crisis lasted until the mid 1920s. It was mitigated slightly by some support from the Academy's *Treitl-Stiftung* (Treitl fund) resources; by the municipal administration of Vienna; and by the state-undersecretary for education, Otto Glöckel, who agreed to the directors' request to take over the institute's salaries and wages.[18] However, in 1920 the Social Democrats left government, with the result that the funding arrangement of 1914 was not adjusted to the new economic situation.

In the following years, the municipal administration of Vienna and the federal Ministry of Education provided additional support for building maintenance, the procurement of fuel, animal feed, and so forth, as well as for the Physiology Department, for expenses such as renovation work and the employment of assistants.[19] From 1921 to 1925, a number of private donations helped to ensure the survival of the research institute, for example by Jerome and Margarethe Stonborough (Ludwig Wittgenstein's sister), Camillo Castiglioni, Louis Rothschild, and Franz Boas, who directed the Emergency Society for German and Austrian Science and Art in New York.[20]

Quantitative experimental research was fully relaunched in March 1931 after the air-conditioning system was finally repaired. It had been installed in 1911 and had broken down in 1914. In 1932, Przibram and Portheim reported that the year 1931, "the thirtieth after the founding of the institute, was characterized by an intense activity which to this extent had not been recorded since 1914, the date of the takeover by the Academy."[21]

But their joy was short-lived: in January 1931, the state subsidy was cut by 25 percent because of the Depression. In fact, the subsidy shrank by almost half, and once again research could be kept going only on a very limited scale. Temperature and humidity regulation was again suspended. In June 1932, on the institute's thirtieth anniversary, a permanent aquarium show was opened at the BVA, the income of which was used to cofinance the upkeep of the institute. However, despite such limited resources, year after year in the annual reports of the 1930s the directors were able to refer to remarkable research results produced by each of the BVA's three departments.

Between 1903 and 1938, more than 250 scientists from Austria and abroad (around a quarter of them women) conducted research at the Biologische Versuchsanstalt, either unpaid or funded by grants (Brauckmann 2014; Taschwer 2014b, 103). However, during the 1920s, Przibram's closest collaborators and highly gifted staff members had to leave the BVA because they had no prospects of pursuing a professional career. In Austria, anti-Semitism intensified enormously at the universities (see Ash 2015; Taschwer 2015; Feichtinger 2001).

In 1919, Paul Kammerer was refused the title of extraordinary professor at the University of Vienna. In 1926, both Paul Weiss, who had succeeded Kammerer as academic adjunct at the BVA, and Leonore Brecher were refused the Habilitation, the qualification to teach courses at the University of Vienna, for anti-Semitic reasons (Taschwer 2016; Taschwer 2012; Taschwer 2014b, 102–103; Brauckmann 2013; Drack, Apfalter, and Pouvreau 2007; Hofer 2002). They left Austria. In 1913, Hans Przibram had been appointed associate professor of zoology at the University of Vienna. In 1925, he was not even considered for one of the two vacant full professorships, despite his academic achievements and his contributions to Austrian science (Taschwer 2012; 2014b, 103).

After the *Anschluss*, Franz Köck, an engineer who had been hired as adjunct after the departure of Paul Weiss in 1927, was appointed commissarial sub-authorized agent (*Unterbevollmächtigter*) for the BVA (Reiter 1999, 610); Köck announced his loyalty to the National Socialist party (Taschwer 2014b, 104). In January 1938, Fritz Knoll, a professor of botany and a member of the Academy of Sciences, was elected to the board of the BVA and appointed its chairman. Following the *Anschluss* in March, Knoll was appointed commissarial rector of the University of Vienna and was also commissioned to safeguard "the interests of the provincial administration of the Nazi party in Austria" at the Academy of Sciences (Taschwer 2014c, 46). On April 4, 1938, Knoll ordered "Party comrade University Professor Dr. Victor Junk, Actuary of the Academy of Sciences, together with Mrs. Marie Repotschnig, accounting assistant of the Academy of Sciences, to take over as signatories for the Fund of the Institute for Experimental Biology and the ordinary account."[22] At the same time, the former signatories Hans Przibram and Leopold von Portheim were "relieved of their signing authority"[23] (see Taschwer 2014b, 105). On April 13, the BVA was closed by Knoll and the designated Academy President, Heinrich Srbik, for two weeks for "cleaning work." After its reopening on April 26, Przibram, Portheim,

Steinach, and fifteen staff members were prevented from entering the BVA building. They were classified as Jews according to the Nuremberg Laws and deprived of their functions (see Taschwer et al. 2016, 50–52).

Hans and Karl Przibram and Leopold von Portheim had invested approximately half a million crowns in the BVA before it was donated to the Academy of Sciences in 1914. In addition, the founders of the BVA had paid the salaries of the scientific staff from their own funds in the first decade of the institute's existence, and after 1914 they paid the wages of their private assistants. For 35 years, Przibram, Portheim, and Figdor had carried out scientific research at the BVA without receiving for this any salary from the public sector.

Wilhelm Figdor died just before the *Anschluss*. Portheim and Steinach were forced to leave the country. Most employees were persecuted on "racial" grounds, some perishing in the National Socialist concentration camps (Gedenkbuch 2015; Taschwer 2014b, 101–111). Leopold von Portheim was able to leave Austria for the United Kingdom in 1938 as a refugee (Metcalfe 1947, 835). Eugen Steinach, who had been lecturing in Switzerland, never returned to Austria and died in 1944 (see chapter 11 of this volume and Walch 2016; Logan 2013; Peczenik 1945). Hans Przibram was forced to remain in Vienna. Tragically for both Przibram and his spouse, although visas had been issued to them in December 1938 by the British Home Office, Przibram had not been able to procure the tax clearance certificate (*Steuerunbedenklichkeit*) required for departure, and the couple were unable to leave the country before war broke out in September 1939. A new Home Office ruling stated that expired visas would be reissued only to refugees who had already been granted them before the outbreak of war and now found themselves in a neutral country. Hans Przibram was still in Vienna.[24] In 1939, Przibram asked the Actuary of the Academy of Sciences, Viktor Junk (1875–1948), for confirmation that he had donated the BVA to the Academy on January 1, 1914, and had directed the research facility on an unsalaried basis. Przibram referred to a conversation he had conducted with Heinrich Srbik (1878–1951) and Egon Schweidler (1873–1948), who served as president and vice president of the Academy of Sciences in Vienna during the Nazi period. In a letter dated August 17, 1939, Junk, a Nazi party member, informed Schweidler that he had "of course" refused to issue the confirmation, as long as an additional demand of 50,000 Reichsmark was not settled by Przibram. In his letter to Schweidler, Junk relied on information supplied by Franz Köck, who had reported Przibram's liabilities to the property registration office (*Vermögens-verkehrsstelle*). The document further stated that the amount demanded by the BVA (under its new Nazi leadership) had been approved by the property registration office and that Przibram's fortune to the amount of 150,000 Reichsmark had been confiscated. Junk noted further: "If the confirmation had been issued, the money would probably have been released. However, the BVA had already taken the initiative to collect Przibram's debt."[25]

These intentional obstacles prevented Hans Przibram's planned emigration in an orderly manner to the United Kingdom and the United States. Eventually, in December 1939, he and his spouse were able to flee to Amsterdam. After German troops invaded the Netherlands in the spring of 1940, Nazi terror struck Hans Przibram once again. In April 1943, the couple was deported to the Theresienstadt ghetto / concentration camp, where Hans died on May 20, 1944. Elisabeth committed suicide the following day (Gedenkbuch 2015; Taschwer 2014a; Taschwer 2014b, 108–109).

After the *Anschluss*, two-thirds of the scientific staff were required to leave the institute immediately on "racial" grounds and were forced to leave the country. Apart from Hans and Elisabeth Przibram, five former staff members died by Nazi persecution: Helene Jacobi and Leonore Brecher at the extermination camp of Maly Trostinec in 1942, Martha Geiringer and Henriette Burchardt at the Auschwitz-Birkenau concentration camp in 1943 and 1944 respectively, and Heinrich Kun in an unknown camp in Yugoslavia (Gedenkbuch 2015; Taschwer et al. 2016, 53–55; Taschwer 2014b, 109).

In the wake of the *Anschluss*, serious scientific work ceased at the BVA. The building was used as a public aquarium until 1943, when it was handed over to the Kaiser Wilhelm Institute for Cultivated Plant Research (KWI für Kulturpflanzenforschung). In the last days of war in Vienna, the building was destroyed in combat operations, and the burned-out ruins were sold by the Academy of Sciences in 1947/48 (see Taschwer et al. 2016, 56–58).[26] In retrospect, it seems bizarre that the revenue from the sale of the BVA's ruins was invested in the Biological Research Station in Lunz am See, whose staff were mostly former Nazi party members. They were denazified after 1945 and continued their research—nobody was replaced (see Feichtinger and Hecht 2014, 184). In 1956, the Biological Research Station in Lunz am See received the remaining funds that had been dedicated in 1914 by the BVA's founders to the Academy of Sciences for running the institute.[27]

Responsibility for the destruction of the BVA and the expulsion of its staff on "racial" grounds was not clarified after the war, even though Fritz Knoll, to whom the BVA had been entrusted in 1938, was subject to registration as a former Nazi party member. Knoll, who had been appointed commissarial rector of the University of Vienna after the *Anschluss* and held that position until 1943, was dismissed in 1945. In 1947, his dismissal was transformed into retirement. At the Academy of Sciences, his membership was deemed terminated in 1945. However, the State Amnesty Law of 1948 permitted his membership to be reinstated. In 1957, Knoll was elected secretary of the Section for Mathematics and the Natural Sciences, and in 1959 he was appointed secretary-general of the Austrian Academy of Sciences, an office he held until 1964.

Summary

This chapter portrays the Biologische Versuchsanstalt (BVA) in its historical context: its rise to become an outstanding research center, its affiliation with the University of Vienna and with the Imperial Academy of Sciences, and its destruction by the National Socialists.

Special attention is given to the wider institutional aspects of "scientific Vienna" at the fin-de-siècle and in the interwar period. In 1902, Hans Przibram, Leopold von Portheim, and Wilhelm Figdor, themselves biologists, privately purchased the former Vivarium building in the Vienna Prater area. After the BVA opened in 1903, they made it into one of the world's foremost research facilities for experimental biology with a leading role in the emergence of a modern culture of experimental biological research. The laboratory in the Prater broke new ground in organizing, funding, and practicing science.

In its first decade, the BVA was affiliated with the University of Vienna, whose chairs of zoology and botany constituted the scientific board. After the building and a large endowment were donated by the founders in 1914, the BVA became affiliated with the Imperial Academy of Sciences. Some outstanding members of its Section for Mathematics and the Natural Sciences acted as senior scientific and administrative members of the board. To a large extent, research at the BVA was financed by its founders. Additional funding was provided by the federal government and, after the BVA became one of the institutes of the Academy of Sciences in Vienna, by the Academy, though only to a small extent. The BVA became a model of collaboration among scientists (who themselves acted as sponsors), the federal government, the University of Vienna, and the Academy of Sciences.

At the fin-de-siècle, following years of expansion in the university system, some observers saw a need for doing experimental research in a non-university setting. The demand was met by affluent Austrian businessmen who invested huge amounts of money for the promotion of science in Austria, at a time when private science funding was still in its infancy internationally. Hans and Karl Przibram and the Austrian entrepreneur Karl Kupelwieser established new research facilities: the BVA (established in 1903), the Biological Research Station at Lunz am See (established in 1906), and the Institute for Radium Research (established in 1910). As a result of these private initiatives, experimental research institutions were already operating in Austria in 1910–11 when the newly established Kaiser-Wilhelm-Gesellschaft began to invest huge amounts of public money in the establishment and operation of various non-university research institutes for the experimental sciences in Germany. In January 1911, Hans Przibram initiated a parallel effort and offered to transfer the BVA to the Imperial Academy of Sciences, at that time the only scientific body in Austria that maintained research institutes outside the universities. By institutionalizing the BVA at the Academy of Sciences, Przibram expected to secure its ongoing existence. The BVA survived both World War I and the postwar economic crisis, the latter only through additional support given by the city of Vienna, the federal government, and foreign donors. However, it would not survive National Socialism. After 1914, the institute could carry out only very limited research for several reasons: the disastrous financial situation, the unrepaired air-conditioning system, and the loss of its most gifted staff scientists. After the *Anschluss* of Austria by Nazi Germany in 1938, the BVA was gradually destroyed: research was stopped, and research objects were deliberately damaged. The directors and most members of the scientific staff, who were classified as Jews according to the Nuremberg Laws and deprived of their functions and rights, were expelled and forced to leave the country. Some died in Nazi concentration camps. In the last days of the war, the BVA building was destroyed by fire and the ruins were sold. After World War II, those who had been responsible for the destruction of the BVA were not brought to account.

Notes

1. Przibram H., Vortrag gehalten von H. Prof. Dr. H. Przibram am 9. Mai 1935. *Archiv der Österreichischen Akademie der Wissenschaften* (Archive of the Austrian Academy of Sciences; hereafter AOeAW). BVA Box 4, Folder 4. The original German was translated by the author if not otherwise noted.

2. Loeb J. an Hans Przribram, 20.11.1910. Äusserungen des Auslandes über die Biologische Versuchsanstalt in Wien. *Österreichisches Staatsarchiv* (Austrian State Archives; hereafter OeSTA)/AVA Unterricht UM Allg. Akten. Biologische Versuchsanstalt 1902–1919 Box 128.

3. These endowments amount to nearly USD 10 million today.

4. Promemoria an das hohe Präsidium der kaiserlichen Akademie der Wissenschaften in Wien, die Angliederung einer Forschungsstätte für experimentelle Biologie betreffend, von Hans Przibram, Leopold von Portheim und Wilhelm Figdor, 5.1.1911. OeStA/AVA Unterricht UM Allg. Akten. Biologische Versuchsanstalt 1902–1919 Box 128.

5. Ibidem.

6. Memorandum betr. die Übernahme der biologischen Versuchsanstalt in Wien (Prater) durch die k. Akademie. In Präsidium der k. Akademie der Wissenschaften in Wien betr. die Übernahme der biologischen Versuchsanstalt in Wien (Prater) durch die k. Akademie, 29.5.1911. OeStA/AVA Unterricht UM Allg. Akten. Biologische Versuchsanstalt 1902–1919 Box 128.

7. Memorandum betr. die Übernahme der biologischen Versuchsanstalt in Wien (Prater) durch die k. Akademie. In Präsidium der k. Akademie der Wissenschaften in Wien betr. die Übernahme der biologischen Versuchsanstalt in Wien (Prater) durch die k. Akademie, 29.5.1911. OeStA/AVA Unterricht UM Allg. Akten. Biologische Versuchsanstalt 1902–1919 Box 128.

8. Bericht über die Tätigkeit der Biologischen Versuchsanstalt in Wien im Jahre 1903. OeStA/AVA Unterricht UM Allg. Akten. Biologische Versuchsanstalt 1902–1919 Box 128.

9. Ausweis über die laufenden Ausgaben der Biologischen Versuchsanstalt in Wien, in den Jahren 1903–1910. In diesem Vorschlage sind die Auslagen für die wissenschaftlichen Kräfte des Institutes nicht enthalten. In: Präsidium der k. Akademie der Wissenschaften in Wien betr. die Übernahme der biologischen Versuchsanstalt in Wien (Prater) durch die k. Akademie, 29.5.1911. OeStA/AVA Unterricht UM Allg. Akten. Biologische Versuchsanstalt 1902–1919 Box 128.

10. Kommissionsbericht 1911. In Präsidium der k. Akademie der Wissenschaften in Wien betr. die Übernahme der biologischen Versuchsanstalt in Wien (Prater) durch die k. Akademie, 29.5.1911. OeStA/AVA Unterricht UM Allg. Akten. Biologische Versuchsanstalt 1902–1919 Box 128.

11. Kommissionsbericht 1911. In Präsidium der k. Akademie der Wissenschaften in Wien betr. die Übernahme der biologischen Versuchsanstalt in Wien (Prater) durch die k. Akademie, 29.5.1911. OeStA/AVA Unterricht UM Allg. Akten. Biologische Versuchsanstalt 1902–1919 Box 128.

12. Protokoll der Sitzung der mathematisch-naturwissenschaftlichen Klasse der kaiserlichen Akademie der Wissenschaften in Wien, 26.3.1914. AOeAW B 1919.

13. Protokoll. Sitzung des Curatoriums der biologischen Versuchsanstalt der kaiserlichen Akademie der Wissenschaften, 17.6.1915. AOeAW BVA Box 2, Folder 3.

14. Protokoll der Sitzung des Curatoriums der biologischen Versuchsanstalt, 5.3.1914. AOeAW BVA Box 2, Folder 3.

15. In 1914, the Academy of Sciences received 130,000 crowns a year of government funding in addition to the funding for both institutes. The maximum salary of a full university professor at the University of Vienna at this time was approximately 7,500 crowns a year (according to the salary law of 1898).

16. Bericht über die Gesamtsitzungen des Kuratoriums der biologischen Versuchsanstalt am 6. und 13. Februar 1919, 18.2.1919. AOeAW BVA Box 2, Folder 4.

17. Vorschlag eines eingeschränkten Betriebes der biologischen Versuchsanstalt für den Winter 1919/20. AOeAW BVA Box 1, Folder 2; Bericht über die Lage der Biologischen Versuchsanstalt im Herbst 1919. AOeAW BVA Box 1, Folder 2.

18. Bericht über die Tätigkeit der biologischen Versuchsanstalt der Akademie der Wissenschaften in Wien für das Jahr 1919 (und Programm für 1920). AOeAW BVA Box 2, Folder 1.

19. Bericht über die Tätigkeit der Biologischen Versuchsanstalt der Akademie der Wissenschaften in Wien, für das Jahr 1920. AOeAW BVA Box 2, Folder 1.

20. For the reports on the years 1921 to 1923 and the program until 1924, see AOeAW BVA Box 2, Folder 1.

21. Bericht über die Tätigkeit der Biologischen Versuchsanstalt der Akademie der Wissenschaften in Wien für das Jahr 1931 und Programm für das Jahr 1932. AOeAW BVA Box 2, Folder 1.

22. AOeAW Papers of Fritz Knoll, Box 1, Folder 2, published in Taschwer 2014b, 105.

23. AOeAW Papers of Fritz Knoll, Box 1, Folder 2.

24. Correspondence E. Simpson to K. Soffel, 11.11.1939. Archive of the Society for the Protection of Science and Learning, Bodleian Library, Oxford. Personal File Hans Przibram Box 538, Folder 2; Correspondence M. Hope to D. C. Thomson, 16.6.1939. Archive of the Society for the Protection of Science and Learning, Bodleian Library, Oxford. Personal File Hans Przibram Box 538, Folder 2.

25. Correspondence V. Junk an E. Schweidler, 17.8.1939. AOeAW BVA Box 3, Folder 4.

26. Albert Weiser. Gedächtnisniederschrift über die Kriegsereignisse im April, 1945 die zur Vernichtung der Biologischen Versuchsanstalt führten, 18.12.1945. AOeAW BVA Box 4, Folder 2.

27. Protokoll der Sitzung der Gesamtakademie, 27.1.1956. AOeAW A 1078. Wissenschaften in Wien, 26.3.1914. AOeAW B 1919.

References

Ash, M. 2015. Die Universität Wien in den politischen Umbrüchen des 19. und 20. Jahrhunderts. In *Universität–Politik–Gesellschaft*, vol. 2 of *650 Jahre Universität Wien–Aufbruch ins neue Jahrhundert*, ed. M. Ash, and J. Ehmer, 29–172. Göttingen: V&R unipress.

Brauckmann, S. 2013. Weiss, Paul Alfred. *eLS* (September 20): 1–7.

Brauckmann, S. 2014. A Laboratory in the Prater: The Biologische Versuchsanstalt in Vienna. Unpublished chronology.

Broda, E. 1979. Warum war es in Österreich um die Naturwissenschaft so schlecht bestellt? *Wiener Geschichtsblatter* 34 (3): 89–107.

Coen, D. R. 2006. Living precisely in fin-de-siècle Vienna. *Journal of the History of Biology* 39 (3): 493–523.

Coen, D. R. 2007. *Vienna in the Age of Uncertainty: Science, Liberalism, and Private Life.* Chicago: University of Chicago Press.

Csáky, M., et al., eds. *1996–2008. Studien zur Moderne.* Vienna: Passagen Verlag.

Drack, M., W. Apfalter, and D. Pouvreau. 2007. On the making of a system theory of life: Paul A. Weiss and Ludwig von Bertalanffy's conceptual connection. *Quarterly Review of Biology* 82 (4): 349–373.

Edwards, C. L. 1911. The Vienna Institution for Experimental Biology. *Popular Science Monthly* 78 (1): 584–601.

Feichtinger, J. 2001. Kulturelle Marginalität und wissenschaftliche Kreativität. Jüdische Intellektuelle im Österreich der Zwischenkriegszeit. In *Das Gewebe der Kultur. Kulturwissenschaftliche Analysen zur Geschichte und Identität Österreichs in der Moderne,* ed. J. Feichtinger and P. Stachel, 311–333. Innsbruck, Vienna, Munich: Studienverlag.

Feichtinger, J. 2010. *Wissenschaft als reflexives Projekt. Von Bolzano über Freud zu Kelsen: Österreichische Wissenschaftsgeschichte 1848–1938.* Bielefeld: Transcript.

Feichtinger, J. 2015. Die verletzte Autonomie. Wissenschaft und ihre Struktur in Wien 1848 bis 1938. In *Universität–Forschung–Lehre,* vol. 1 of *650 Jahre Universität Wien–Aufbruch ins neue Jahrhundert,* ed. F. Stadler, K. Kniefacz, E. Nemeth, and H. Posch, 259–290. Göttingen: V&R unipress.

Feichtinger, J., and D. Hecht. 2014. Denazification at the Academy of Sciences. In *The Academy of Sciences in Vienna 1938 to 1945,* ed. J. Feichtinger, H. Matis, S. Sienell and H. Uhl, 163–179. Vienna: Austrian Academy of Sciences Press.

Gedenkbuch 2015. Gedenkbuch für die Opfer des Nationalsozialismus an der Österreichischen Akademie der Wissenschaften. www.oeaw.ac.at/gedenkbuch/

Gliboff, S. 2006. The case of Paul Kammerer. Evolution and experimentation in the early 20th century. *Journal of the History of Biology* 39:525–563.

Hatschek, B. 1902. Diskussionsbeitrag. In *Vorträge und Besprechungen über Die Krisis des Darwinismus (Dr. Max Kassowitz, Dr. Richard v. Wettstein, Dr. Berthold Hatschek, Dr. Christian Freih. von Ehrenfels, Dr. Josef Breuer),* 35–38. Leipzig: Johann Ambrosius Barth.

Hirschmüller, A. 1991. Paul Kammerer und die Vererbung erworbener Eigenschaften. *Medizinhistorisches Journal. Internationale Vierteljahresschrift für Wissenschaftsgeschichte* 26 (1/2): 26–77.

Hofer, V. 2002. Rudolf Goldscheid, Paul Kammerer und die Biologen des Prater-Vivariums in der liberalen Volksbildung der Wiener Moderne. In *Wissenschaft, Politik und Öffentlichkeit. Von der Wiener Moderne bis zur Gegenwart,* ed. M. G. Ash and C. H. Stifter, 149–184. Vienna: Wiener Universitätsverlag.

Höflechner, W. 1990. Zur nichtstaatlichen Wissenschaftsförderung in Österreich in der Zeit von 1848 bis 1938 am Beispiel der Akademie der Wissenschaften in Wien. In *Formen außerstaatlicher Wissenschaftsförderung im 19. und 20. Jahrhundert,* ed. R. vom Bruch and R. A. Müller, 211–225. Stuttgart: Steiner.

Jahrhundertfeier der Königlichen Friedrich-Wilhelms-Universität zu Berlin. 10–12 Oktober 1910. 1911. Bericht im Auftrag des Akademischen Senats erstattet von dem Prorektor Erich Schmidt. Berlin: Schade.

Koestler, A. 2010. *Der Krötenküsser. Der Fall des Biologen Paul Kammerer. Mit einem Nachwort von Peter Berz und Klaus Taschwer.* Vienna: Czernin Verlag.

Kofoid, C. A. 1910. *The Biological Stations of Europe. United States Bureau of Education Bulletin 440.* Washington, DC: Government Printing Office.

Kühnelt, W. 1985. Zoologische Forschung im Bereich der Wiener Universität. *Archiv der Naturwissenschaften* 14/15:663–679.

Leitungsordnung für die Biologische Versuchsanstalt der Kaiserlichen Akademie der Wissenschaften in Wien. 1914. *Almanach der kaiserlichen Akademie der Wissenschaften* 64: 242–248. Vienna: Alfred Hölder.

Logan, C. A. 2013. *Hormones, Heredity, and Race: Spectacular Failure in Interwar Vienna.* New Brunswick, NJ: Rutgers University Press.

Logan, C. A., and S. Brauckmann. 2015. Controlling and culturing diversity: Experimental zoology before World War II and Vienna's *Biologische Versuchsanstalt. Journal of Experimental Zoology. Part A, Ecological Genetics and Physiology* 323 (4): 211–226.

Metcalfe, C. R. 1947. Dr. Leopold von Portheim. *Nature* 159 (June 21): 835.

Müller, G. B., and H. Nemeschkal. 2015. Zoologie im Hauch der Moderne: Vom Typus zum offenen System. In *Reflexive Innensichten aus der Universität. Disziplinengeschichten zwischen Wissenschaft, Gesellschaft und Politik*, vol. 4 of *650 Jahre Universität Wien–Aufbruch ins neue Jahrhundert*, ed. K. A. Fröschl, G. B. Müller, T. Olechowski, and B. Schmidt-Lauber, 355–369. Vienna: VR unipress.

Peczenik, O. 1945. Eduard [sic] Steinach. In *Oesterreichische Wissenschaft. Essays, Biographien, Betrachtunge*, ed. H. Ullrich, with a preface by Prof. Dr. Erwin Schroedinger, 29–32. London: Free Austrian Movement.

Philippovich, E. 1906. Bericht über das Studienjahr 1905/06, dz. Prorektor der k. k. Universität. In *Die feierliche Inauguration des Rektors der Wiener Universität für das Studienjahr 1906/07*, 3–29. Vienna: Selbstverlag der Universität Wien.

Przibram, H. 1903. Die neue Anstalt für experimentelle Biologie in Wien. In *Verhandlungen der Gesellschaft deutscher Naturforscher und Ärzte. 74. Versammlung zu Karlsbad. 21–27 September 1902. 2.1: Naturwissenschaftliche Abtheilungen*, ed. A. Wangerin, 152–155. Leipzig: F. C. W. Vogel.

Przibram, H. 1908/09. Die Biologische Versuchsanstalt in Wien. Zweck, Einrichtung und Tätigkeit während der ersten fünf Jahre ihres Bestandes (1902–1907). Bericht der zoologischen, botanischen und physikalisch-chemischen Abteilung. *Zeitschrift für biologische Technik und Methodik* 1: 234–264; 1. Fortsetzung: 329–362; 2. Fortsetzung: 409–433.

Przibram, H. 1913. Die Biologische Versuchsanstalt in Wien. Ausgestaltung und Tätigkeit während des zweiten Quinquenniums (1908–1912). Bericht der zoologischen, botanischen und physikalisch chemischen Abteilung. *Zeitschrift für biologische Technik und Methodik* 3: 163–245.

Przibram, H. 1926. Paul Kammerer als Biologe. *Monistische Monatshefte* 11:401–405.

Przibram, K. 1959. Hans Przibram (1874–1944). In *Grosse Österreicher. Neue Österreichische Biographie ab 1815*. vol. 13., 184–191. Zurich, Leipzig, Vienna: Amalthea.

Reiter, W. L. 1999. Zerstört und Vergessen. Die Biologische Versuchsanstalt und ihre Wissenschaftler/innen. *Österreichische Zeitschrift für Geschichtswissenschaften* 10 (4): 585–614.

Reiter, W. L. 2014. Mäzenatentum, Wissenschaft und Politik in Österreich um 1900. *Österreichische Zeitschrift für Geschichtswissenschaften* 25 (3): 212–247.

Sandgruber, R. 2013. *Traumzeit für Millionäre. Die 929 reichsten Wienerinnen und Wiener im Jahr 1910*. Vienna, Graz, Klagenfurt: Styria.

Schorske, C. E. 1980. *Fin-de-Siècle Vienna. Politics and Culture*. London: Weidenfeld and Nicolson.

Statut. 1914. In *Almanach der kaiserlichen Akademie der Wissenschaften* 64: 231–236. Vienna: Alfred Hölder.

Stock, A. 1945. Das Wiener Vivarium. In *Oesterreichische Wissenschaft. Essays, Biographien, Betrachtungen*, ed. H. Ullrich, with a preface by Prof. Dr. Erwin Schroedinger, 25–28. London: Free Austrian Movement.

Taschwer, K. 2012. Andenken an eine völlig vergessene Forscherin. Vor genau siebzig Jahren wurde die österreichische Biologin Leonore Brecher ermordet. Eine späte Würdigung der tragisch Gescheiterten. *Der Standard* (September 23): derstandard.at/1348283731761/Andenken-an-eine-voellig-vergessene-Forscherin

Taschwer, K. 2014a. Ein tragischer Held der österreichischen Wissenschaft. *Der Standard* (February 5): 16–17.

Taschwer, K. 2014b. Expelled, burnt, sold, forgotten, and suppressed: The permanent destruction of the Institute for Experimental Biology and its academic staff. In *The Academy of Sciences in Vienna 1938 to 1945*, ed. J. Feichtinger, H. Matis, S. Sienell and H. Uhl, 101–111. Vienna: Austrian Academy of Sciences Press.

Taschwer, K. 2014c. The two careers of Fritz Knoll: How a botanist furthered the Nazi Party's interests after 1938—and successfully lived it down after 1945. In *The Academy of Sciences in Vienna 1938 to 1945*, ed. J. Feichtinger, H. Matis, S. Sienell and H. Uhl, 45–52. Vienna: Austrian Academy of Sciences Press.

Taschwer, K. 2015. *Hochburg des Antisemitismus. Der Niedergang der Universität Wien im 20. Jahrhundert.* Vienna: Czernin Verlag.

Taschwer, K. 2016. *Der Fall Paul Kammerer. Das abenteuerliche Leben des umstrittensten Biologen seiner Zeit.* Munich: Hanser.

Taschwer, K., J. Feichtinger, S. Sienell, and H. Uhl, eds. 2016. *Experimental Biology in the Vienna Prater: On the History of the Institute for Experimental Biology 1902 to 1945.* Vienna: Verlag der Österreichischen Akademie der Wissenschaften.

Übergabsdokument. 1914. Biologische Versuchsanstalt der Kaiserlichen Akademie der Wissenschaften in Wien. In *Almanach der kaiserlichen Akademie der Wissenschaften* 64: 229–230. Vienna: Alfred Hölder.

Universität Wien. Philosophische Fakultät. 1902. *Denkschrift über die gegenwärtige Lage der Philosophischen Fakultät der Universität Wien.* Vienna: Adolf Holzhausen.

Vom Brocke, B., and H. Laitko eds. 1996. *Die Kaiser-Wilhelm-/Max-Planck-Gesellschaft und ihre Institute. Das Harnack-Prinzip.* Berlin: de Gruyter.

Walch, S. 2016. *Triebe, Reize und Signale. Eugen Steinachs Physiologie der Sexualhormone. Vom biologischen Konzept zum Pharmapräparat, 1894–1938.* Vienna, Cologne, Weimar: Böhlau.

Wettstein, R. 1902a. Oesterreichische biologische Stationen. *Neue Freie Presse* (August 21): 14–15.

Wettstein, R. 1902b. Über directe Anpasssung. Vortrag gehalten in der Feierlichen Sitzung der Kaiserlichen Akademie der Wissenschaften am 28. Mai 1902. In *Almanach der kaiserlichen Akademie der Wissenschaften* 52: 311–337. Vienna: Alfred Hölder.

Wettstein, R. 1902c. Die Stellung der modernen Botanik zum Darwinismus. Vortrag, gehalten am 20. Jänner 1902. In *Vorträge und Besprechungen über Die Krisis des Darwinismus (Dr. Max Kassowitz, Dr. Richard v. Wettstein, Dr. Berthold Hatschek, Dr. Christian Freih. von Ehrenfels, Dr. Josef Breuer)*, 19–32. Leipzig: Johann Ambrosius Barth.

Wettstein, R. 1903. *Der Neo-Lamarckismus und seine Beziehungen zum Darwinismus. Vortrag gehalten in der allgemeinen Sitzung der 74. Versammlung deutscher Naturforscher und Aerzte in Karlsbad am 26.8.1902.* Jena: Gustav Fischer.

Wettstein, R. 1912. Hochschule und selbständiges Forschungsinstitut. *Neue Freie Presse* (February 8): 22–24.

W.T.C. 1932. Connecting Laws in Animal Morphology. Four Lectures held at the University of London, March 1929. *Nature* 129 (February 27): 298–299.

III   SCIENTIFIC ZEITGEIST

5 Experimental Biology and the Biomedical Ideal around the Year 1900

Heiner Fangerau

In 1949, the French biologist and writer Jean Rostand published a short popular science book, entitled *La biologie et l'avenir humain* (Biology and the human future), in which he summarized the current state of experimental biological research. His publication begins with a commentary from 1890 by Anatole France, recipient of the Nobel Prize in literature, who stated: "There is not yet any biology," but perhaps it will exist "in a few million years" (France 1892; Rostand 1950, preface).[1] Rostand stands up to contradict this pessimistic prediction. Sixty years after France made his statement, Rostand proudly notes that not only does a "science of life" exist now, but that it also plays an important role in society, especially through its influence on medicine. In his analyses of the future of man, an enthusiastic Rostand goes on to sketch the prospect of biological research work: rather than an impossible dream, experimental genetics and embryology would enable human abilities to be improved, the human life span to be increased, and human nature to be surpassed. Establishing his own narrative of modern biology, Rostand describes a "still young science" that originated in the late nineteenth century and successfully aided scientists in understanding life processes. He claims that this science was launched with the beginnings of experimental biology and that it would now progress from a phase of analyzing living matter to a phase of targeted manipulation of life's processes.

Rostand repeats a topos that can be traced back to the nineteenth century. Biologists like Jacques Loeb at the end of the nineteenth century referred to the development of chemistry when they tried to explain the prospects of successful biological research. Morphological analysis should be replaced by a synthetic phase, just as in chemistry analytical approaches had been replaced by the artificial synthesis of organic material, such as urea, synthesized in the laboratory by Friedrich Wöhler in 1828. In the end this should result in the biologist engineering nature the way the engineer designs technology (Fangerau 2012; Pauly 1987). The laboratory as a place of experimental research and technological manipulation played a crucial role in this concept of biology. Similar to chemical laboratories of the eighteenth century, where innovation and research occurred alongside activities like the craft-style production of materials such as gunpowder, or artisan metallurgical testing (Klein 2008a, 10; 2008b), the biological laboratories established during the late nineteenth century were designed to investigate, to control, and to synthesize nature. As a desired consequence, Rostand and earlier biologists stated, by following the laboratory path of synthesizing life, biology would in the end help to transform medicine from a curative focus to an approach for enhancing health and optimizing human existence.

While Rostand's foundational narrative of modern biology and medicine is understandable in light of the scientific enthusiasm that still existed in the late 1940s and early 1950s, even for his period he was smoothing over many details of the history of biology and its interconnections with medicine. Even his basic concept of the history of biology—namely that biology had started becoming a science with the advent of experimental biology in the late nineteenth century—might have been disputed around 1950. First, his version of the beginnings of experimentalization and how it served as the new key for biological research in the 1880s and 1890s is at least as problematic as many other attempts at historical periodization and structuring. All efforts at periodization (political caesura, battles, shifting ideas, etc.) have their limits and rely on contingent factors (Bentley 1996; Osterhammel 2006). The validity of Rostand's argument that a genuinely new approach had emerged at the end of the nineteenth century depended on the depth of the analysis, on the perspective, and on the definition of "experimentalization." Second, the idea that a new era of biology had started with what was perceived and defined as a new experimental biology had already been disputed around 1900.

In the following pages, I would like to challenge Rostand's narrative by examining two discourses about experimentation in the field of biology, one written around 1900 and the other 20 years later. Whereas the first dealt with the foundational narrative of experimentalization, the second treated the influence of experimental biology on medicine, an influence Rostand regarded as so great that it would in the near future change the prospect of health care from cure to enhancement. My hypothesis is that both discourses not only reflect methodological debates in these two life sciences but also touch the professional self-image of both biologists and physicians. They are also relevant to the conceptualization of biology and medicine as academic and practical disciplines: Both discourses helped to shape the self-concept of these disciplines in a later phase of specialization and differentiation, processes that had started in the nineteenth century. The earlier discourse was important for biology, the second influenced the self-image of the medical discipline.

Several great historical studies have already described the process by which the experimental approach as a specific branch of biological research was formulated, contoured, and transferred into practice.[2] My aim is to add to that history some minor facets concerning the roles of ideas and the sociology of knowledge. Around the year 1900, a supposedly new form of experimental research was established; and, around 1920, some researchers advocated the benefits of this particular biological approach for clinical and practical medicine more vehemently than before. Here, in a first step I will analyze whether a new form of biology, in the guise of "developmental mechanics," truly emerged around 1900. This question was raised by contemporaries as well as by historians of biology, and I will reconstruct some of the main arguments of the time. In a second step, I will highlight the debate surrounding the transfer of a biological perspective into medicine in the form of an early "biomedical program" in the 1920s. I will examine its arguments, intentions, and effects on the lasting narrative offered by later biologists like Rostand.

Developmental Mechanics—A New Discipline?

In 1911, the Austrian Academy of Sciences discussed the acquisition of the Institute for Experimental Biology (Biologische Versuchsanstalt, or BVA) in Vienna, which had been founded by Hans Przibram, Wilhelm Figdor, and Leopold von Portheim in 1903 (Reiter 1999).[3] According to a commission report regarding the takeover, the purpose of the institution "should be to cultivate the experimental branches of biology, particularly in a direction that examines the function of the morphological development of various kinds of conditions and which is ... in more recent times called developmental mechanics according to Roux and experimental morphology according to Davenport."[4]

Hans Przibram himself had presented the program of this experimental institution at the 74th meeting of the Society of German Natural Scientists and Physicians in Karlsbad in the year 1902, shortly before the official inauguration of the biological station. At the same time, he delimited the purpose of the institution from the objectives followed by a great number of already existing institutions that, he stated, "primarily served the purpose of descriptive research, of systematic description."[5] In contrast, his research institution focused on the experimental method, which according to him only "during the last third of the nineteenth century" had become a prominent tool of biological research. Przibram also referred explicitly to the developmental mechanics program formulated by Wilhelm Roux in the late 1880s as a countermodel to descriptive and morphological studies in biology.

Roux's program was part of a longer process to make the sciences experimental, which had been ongoing since at least the seventeenth century (Cohen 1994, 2010). Proceeding from physics and chemistry, this process had later reached medicine and the life sciences and, according to the prevalent narrative, had undergone a decisive change around the year 1850. At that time the field of physiology, working with the natural sciences and experimentation, had advanced to become the guiding discipline of scientifically oriented medicine, in terms of the number of institutions and personnel. It competed with anatomy (including histology and cytology) for influence and power in the medical faculties.[6] After the 1880s, Wilhelm Roux and others sought to transfer the physics- and chemistry-oriented model that had been so successful in physiology to explore the development of organisms and the processes of life. In this context, they explicitly endeavored to distinguish their method from the dominant science of comparative anatomy/embryology, as personified by such charismatic personalities as Ernst Haeckel. As Roux had written in 1885, and repeated in 1897 in a reply to criticism by Oscar Hertwig (1897) regarding the novelty of Roux's ideas, while descriptive embryology had succeeded in describing the changes in form shown by various species during their development, it was now time to consider the processes that functionally determined these anatomical changes (Roux 1897a, 4). Using his approach, Roux wanted to understand and to prove experimentally the causality of the processes that determined changes in form (Roux 1897a, 23). His aim was to use a causal-analytical process to explore whether "physical-chemical forces" (*physikalisch-chemischen Kräfte*) actually influenced the growth, division, and movement of cells, aspects of development that, from Roux's perspective, the so-far-dominant comparative anatomy/morphology approach had been satisfied to simply describe (Roux 1897a, 52). He wanted to investigate the causal mechanics of development. Accordingly Roux coined his approach "developmental mechanics" (*Entwicklungs-mechanik*). With this term he also aimed to demarcate his approach from that of William Preyer, who had created the term *Entwicklungsphysiologie* (developmental physiology) in 1880 to describe the investigation of the functions of preservation during developmental processes. Roux, for his part, intended to stress the functions of configuration and arrangement (Sucker 2002, 22). Authors like Thomas Morgan preferred to call the approach "experimental physiology" (Morgan 1898, 158), and Hans Driesch, who had worked together with Morgan in the Marine Biological Station in Naples in 1894/95 (Müller 1976) and studied with Preyer in Jena, again used the terms *Entwickelungsphysi-ologie* or *Entwickelungsanalytik* in order to avoid the machine-like connotations of "mechanics" (Driesch 1897).

In 1922, the physiologist Georg Ettisch[7] expressed the basic idea in a nutshell, stating that developmental mechanics does not ask for the "how" of the form or the construction, but principally for the "why" (Ettisch 1922a, 632). In a retrospective attribution of meaning, Ettisch stated that the "discipline" founded by Roux was of the highest importance for the theory and practice of medicine, "but also in relation to the history of science, although it has not acquired in Germany a propagation and recognition that corresponded to this importance" (Ettisch 1922a, 631).[8]

Claims and Counterclaims

From the very beginning, the repeated claim that developmental mechanics represented a new experimental approach to biology, and the parallel desire of its proponents for its recognition and incorporation by institutions, resulted in a heated debate, which finally after a climax seemed to bore even the contemporary observer. The biologist Oscar Hertwig questioned above all the novelty of Roux's approach (Hertwig 1897). Roux

answered with a long attack on Hertwig (Roux 1897a).[9] Both works were reviewed anonymously in *Nature* with expressions of some disappointment that Hertwig and Roux spent
so many pages arguing against and for the idea of a new program instead of concentrating
on the research itself (Anonymous 1897b, Anonymous 1897a). In his book *Zeit- und
Streitfragen der Biologie* (Current issues of biology), Hertwig sought to deconstruct
Roux's claim to novelty. He asked, polemically, what was the "aim or task of the new
future science," and what new ways, resources, or methods did it bring with it? (Hertwig
1897, 9)[10]? He argued that its aims were unclear (p. 63), its methods not new, and that the
placement of experiments in the foreground was not a new approach. Hertwig wrote that
he too had done experiments, like Roux, "in the most varied modifications" (p. 78).

Retrospectively it might be assessed that some of the disagreement between Hertwig
and Roux could have been due to different concepts of methods and methodologies. In
his reply Roux stated that regarding the "technical methodology," developmental mechanics did not require previous experimental methods to be improved. Only the prevailing
microscopy techniques (e.g., volume meters, goniometers, aerometers, etc.) needed to be
updated with methods from physics and chemistry laboratories (Roux 1897a, 220). On an
institutional level the novelty of a biological laboratory as envisioned by Roux can also
be questioned. The foundation of the BVA in Vienna, for example, followed a tradition
that suggests a process-like narrative to experimental biology, rather than a sharp turning
point at the moment that developmental mechanics was introduced. Other non-university
institutions also followed the growing trend of adopting an experimental approach against
the background of descriptive biology. These institutions included several marine biological stations, for example, in Concarneau (founded in 1859), Roscoff (1872), Sebastopol
(1871), Naples (1874), Trieste (1875), Kristineberg (1877), Villefranche-sur-Mer (1880),
Woods Hole (1884), Plymouth (1887), Bergen (1891), and Helgoland (1892) (Penzlin
2004; Kofoid 1910; de Bont 2009; Maienschein 1989). Research equipment and materials
at these stations consisted of aquariums, various test animals, and chemicals, which
favored the formation of a developmental mechanics program, in contrast with other
institutions where the foundation of the program had to be enforced. Moreover, the equipment at the research institution in Vienna did not seem to differ from those at other stations.
It also included aquariums, terrariums, an electric light, chemicals, electric current, burners,
etc.[11] Hence, at the technical and institutional level of experimentation, Hertwig's objections seem to have had some justification.

Regarding the "method," however, even Hertwig admitted a central difference in his
and Roux's views about the fundamental approach of experimental work. Hertwig insisted
that experiments could clarify developmental pathologies but were not suitable for obtaining insights into normal development. For him, nature was a better teacher than "the
experimenting anatomist" (Hertwig 1897, 74, 76, 79). On the other hand, Roux not only
analyzed normal development, but also experimentally influenced and controlled variations from normal (Roux 1897a, 227). Since it was impossible to observe developmental

forces within organisms, Roux claimed that only "through isolation, displacement, destruction, weakening, irritation, wrong connections, passive deformation, changing the nutritional and functional variables, as well as by the effects of special agents" on development was it possible to reveal the mutual "effects of parts of the organism on one another" (Roux 1897a, 240).[12] The interference with normal development to see abnormal results was what was new in the understanding of the experimenting biologist. Through this concept of targeted influence and invasive experimentation, Roux and many of his contemporaries wanted to let biology enter a new phase of experimental understanding, in which observation was complemented by manipulation, deliberate design, and the "conscious combination of events" (Driesch 1897, 138).

Apart from the institutional connections of biologists who were working experimentally at medical departments, research involving the production of various malformations also had a direct connection to medicine itself. On the one hand, the production of malformation was linked to cancer research and medicine; on the other, questions of growth, form, and (tissue) regeneration puzzled both anatomists and surgeons (Cooper 2004, 2009). Tissue cultivation and regeneration were considered as ultimately practical fields of application of modern biology. Thus, medical research seemed to be very receptive to a causal-analytical approach. Experimentalizing, understood as an enlargement of experimental laboratory methods and an attempt to understand pathologies as variations of normal, had been far advanced in medicine via physiological and pathological research. In 1896, for example, the first issue of the *Journal of Experimental Medicine* appeared. Its founder and editor, the first dean of the Johns Hopkins University School of Medicine, the pathologist William Welch, stated in his introduction: "In medical, no less than in other sciences the great advances of modern times have been owing to the method of experiment combined with that of exact observation" (Welch 1896, 3). With his journal he wanted to offer American scholars who were already doing research in the field of experimental medicine and the allied sciences a more specified platform than clinical journals could provide. According to him, in European countries such journals already existed and American scholars had felt a keen need for such an organ. He explicitly excluded "purely clinical observations," meaning bedside research or case reports, but opened the journal to the publication of investigations based on "methods of the laboratory, microscopical, chemical, and bacteriological." He wanted to foster what he called "scientific medicine." Experimental biology responded to this in 1904 with the founding of the *Journal of Experimental Zoology*. Retrospectively, Ross Harrison directly linked the journal back to the journal Roux had founded to propagate *Entwicklungsmechanik* (*Wilhelm Roux' Archiv für Entwicklungsmechanik der Organismen*) when he remembered, "There was no avenue of publication to which those working in experimental zoology might look, so that many papers by American investigators were sent to Roux' Archiv. The need for a medium of publication in this country was obvious and steps were taken to establish it" (Harrison 1945, xi). Nevertheless, the links between experimental biology and medicine remained obvious in this endeavor. The journal was hosted in the anatomical laboratory of the Johns Hopkins Medical School, and, according to Harrison, "the distinguished physician and physiologist" Samuel Meltzer (note the order of the characterization), the founder of the "Society for Experimental Biology and Medicine," was the first subscriber (Harrison 1945, xix). This society has existed since 1903 and has issued the *Journal for Experimental Biology and Medicine* since the same year (Howell 1923, 10).

However, the strict applicability to practical medicine of the physical-chemical approach in the (sub)microscopic area that was propagated in these journals was not completely uncontested. When, for example, the German physiologist Nathan Zuntz reported on a local level in Berlin about experimental biological work he had performed in California with the aforementioned experimental biologist Jacques Loeb, he could state that clinicians had been vividly interested in the research but that the relevance of Loeb's research (on regeneration, growth, and development) had been the topic of a long discussion.[13] Similarly, when the Rockefeller Institute for Medical Research wanted to hire Jacques Loeb as director of the Department of Experimental Biology in 1909, Loeb, on the basis of his prior experiences, expressed his skepticism about the physicians' receptiveness to the application of the experimental biological approach in medicine, writing: "In my opinion, experimental biology ... will have to form the basis not only of Physiology but also of General Pathology and Therapeutics. I do not think that Medical Schools in this country are ready for this new departure ... The medical public at large does not yet fully see the bearing of the new Science of Experimental Biology ... on Medicine."[14]

Practical Medicine

The same tension seemed to exist in the Central European context, especially with regard to the experimental method and the integration of the ideas put forward by the program of developmental mechanics. In the first edition of the journal *Die Naturwissenschaften* (The natural sciences) in 1913, the internal medicine specialist Wilhelm His Jr., expressed hopeful sentiments about the future of medicine that very much resemble the statements of his successors like Rostand: "Our thinking, our methodology has become scientific, and there is no amount of abstract research in chemistry, physics, or biology that could not one day enrich our knowledge and abilities in a revolutionary manner ... our relationship to technology is similarly close" (His 1913, 3).[15] However, fourteen days later in the same journal, an internal medicine specialist from Tübingen, Otfried Müller, penned an article entitled "Medizinische Wissenschaft und Ärztliche Kunst" (Medical science and the art of the physician), that concluded: "Indeed, we have a biological science, the special pathology; in certain branches, we also obtain a rational therapy based on biology. However, a great part of the therapeutic field itself is ... a matter of the personal, the human, an artistic field" (Müller 1913, 90).[16]

Thus, even by about 1900, practical medicine was having difficulty with the application of experimental biological methods. Debates that had already accompanied the technization of biology and the interpretation of the experimental results from developmental mechanics erupted on the practical medicine front. Views on the question of experimentalization, in particular, divided along the lines of vitalism vs. mechanism, causality vs. conditionality, and holism vs. reductionism. Whereas researchers such as Jacques Loeb tried to explain life processes according to purely physical-chemical points of reference, other experimentally oriented scientists, such as Hans Driesch, advocated or formulated vitalist positions, considering forces other than physical and chemical ones to be influential. Still others, like Max Verworn, held conditionalist viewpoints that opposed the causal interpretations of development. In this context, if medical sciences were understood causally and in a physical-chemical manner, then conditional and vitalist elements would be placed on the side of the physician's "art."

The inconclusiveness and obscurity of the debate about the art and science of medicine that is reflected in His's and Müller's statements mirrors the indecisiveness of many biologists themselves with regard to the examination and explanation of living phenomena. In retrospect it is difficult to definitively locate the position of several actors along this matrix between vitalistic, mechanistic, causal, and conditional concepts. Not all actors could be classified along the lines of these extreme positions (Diepgen 1931, 1437). Exemplarily, the botanist Johannes Reinke declared in his comments on a monograph by Otto Bütschli about "mechanism and vitalism" that he considers himself to be a mechanist erroneously marked as a vitalist by Bütschli (Reinke 1902, 24; Diepgen 1931).[17]

This debate, at least in Germany, broke out anew in the 1920s, in the context of the incipient dispute about a crisis in medicine. Some physicians initiated a discourse about such a crisis in the Weimar Republic, asserting that medical doctors had moved too far in the direction of a mechanical and technical understanding of healing and had forgotten to care for the whole patient, including his or her soul and feelings. According to authors like the notorious Erwin Liek, not mysticism but a return to caring should become the guideline of medicine.[18] Regarding the relationship of experimental biology and medicine in 1922, several articles appeared in the *Klinische Wochenschrift* (Clinical weekly), such as "Einige Bemerkungen über die Grundlagen des ärztlichen Denkens von heute" (Some remarks on the foundation of contemporary medical thinking), by Friedrich Martius (1922), "Die Sinnfindung als Kategorie des ärztlichen Denkens" (The search for meaning as a category of medical thinking), by Oswald Schwarz (Schwarz 1923), and "Vitalismus und ärztliches Denken" (Vitalism and medical thinking), by Helmuth Plessner (1922; Plessner had been a pupil of Hans Driesch, who had been an ally of Wilhelm Roux), which considered the subject of the relation between experimental biology and medicine from slightly different perspectives. They all discussed the tension between medical science and medical practice and the problems of applying chemical and physical approaches used in the laboratory to individual persons who are seeking help. These contributions, which try to offer syntheses between laboratory bench and bedside, demonstrate the ambiguity of the position of medicine, which around 1920 still oscillated between systematic science-oriented medical researches and clinical practice oriented to different norms.

Self-Images and Resources

For an author like Jacques Loeb, his position was clear: he considered himself a strictly reductionist experimenter. He actively sought to disseminate and integrate the biochemical and physiological mindset into clinical medicine. Moreover, he wanted to move toward a synthetic phase of biological creation. In 1923, *Dorland's Medical Dictionary* added this approach to its inventory of medical concepts as "biomedicine." It defined biomedicine as "clinical medicine based on the principles of physiology and biochemistry" (Dorland 1923, 172). This definition did not specifically include the terms "synthesis" or "engineering," but the reference to physiology and biochemistry implied not only the observation and therapy of diseases but also the attempt to come to an experimental understanding of the development of disorder and disease. Although the concept of biomedicine as well as the term are usually attributed to a molecular understanding of health and disease on a (sub)cellular level, the basics of this concept had been well formulated by around 1920. In contrast to later approaches, the Dorland definition still included the clinical perspective of medicine, a dimension that, according to Ilana Löwy, seems to have been neglected somewhat in the historiography of biomedicine (Löwy 2011).

The Biologische Versuchsanstalt in Vienna, at least in the surgical and endocrinological works of Eugen Steinach, embodied the classical concept of biomedicine by its attention to possible clinical applications (see chapter 11). A concept like this could have had the power to reconcile medical science with medical practice at the beginning of the twentieth century; however, it can be argued that contemporary physicians and biologists invested much of their scientific effort in a strong positioning along the morphology–experiment, vitalism–mechanism, or causality–conditionality dichotomies. Such positioning was integral to their self-representation, and facilitated their ability to compete for academic relevance, resources, and positions. Therefore, they were not interested in harmony. Letters exchanged between Jacques Loeb and several of his colleagues provide ample evidence to support this argument. For instance, in his letter to Loeb regarding positions at the Kaiser Wilhelm Institute for Biology (which was in the process of being founded in 1912), Hans Driesch wrote that Oscar Hertwig, as a representative of the old morphology approach, was working against the appointment of his friend the experimental biologist Curt Herbst. Not only was Hertwig against Herbst getting this position, but "he was, in general, against an institute for developmental physiology because it would make for competition against his laboratory." Driesch added, "What an interesting perspective! The fact that Herbst was passed over in Freiburg again and that they gave Weismann's position to the totally mediocre Doflein was another sign that the German zoologists are really 'hopelessly reactionary'!"[19] After Herbst was not considered, Loeb wrote a letter of regret to Herbst using the same line of attack. He asserted that the American biology field was more open to experimentation than the German field. Loeb had just been appointed in 1910 to the Rockefeller Institute for Medical Research, which, under the guidance of Flexner and in combination with the Rockefeller Hospital, followed an ideal of basic research in medicine (Amsterdamska 2004). Loeb had his own department and felt rather free in this research environment to follow his experimental approach; he contrasted his own experimental freedom with the German situation as it was reported to him by his friends. He framed the issue not as a question of mentality, but as one of power relations within (medical, zoological, philosophical, etc.) institutions. Thus, he wrote to Herbst: "As far as the biological situation in Germany is concerned, I consider it almost as a shame. It seems that Oskar Hertwig, who has long ago ceased to be a modern biologist, can cause all kinds of mischief without any opposition whatsoever. If possible, he is worse than the pure zoologists."[20]

It should be noted that Hertwig, like Driesch, Herbst, and Loeb, was himself an active and conscious player in this intergenerational conflict about university positions. He also welcomed the (institutional) expansion of biology and embraced the idea that such expansion would result in new fields of action and job prospect for young biologists. Explicitly, he also praised the promotion biology had enjoyed in the United States since the 1890s (Hertwig 1913).

Loeb's us-versus-them attitude (supported by Driesch, Roux, and Herbst) was again reinforced when he had to read from one of his pupils that almost all zoological institutions in Germany were "wrapped up in morphology."[21] Similar messages had reached him from Austria and the BVA, where in 1908 Przibram complained that "Kammerer's postdoctorate had been delayed by those who were not pleased with the experimental direction."[22] Unsurprisingly, in his exchange of letters with Roux, "descriptive biologists," envy, and malevolence were recurrent topoi, too.[23] Along those same lines, the still-young Wolfgang Ostwald—who had been a student of Loeb between 1904 and 1906—complained in 1911 that "people [in Germany] know little about the independence of experimental biology as a science and as a taught discipline … and there is not one single assistant professor of experimental biology. Such works are only performed by zoologists or physicians as a secondary business."[24] This perspective on Germany was combined with Loeb's interpretation of the First World War, whose causation he attributed to the German Empire's claim to hegemony, and was fueled by the biological theories of the antiexperimentalists. All of these considerations induced Loeb after the war to exert influence more vigorously than ever before in favor of a scientific direction for biology and medicine (Fangerau 2007, 2009, 2010).

To accelerate the integration of the physical-chemical orientation into biology and medicine, Loeb created a series of books (Monographs on Experimental Biology, together with Thomas Morgan), founded the *Journal of General Physiology*, promoted the aforementioned *Journal of Experimental Zoology* (founded in 1904), and fostered research funding in Europe. In a draft for an announcement of the journal, Loeb reiterated his earlier position that medicine as a science did not adequately consider experimental biology. He further argued that physiology, as a fundamental subject of medicine, had developed under the pressures of practical medicine to become an applied science. Consequently (he argued), there was a demand for a journal that was interested in basic research that was quantitative and experimental.[25]

The scientific experimental approach per se—at least at the level of laboratory research—was hardly questioned in medicine at that time. Clinical observation at the bedside was no longer considered the only sufficient epistemological tool. Special locations related to diagnosis, therapy, teaching, and research were created. By the turn of the century, only very few clinical fields could work on an institutional level without at least a basic laboratory. For example, in 1894 the German psychiatrist Ernst Siemerling and the neurologist Alfred Goldscheider, in a clinical journal (*Klinisches Jahrbuch*), gave an overview of the various links between the laboratory and the bedside. Referring to the example of psychiatry, they expressed the clinical value of laboratory research in order to state in the end that laboratory research was a necessity for clinicians and that it stimulated clinical studies (Siemerling and Goldscheider 1894, 45). Consequently, they claimed that rooms for the laboratory should be integrated in hospitals and not be separated from the clinic (a concept later followed, for example, by the Rockefeller Hospital).

The role of experimental biology for medicine, however, remained a topic of discussion. On a practical level, just the example of Wolfgang Pauli's work at the BVA[26] shows how difficult the balancing act between work in a polyclinic and experimental work could be. In a letter to Jacques Loeb he complained that clinical work distracted him from experimenting,[27] and when he had the prospect of being promoted to director of a polyclinic he wrote to Loeb that the latter should not be puzzled if he accepted the position. Pauli's real interest was and would always remain scientific work.[28] On a theoretical level, clinical journals diligently promoted experimental science, especially with the help of pupils and friends of Roux, who wanted to foster the legacy of Roux's developmental mechanics. In 1920 and 1925—on Roux's 70th birthday and one year after his death—some articles appeared in the *Archiv für Orthopädische und Unfall-Chirurgie* (Archives of orthopedic and casualty surgery) that clarified the influence of Roux's approach and its lasting effect on practical medicine (Anton 1920; Wetzel 1925). In 1922, Ettisch published a series of articles on the effects of developmental mechanics on practical medicine in the *Deutsche Medizinische Wochenschrift* (German medical weekly) (Ettisch 1922a, 1922b, 1922c, 1922d). All of these articles emphasized the four periods of gestalt-determined life suggested by Roux, including phases of (1) prefunctional, inherited design, (2) a mixed intermediate period, (3) a phase marked by the predominance of the functional stimulus, and (4) a phase of the predominance of pure, "disease-free" senility (Wetzel 1925, 8). The authors argued that practical medicine could gain useful insights from developmental mechanics by incorporating the concept of these periods, with many applications to genetics, fertilization, tumor research, transplantation medicine, and tissue regeneration. At the end of his article series, Ettisch again made a connection to the physical-chemical consideration of life. In the spirit of the mechanistic-synthetic approaches propagated by Loeb, Ettisch emphasized that "it is only a matter of time" before the precipice between the inorganic and organic sciences would be bridged (Ettisch 1922d, 776).

Conclusion

Possible applications of experimental biology to medicine that were proposed as new around the year 1900 and mentioned again in the 1920s concerned the fields of regeneration, transplantation, control of fertilization, and hereditary transmission. Almost 30 years later, Jean Rostand outlined exactly these fields in describing his future vision of optimizing the human experience. Within half a century, the biomedical perspective initiated by, for example, the physiologist Jacques Loeb, on the basis of Roux's program, had set in place the conceptual framework that Roux and Loeb had envisioned in 1900. Similar to Anatole France, Loeb expressed the following idea in a letter to Ernst Mach in 1890: Only with the successful development of experimental abiogenesis would biology begin to be a science. In later years, he campaigned intensively for implementing a synthetic vision of biological work, meaning that he did not want only to observe results of experimentation but to synthetically create living matter as well (Fangerau 2010, 2012).

While for biologists the shaping of biology as an experimental science seems to have been crucial for their self-understanding as life scientists,[29] medical faculties in Central Europe at the same time discussed their relation to biology. For both professions it was necessary to clarify their relationship and their role in academic organizations. Due to specialization, neither medicine nor biology as a whole had anything to do with the disciplines they had been at the beginning of the nineteenth century. Consequently, the biologist and hygienist Vladislav Růžička argued for an alliance of medicine and biology in the medical faculties in order to (a) institutionalize experimental biology and (b) work for the unifying principles of medicine as a life science (Růžička 1922). According to a summarizing analysis from 1922 published by Růžička, who had been a pupil of Oscar Hertwig's brother Richard in Munich, in Roux's monograph series the specialization of medicine went along with a growing study load for medical students. As a consequence, Růžička asked which kind of biology should be taught in medical faculties. Not surprisingly, he reminded the reader that the experimental reform of biology had been promoted by trained physicians. He saw a great divide between morphological zoology and medicine, and thus argued for the integration of what he and others called general biology (Allgemeine Biologie) in the medical curriculum (instead of botany and zoology).[30] According to his views, general biology had become a legitimate, separate discipline through the works of Roux, Driesch, Schaxel, and others, and the medical faculties should be the place to establish an institutionalized general biology (zoology often had been located in philosophical faculties). In the medical faculties an institutionalized general biology should complement other basic medical sciences like anatomy and physiology. In this environment it could serve to find answers to basic questions regarding living matter, and it could be of good use for applied medical fields such as "regeneration, transplantation, tissue cultures, ageing and so on" (Růžička 1922, 17). This general biology, however, should be based on the causal principles established by Roux for his developmental mechanics (23).

In the face of statements like Růžička's, which were taken up by later authors like Rostand, a hypothetical question emerges (which is perhaps superfluous because it is counterfactual): Biology and physiology were already, in the middle of nineteenth century, well on the way to becoming experimental sciences, and, as such, basic research in medicine (e.g., in physiology) also was already deeply anchored in the laboratory. However, would these sciences have developed in the same manner without the express formulation of a developmental mechanics program and its continuation in the form of "synthetic morphology"? It seems indisputable that scientists such as Roux and Loeb provided essential catalyzing impulses for the use of manipulative and formative approaches in the life sciences laboratory. They inspired colleagues like Hans Przibram, who, explicitly following the way they had paved, tried to institutionalize this approach in the BVA. Furthermore, obvious successful representations of the link between experimental biology and medicine—personified, for example, in Eugen Steinach's or Wolfgang Pauli's work at the BVA—proved the validity and relevance of their approaches. Thus, the rhetoric and conceptual impulses of developmental mechanics and synthetic morphology were so powerful that they were brought into practical medicine not only by pure propagation, but also by compelling practical medicine to join a sustained debate about their effects that changed the practice of physicians. Debates about mechanization and experimentalization and their effects on clinical work last until today.

Notes

1. If not otherwise stated, all translations are by the author.

2. Garland Allen, for example, confronted a morphological with an experimental approach, the latter of which eclipsed the first at the end of the nineteenth century (see, e.g., Allen 1975). Several other authors differentiated this narrative—see among others Benson (1985), Maienschein (1987, 1994, 1999), Müller (1976), Pauly (1984, 1987, 1988), Penzlin (2004), and Laubichler and Maienschein (2007).

3. See also the respective articles in this volume.

4. Der Zweck "sollte die Pflege der experimentellen Zweige der Biologie, insbesondere jener Richtung sein, welche die Abhängigkeit der morphologischen Gestaltung von Bedingungen verschiedenster Art prüft und die in neuerer Zeit nach Roux als Entwicklungsmechanik … bezeichnet wird." Bericht der Kommission betr. Übergabedokument und Statut (ad No. 340/1913), Österreichische Akademie der Wissenschaften, Archiv, Akten zur Biologischen Versuchsanstalt, Karton 1, Mappe 1.

5. "vorwiegend der descriptiven Forschung, der systematischen Beschreibung" (Przibram 1903, 152).

6. Regarding the historical caesura, see Rheinberger and Hagner (1993, 10). A discussion of this caesura can be found in Hagner (2003) or Cunningham (2002, 2003).

7. On Ettisch, who later had to leave Germany due to the Nazi terror, see Rürup and Schüring (2008, 187–188).

8. "… aber auch in wissenschaftshistorischer Beziehung von höchster Bedeutung ist, ohne daß ihr indessen bisher in Deutschland eine dieser Bedeutung entsprechende Verbreitung und Anerkennung beschieden war."

9. Also printed as a book (Roux 1897b).

10. "das neue Ziel oder die Aufgabe der neuen Zukunftswissenschaft."

11. Vorschlag für ein Statut der Biologischen Versuchsanstalt, Ausstattung der Arbeitsplätze, Österreichische Akademie der Wissenschaftem, Archiv, Akten zur Biologischen Versuchsanstalt, Karton 1, Mappe 1, p. 19a.

12. "Durch Isolation, Verlagerung, Zerstörung, Schwächung, Reizung, falsche Verbindung, passive Deformation, Änderung der Ernährung und der Funktionsgröße von Theilen des Eies, Embryos oder weiter ausgebildeten Individuums, sowie durch besondere Einwirkung von Agentien … wird es uns möglich sein vielerlei gestaltende Wirkungen der Theile der Organismen auf einander kennen zu lernen."

13. Nathan Zuntz to Jacques Loeb 04.12.1908, Loeb Papers, Library of Congress, Washington, DC.

14. Jacques Loeb to Simon Flexner 1910, lost manuscript (Osterhout 1928, xvii–xix).

15. "… unser ganzes Denken, unsere Methodik, ist naturwissenschaftlich geworden, und es gibt keine noch so abstrakte Forschung der Chemie, Physik und Biologie, die nicht eines Tages unser Wissen und Können umwälzend bereichern könnte … Nicht minder eng sind unsere Beziehungen zur Technik."

16. "Wir haben in der Tat eine biologische Wissenschaft, die spezielle Pathologie, wir gewinnen in einzelnen Zweigen auch eine rationelle, auf biologischer Grundlage fußende Therapie. Ein großer Teil des therapeutischen Feldes selbst aber … ist Sache des Persönlichen, des Menschlichen, des Künstlerischen."

17. Bütschli also commented on the dispute of Roux and Hertwig (Bütschli 1901, 68ff.).

18. At the same time health insurance and the health care system were seen both as responsible for the crisis and in a crisis as well. On the perception of a "crisis in medicine" in the Weimar Republic see, e.g., Geiger (2010) and Schmiedebach (1989).

19. Hans Driesch to Jacques Loeb 25.05.1912, Loeb Papers, Library of Congress, Washington, DC. ("Es heisst, er [Hertwig] sei überhaupt gegen ein Institut für Enwickl. physiologie, weil es seinem Laboratorium Concurrenz machen werde! Ein hübscher Gesichtspunkt! Daß man in Freiburg Herbst wieder übergangen und den ganz unbedeutenden Doflein an Weismanns' Stelle gesetzt hat, war auch wieder ein Zeichen dafür, daß die deutschen Zoologen wirklich 'hopelessly reactionary' sind!").
   This letter has been cited by Reinhard Mocek (1987) as well, who referred also to the dispute between Hertwig and Roux from the perspective of Roux, Driesch, and Jacques Loeb. The referenced article by Mocek uses archival material, which I have used in my book on Jacques Loeb (Fangerau 2010). However, I did not know Mocek's paper before 2014 when Kai Torsten Kanz, whom I thank for that, sent a copy of it to me.

20. Jacques Loeb to Curt Herbst 29.12.1913, Loeb Papers, Library of Congress, Washington, DC. ("Was die biologische Situation in Deutschland angeht, betrachte ich es nahezu als Schande. Es scheint, dass Oskar Hertwig, der schon lange aufgehört hat, ein moderner Biologe zu sein, den ganzen Unfug machen kann ohne irgendeine Opposition. Er ist wirklich, wenn das möglich ist, schlimmer als die reinen Zoologen").

21. Wolfgang Felix Ewald to Jacques Loeb 22.01.1912. Loeb Papers, Library of Congress, Washington, DC.

22. Hans Przibram to Jacques Loeb 24.07.1908. Loeb Papers, Library of Congress, Washington, DC ("wenig über die experimentelle Richtung erbauten Professoren").

23. Wilhelm Roux to Jacques Loeb 01.10.1906, Wilhelm Roux to Jacques Loeb 10.12.1910, Wilhelm Roux to Jacques Loeb 06.04.1913, Jacques Loeb to Heinrich Schmidt 08.12.1913, Loeb Papers, Library of Congress, Washington, DC.

24. Wolfgang Ostwald to Jacques Loeb 06.02.1911. Loeb Papers, Library of Congress, Washington, DC. ("Obgleich ich ja selbst nicht mehr direkt daran beteiligt bin, so spüre ich deutlich, wie wenig die Leute hier in Deutschland über die Selbständigkeit der experimentellen Biologie als Wissenschaft und Lehrfach wissen. Wir haben in Deutschland ja noch nicht einen einzigen Privatdozenten direkt für experimentelle Biologie. Derartige Arbeiten werden nur von Zoologen und Medizinern als Nebenbeschäftigung betrieben").

25. Draft by Jacques Loeb, commented by Simon Flexner enclosed to a letter by Simon Flexner to Jacques Loeb 25.05.1918. American Philosophical Society, Flexner Papers.

26. See chapter 10 in this volume.

27. Wolfgang Pauli to Jacques Loeb (without date, prior to 1903), Loeb Papers, Library of Congress, Washington, DC. Paul Ehrlich complained to Loeb in a similar way when he remembered that when he was younger he had worked in a clinic with only limited space and limited time for laboratory work. Paul Ehrlich to Jacques Loeb (03.07.1906), Loeb Papers, Library of Congress, Washington, DC.

28. Wolfgang Pauli to Jacques Loeb (01.01.1906), Loeb Papers, Library of Congress, Washington, DC.

29. See also, for example, Driesch (1905), and von Uexküll (1913).

30. On general biology see chapter 6 in this volume.

References

Allen, Garland E. 1975. *Life Science in the Twentieth Century*. New York: Wiley.

Amsterdamska, Olga. 2004. Research at the hospital of the Rockefeller Institute for Medical Research. In *Creating a Tradition of Biomedical Research: Contributions to the History of the Rockefeller University*, ed. Darwin H. Stapleton, 111–126. New York: Rockefeller University Press.

Anonymous. 1897a. Developmental mechanics: Review on Wilhelm Roux's Programm und Forschungsmethoden der Entwicklungsmechanik der Organismen, leichtverständlich dargestellt. *Nature* 57 (1484): 531–533.

Anonymous. 1897b. Mechanism and biology: Review on Oscar Hertwig: Zeit- und Streitfragen der Biologie. *Nature* 56 (1440): 98–100.

Anton, Gabriel. 1920. Was bedeutet die Entwicklungsmechanik von W. Roux für den Arzt? *Archiv für orthopädische und Unfall-Chirurgie* 28:551–558.

Benson, Keith R. 1985. American morphology in the late 19th century: The Biology Department at Johns Hopkins University. *Journal of the History of Biology* 18 (2): 163–205.

Bentley, Jerry H. 1996. Cross-cultural interaction and periodization in world history. *American Historical Review* 101 (3): 749–770.

Bütschli, Otto. 1901. *Mechanismus und Vitalismus*. Leipzig: Engelmann.

Cohen, H. Floris. 1994. *The Scientific Revolution: A Historiographical Inquiry*. Chicago: University of Chicago Press.

Cohen, H. Floris. 2010. *How Modern Science Came into the World*. Amsterdam: Amsterdam University Press.

Cooper, Melinda. 2004. Regenerative medicine: Stem cells and the science of monstrosity. *Medical Humanities* 30:12–22.

Cooper, Melinda. 2009. Regenerative pathologies: Stem cells, teratomas and theories of cancer. *Medicine Studies* 1 (1): 55–66.

Cunningham, Andrew. 2002. The pen and the sword: Recovering the disciplinary identity of physiology and anatomy before 1800. I: Old physiology—the pen. *Studies in History and Philosophy of Science. Part C, Studies in History and Philosophy of Biological and Biomedical Sciences* 33 (4): 631–665.

Cunningham, Andrew. 2003. The pen and the sword: Recovering the disciplinary identity of physiology and anatomy before 1800: II, Old anatomy—the sword. *Studies in History and Philosophy of Science. Part C, Studies in History and Philosophy of Biological and Biomedical Sciences* 34 (1): 51–76.

de Bont, Raf. 2009. Between the laboratory and the deep blue sea: Space issues in the marine stations of Naples and Wimereux. *Social Studies of Science* 39 (2): 199–227.

Diepgen, Paul. 1931. Vitalismus und Medizin im Wandel der Zeiten. *Klinische Wochenschrift* 10:1433–1438.

Dorland, William Alexander Newman. 1923. *The American illustrated Medical Dictionary*. 12th ed. Philadelphia: W. B. Saunders.

Driesch, Hans. 1897. Über den Werth des biologischen Experiments. *Archiv für Entwicklungsmechanik der Organismen* 5 (1): 133–142.

Driesch, Hans. 1905. Das System der Biologie. *Süddeutsche Monatshefte* 6:1–16.

Ettisch, Georg. 1922a. Entwicklungsmechanik und praktische Medizin I. *Deutsche Medizinische Wochenschrift* 48 (19): 631–633.

Ettisch, Georg. 1922b. Entwicklungsmechanik und praktische Medizin II. *Deutsche Medizinische Wochenschrift* 48 (20): 662–664.

Ettisch, Georg. 1922c. Entwicklungsmechanik und praktische Medizin III. *Deutsche Medizinische Wochenschrift* 48 (22): 731–733.

Ettisch, Georg. 1922d. Entwicklungsmechanik und praktische Medizin IV. *Deutsche Medizinische Wochenschrift* 48 (23): 773–776.

Fangerau, Heiner. 2007. Biology and war: American biology and international science. *History and Philosophy of the Life Sciences* 29:395–428.

Fangerau, Heiner. 2009. From Mephistopheles to Isaiah: Jacques Loeb, technical biology and war. *Social Studies of Science* 39 (2): 229–256. doi: 10.1177/0306312708101045.

Fangerau, Heiner. 2010. *Spinning the Scientific Web. Jacques Loeb (1859–1924) und sein Programm einer internationalen biomedizinischen Grundlagenforschung*. Berlin: Akademie Verlag.

Fangerau, Heiner. 2012. Zur Geschichte der Synthetischen Biologie. In *Synthetische Biologie. Entwicklung einer neuen Ingenieurbiologie? Themenband der interdisziplinären Arbeitsgruppe Gentechnologiebericht*, ed. Kristian Köchy and Anja Hümpel, 61–84. Dornburg: Forum W-Wissenschaftlicher Verlag.

France, Anatole. 1892. *La vie littéraire. Quatrième série*. Paris: Calmann-Lévy.

Geiger, Karin. 2010. "Krise"–zwischen Schlüsselbegriff und Schlagwort. Zum Diskurs über eine "Krise der Medizin" in der Weimarer Republik. *Medizinhistorisches Journal* 45 (3/4): 368–410.

Hagner, Michael. 2003. Scientific medicine. In *From Natural Philosophy to the Sciences: Writing the History of Nineteenth-Century Science*, ed. David Cahan, 49–87. Chicago: University of Chicago Press.

Harrison, Ross G. 1945. Retrospect. *Journal of Experimental Zoology* 100 (3): xi–xxxi.

Hertwig, Oscar. 1897. *Mechanik und Biologie. Mit einem Anhang: Kritische Bemerkungen zu den entwicklungsmechanischen Naturgesetzen von Roux*. Volume 2, Zeit- und Streitfragen der Biologie. Jena: G. Fischer.

Hertwig, Oscar. 1913. Naturwissenschaft und Biologie. *Naturwissenschaften* 1 (1): 2–3.

His, Wilhelm, Jr. 1913. Arzt und Naturwissenschaften. *Naturwissenschaften* 1 (1): 3–4.

Howell, William H. 1923. Biographical memoir: Samuel James Meltzer. *Biographical Memoirs, National Academy of Sciences (USA)* 21:1–23.

Klein, Ursula. 2008a. Die technowissenschaftlichen Laboratorien der Frühen Neuzeit. *NTM Zeitschrift für Geschichte der Wissenschaften, Technik und Medizin* 16:5–38.

Klein, Ursula. 2008b. The laboratory challenge: Some revisions of the standard view of early modern experimentation. *Isis* 99 (4): 769–782.

Kofoid, Charles Atwood. 1910. *The Biological Stations of Europe*. Washington, DC: Government Printing Office.

Laubichler, Manfred D., and Jane Maienschein. 2007. *From Embryology to Evo-Devo: A History of Developmental Evolution*. Cambridge, Mass.: MIT Press.

Löwy, Ilana. 2011. Historiography of biomedicine: "Bio," "medicine," and in between. *Isis* 102 (1): 116–122.

Maienschein, Jane. 1987. Physiology, biology, and the advent of physiological morphology. In *Physiology in the American Context 1850–1940*, ed. Gerald L. Geison, 177–193. Bethesda: American Philosophical Society.

Maienschein, Jane. 1989. *100 Years Exploring Life, 1888–1988: The Marine Biological Laboratory at Woods Hole*. Boston: Jones and Bartlett.

Maienschein, Jane. 1994. The origins of Entwicklungsmechanik. In *A Conceptual History of Modern Embryology*, ed. Scott F. Gilbert, 43–61. Baltimore: Johns Hopkins University Press.

Maienschein, Jane. 1999. Diversity in American biology, 1900–1940. *History and Philosophy of the Life Sciences* 21 (1): 35–52.

Martius, Friedrich. 1922. Einige Bemerkungen über die Grundlagen des ärztlichen Denkens von heute. *Klinische Wochenschrift* 1 (2): 49–53.

Mocek, Reinhard. 1987. 100 Jahre Hemiembryonen: Aus Briefen deutscher Entwicklungsphysiologen. *Arbeitsblätter zur Wissenschaftsgeschichte* 19:91–103.

Morgan, Thomas H. 1898. The biological problems of today: Developmental mechanics. *Science* 7 (162): 156–158.

Müller, Irmgard. 1976. *Die Geschichte der zoologischen Station in Neapel von der Gründung durch Anton Dohrn (1872) bis zum ersten Weltkrieg und ihre Bedeutung für die Entwicklung der modernen biologischen Wissenschaften. Univ. Habil.-Schr., Institut für Geschichte der Medizin*. Düsseldorf: HHU.

Müller, Otfried. 1913. Medizinische Wissenschaft und ärztliche Kunst. *Die Naturwissenschaften* 1 (3, 4): 69–72, 87–90.

Osterhammel, Jürgen. 2006. Über die Periodisierung der neueren Geschichte. *Berlin-Brandenburgische Akademie der Wissenschaften, Berichte und Abhandlungen* 10:45–64.

Osterhout, W. J. V. 1928. Jacques Loeb. *Journal of General Physiology* 8:ix–lix.

Pauly, Philip J. 1984. The appearance of academic biology in late nineteenth century America. *Journal of the History of Biology* 17 (3): 369–397.

Pauly, Philip J. 1987. *Controlling Life: Jacques Loeb and the Engineering Ideal in Biology*. New York: Oxford University Press.

Pauly, Philip J. 1988. Summer resort and scientific discipline: Woods Hole and the structure of American biology, 1882–1925. In *The American Development of Biology*, ed. Ronald Rainger, Keth R. Benson, and Jane Maienschein. New Brunswick, London: Rutgers University Press.

Penzlin, Heinz. 2004. Die theoretische und institutionelle Situation in der Biologie an der Wende vom 19. zum 20. Jh. In *Geschichte der Biologie*, ed. Ilse Jahn, 431–440. Hamburg: Nikol Verlag.

Plessner, Helmuth. 1922. Vitalismus und ärztliches Denken. *Klinische Wochenschrift* 1 (39): 1956–1961.

Przibram, Hans. 1903. Die neue Anstalt für experimentelle Biologie in Wien. *Verhandlungen der deutschen Gesellschaft für Naturforscher und Ärzte* 2. Teil, 1. Hälfte: 153–155.

Reinke, Johannes. 1902. Bemerkungen zu O. Bütschli's "Mechanismus und Vitalismus." *Biologisches Zentralblatt* 22:23–29, 52–60.

Reiter, Wolfgang L. 1999. Zerstört und vergessen: Die Biologische Versuchsanstalt und ihre Wissenschaftler/innen. *Österreichische Zeitschrift für Geschichtswissenschaften* 10 (4): 585–614.

Rheinberger, Hans-Jörg, and Michael Hagner. 1993. Experimentalsysteme. In *Die Experimentalisierung des Lebens. Experimentalsysteme in den biologischen Wissenschaften 1850/1950*, ed. Hans-Jörg Rheinberger and Michael Hagner. Berlin: Akademie Verlag.

Rostand, Jean. 1950. *La biologie et l'avenir humain: Uchronie scientifique, Collection Descartes pour la vérité*. Paris: Michel.

Roux, Wilhelm. 1897a. Für unser Programm und seine Verwirklichung. *Archiv für Entwicklungsmechanik der Organismen* 5 (1): 1–80, 219–342.

Roux, Wilhelm. 1897b. *Programm und Forschungsmethoden der Entwicklungsmechanik der Organismen. Leichtverständlich dargestellt; zugleich eine Erwiderung auf O. Hertwig's Schrift: Biologie und Mechanik*. Trans. O. Hertwig. Leipzig: Engelmann.

Rürup, Reinhard, and Michael Schüring. 2008. *Schicksale und Karrieren: Gedenkbuch für die von den Nationalsozialisten aus der Kaiser-Wilhelm-Gesellschaft vertriebenen Forscherinnen und Forscher*. Göttingen: Wallstein Verlag.

Růžička, Vladislav. 1922. *Die allgemeine Biologie als Lehrgegenstand des medizinischen Studiums. Ein Gutachten vorgelegt den Regierungen Mitteleuropas*. In *Vorträge und Aufsätze über Entwicklungsmechanik der Organismen*, vol. 29, ed. Wilhelm Roux, Berlin: Springer.

Schmiedebach, Heinz-Peter. 1989. Der wahre Arzt und das Wunder der Heilkunde. Erwin Lieks ärztlich-heilkundliche Ganzheitsideen. *Das Argument* 162 (Der ganze Mensch und die Medizin): 33–53.

Schwarz, Oswald. 1923. Die Sinnfindung als Kategorie des ärztlichen Denkens. *Klinische Wochenschrift* 2 (24): 1129–1132.

Siemerling, Ernst, and Alfred Goldscheider. 1894. Die Notwendigkeit der Laboratorien bei klinischen Instituten. *Klinisches Jahrbuch* 5:29–47.

Sucker, Ulrich. 2002. *Das Kaiser-Wilhelm-Institut für Biologie. Seine Gründungsgeschichte, seine problemgeschichtlichen und wissenschaftstheoretischen Voraussetzungen (1911–1916)*. Stuttgart: Steiner.

von Uexküll, Jakob. 1913. Der heutige Stand der Biologie in Amerika. *Naturwissenschaften* 1 (34): 801–805.

Welch, William. 1896. Introduction. *Journal of Experimental Medicine* 1 (1): 1–3.

Wetzel, Georg. 1925. Stellung und Bedeutung Wilhelm Rouxs in der Morphologie und in der Heilkunde. *Archiv für orthopädische und Unfall-Chirurgie* 24 (1): 1–13.

6 The Emergence of Theoretical and General Biology: The Broader Scientific Context for the Biologische Versuchsanstalt

Manfred D. Laubichler

Around 1900 the concept of general biology began catching on as a theoretically driven objective of science, a philosophically inspired view of life, an organizing principle for research and teaching, and a way of presenting biology to lay audiences.[1] This development took place primarily in German-speaking countries. Various factors contributed to the emergence of general biology, including the nascent discourse on "theoretical biology" and above all the focus on the autonomy of biology and its status as a fundamental, independent branch of science; the discussions on reforming medical studies, once again being conducted quite intensely at the time; the related debate about the purpose of preclinical subject areas and how much medical students need to know about the basic principles of living systems; and, finally, the popularization of science, which was already well under way around 1900.[2] In this sense, the crystallization of the general perspective in biology followed a pattern consistent with similar developments in other disciplines in the natural sciences and humanities.[3] Furthermore, the idea of the "general" had a constitutive function for biology, because it was within the framework of these discussions that the field emerged as a coherent scientific discipline in the first place. As a result the "general" achieved a special status. This chapter aims to examine the constituent components of this general perspective in biology, which provided an essential context for the founding of the Biologische Versuchsanstalt (BVA) in Vienna, in 1902.

The term *theoretische Biologie* (theoretical biology) was first introduced by the Kiel-based botanist Johannes Reinke in *Einleitung in die Theoretische Biologie*, published in 1901 (Reinke 1901). However, the question of the extent to which biology could be considered a fundamental independent branch of science was discussed as early as 1893 in a publication by Hans Driesch (Driesch 1893). These two works marked the start of an intense debate on the conceptual foundations of biology that in the following decades proved influential in German-language biology. The debate was essentially focused on (1) the interpretation of empirical and experimental findings, to some extent as a reaction to the data crisis becoming evident around 1900, as an increasingly unmanageable number of findings and observations confronted an inadequate conceptual and theoretical framework; (2) building on this interpretation of findings, an attempt to establish the foundations of biology in the form of a system of genuine biological processes and objects; and (3) a critical evaluation of the epistemological and methodological premises of biological research. In this sense Reinke's provisional definition of theoretical biology has lost none of its relevance today:[4]

The results of empirical biology are the subject of theoretical biology. But the task of theoretical biology is not only to determine the basic principles of biological processes, but also to examine the principles on which our biological views rest. The value of theoretical biological discussions must be judged against the notion that knowledge is all the more important, the more general it is, the broader its ramifications, and the more details it encompasses. (Reinke 1901, iii)

Reinke refers explicitly to an epistemological position in which the general has a special value. Here "general" is understood to mean "universally valid," and at the same time it is assigned to "theoretical biological discussions." The related goal of theoretical biology—to explore the fundamental problems of life and thus to assume a special position within the system of the biological sciences—was difficult to achieve at the start. Reinke was fully aware of this problem:

Nothing is further from my thoughts than a comprehensive theory of life or a theoretical biology textbook. The scope of such a work would be quite substantial and a large number of individuals would have to join forces for its publication, not only physiologists and anatomists, but also taxonomists, zoologists and botanists … This is why I have called this book an introduction, setting limitations on its objectives, since it may in the future inspire researchers to create a theoretical biology. (Reinke 1901, iii–iv)

Whereas in the early twentieth century theoretical biology mainly represented a promise or a vision of the future, and as such provoked a debate lasting for decades, in one of its forms it also had far-reaching and very practical effects. After all, even though in the early nineteenth century a concept of biology was circulating that from today's perspective is clearly recognizable as an interpretation of the natural world (Jahn 1998), throughout the nineteenth century biology did not exist as either a research program or the subject of teaching. It is therefore no exaggeration to say that biology at this time was a concept in search of a science. While it is true that the idea of biology as the science of life was intuitively graspable for most researchers and intellectuals, it was anything but clear how "life" was to be understood as a research subject, what the elements of a science of life were, and what methodology scientists should use to investigate life. As a first comprehensive answer to such questions, which in the final years of the nineteenth century found expression in an increasingly intense debate on the foundations of biology and its philosophical consequences (as in Reinke's *Theoretische Biologie*), general biology arose as the first visible sign of the consolidation of the life sciences (Daum 1998; Ziche 2000). The formulation of general biology with elements of a research program, curricular innovations, and small steps of institutional reform was therefore one of the first accomplishments of theoretical biology.

However, the significance of "general biology" was not confined to these debates, which primarily addressed theoretical and philosophical questions. Many of the early treatises on general biology were published in the form of textbooks; these had a far greater influence than most other writings on theoretical biology because they helped to constitute biology as an integrated science in the first place.[5] In other words, biology as an independent branch of science first emerged within the context of teaching. Specific research programs followed over the next few years. The establishment of new research facilities such as the Kaiser Wilhelm Institute for Biology in Dahlem, Berlin, and the Biologische Versuchsanstalt in Vienna, better known as the Prater Vivarium, also played an important role in the establishment of both general and theoretical biology. These were private (Vivarium) and semiprivate (Kaiser Wilhelm) institutions.[6] They demonstrate, on the one hand, that influential circles in society recognized the importance of biology for both the advancement of science and the surrounding philosophical discourse and supported the discipline; and, on the other, that despite that widespread support, university structures adapted only slowly to the new conditions. Researchers at these private institutes were concerned primarily with topics in experimental biology and its theoretical foundations. They focused on problems related to development, heredity, physiology, and behavior. The investigation of these comprehensive general biological problems thus made a significant contribution to the discourse in both theoretical and general biology. It must also be emphasized that many of the representatives of these disciplines did not initially (or ever) have permanent academic positions. If they were not independently wealthy, they were forced to earn an often precarious living mainly through writing and journalism. As a result, the ideas associated with general biology were made accessible to a broad, scientifically interested public. This transfer took place at many different levels, ranging from the encyclopedic conceptual treatment of biology in the monumental work "Die Kultur der Gegenwart" to small inexpensive volumes in popular series such as "Aus Natur und Geisteswelt." Books on general biology were part of various popular book series. Examples include Carl Chun and Wilhelm Johannsen's (1915) edited volume *Allgemeine Biologie* as part of the Kultur der Gegenwart series, Richard Goldschmidt's highly successful popular account *Ascaris: Eine Einführung in die Wissenschaft vom Leben für Jedermann* (Goldschmidt 1922); Hugo Miehe's *Allgemeine Biologie* (Miehe 1915); and, in the context of the BVA, Paul Kammerer's various popular writings, including his *Allgemeine Biologie* (Kammerer 1915).

The dimensions of general biology discussed here clearly reveal the various aspects of "general" that played a role in other scientific disciplines at the turn of the century. At that time, "general" meant comprehensive and universal at the conceptual level, authoritative and fundamental at the institutional and educational level, and comprehensible at the popular level. In this sense, general biology exemplified the concept of general in the sciences.

Biology as a Fundamental Independent Branch of Science and the Formulation of General Biology

In 1893 Hans Driesch published a brief programmatic analysis titled *Die Biologie als selbständige Grundwissenschaft: Eine kritische Studie* (Driesch 1893). The question of how biology could be established as a fundamental independent scientific discipline was in itself not new. Auguste Comte and Herbert Spencer had—each in his own way— assigned biology a place in the overall system of the sciences. Earlier, Ernst Haeckel (1866) had also commented on this question, though not too explicitly, as part of a comprehensive systematic discussion of biological concepts, disciplines, and issues. Compared to these analytical discussions, Thomas Henry Huxley's approach to biology was much more pragmatic. The laboratory course he created on general biology combined zoological, botanical, morphological, and physiological methods in the scientific investigation of life. Of all these definitions and discussions of biology in the nineteenth century, Huxley's course, which served primarily to educate teachers, was by far the most influential (Desmond 1997; Geison 1978).

However, by the end of the nineteenth century, when Driesch began grappling with these issues, the conceptual landscape of the life sciences had changed fundamentally (Driesch 1893). All the elements of what would later be understood as biology now existed, at least in rudimentary form. Whether it was a catalog of the building blocks of life—tissues, cells, and molecules—or a basic understanding of the fundamental life processes—development, heredity, regulation, metabolism, behavior, and evolution—the basic elements of (general) biology were already the subject of intense debate, even if the question of what exactly was to be understood by biology as a science had not yet been clarified. Equally unclear, and also the subject of intense philosophical and ideological debate, was the relationship between the phenomena of life and their underlying physicochemical processes. Here a materialist, mechanistic, and monistic position stood in opposition to different vitalist and life-philosophical views that drew primarily on the diverse regulatory phenomena within organisms. Vitalists argued that it was possible to explain these phenomena only on the basis of "vital forces." The materialist counterposition was that this view endangered the unity of nature and the natural sciences.

In his book Driesch analyzed, from a Kantian perspective, the problems associated with the explanation of the life processes and the autonomy of biology. Most exponents of materialism granted biology a pragmatic autonomy. They believed that, due to their complexity, the life processes could be analyzed from independent viewpoints, but this did not change the basic physicochemical determinism of organic events. By contrast, Driesch introduced the idea of a strict ontological separation of the various natural phenomena. In his view, organic phenomena were characterized by the principles of change, development, teleology, goal-directedness, regulation and self-preservation. The existence of these specific properties permitted an a priori differentiation into various natural fields, which were determined by independent laws of nature. Such a strict separation of nature into different fields guaranteed the autonomy of biology as a fundamental scientific discipline, while within the field of biology the various biological disciplines could be combined into a coherent system of biological theories.

The arguments presented in this discussion by Driesch and a number of other representatives of the "autonomy of the organic" were largely conceptual, but also had empirical foundations. However, by the late nineteenth century a number of genuine biological theories existed that were oriented not toward philosophical considerations but primarily toward empirical results. They were focused on the problems of development and above all the question of the differentiation and determination of cells in the course of ontogenesis. As early as the 1880s, Wilhelm Roux (1881) had attempted to explain ontogenesis both mechanistically and using the Darwinian idea of selection (here with respect to the cells in the embryo). August Weismann, who in the late nineteenth century was rightfully considered one of the most important exponents of theoretical biology, also attempted to arrive at a mechanistic understanding of differentiation and determination, which he regarded as genuine biological properties (Sander 1985; Churchill 2015). In other words, at this time there was a broad consensus on the main phenomena and problems of biology, above all on the problem of development. The only disagreement concerned the details of the conceptual structure within which these phenomena were to be understood. It was these very questions that—following Driesch's famous experiments with sea urchin embryos—inspired his ideas about the autonomy of biology and his vitalist conception of the organic world (Driesch 1894, 1911; Mocek 1974, 1998). And it was these very questions that, as we have seen, formed the heart of the discourse in theoretical biology.

For studies of the origins and foundation of general biology as part of theoretical biology, Julius Schaxel, the "red professor" from Jena, is of great importance (Fricke 1964; Hopwood 1997; Reiß 2007). One of the most controversial figures in biology in the early twentieth century, Schaxel treated general biology from each of the three important perspectives described here: the conceptual, the educational, and the popular. In his own writings—first and foremost the two 1919 works *Über die Darstellung allgemeiner Biologie* and *Grundzüge der Theorienbildung in der Biologie*—he devoted himself to biology's conceptual problems (Schaxel 1919a, 1919b). In addition, as editor of *Abhandlungen zur Theoretischen Biologie*, he played a central role in further developing theoretical biology after the First World War. More than thirty volumes of this series were published between 1919 and 1930.

Schaxel regarded the clarification of theoretical issues as one of the central conditions for further advances in the life sciences, the goal of which was the synthesis of general biology. The attainment of this goal required a historical understanding of key developments in biology. On the basis of that understanding, scientists could then create a system of biological concepts that would allow them to bring together all relevant biological knowledge. For Schaxel the essence of general biology lay not in a more or less accidental stringing together of biological facts, but in their integration through theoretical concepts. Within this framework, he distinguished three basic properties of life: (1) life is based on matter, motion, and energy, (2) organisms are historical beings that have achieved their present state as a result of an ongoing development process, and (3) organisms are distinguished by unity, totality, and individuality. On the one hand, the properties of life addressed in these concepts guaranteed the independence of biology as a science. On the other, the concepts' further elaboration, together with the empirical studies they inspired, formed the basis of a genuine general biology.

For Schaxel, general biology was of great scientific value, but he was well aware that the usual education in biological and medical subjects was not well suited for achieving progress in this type of science. Biological knowledge was for the most part taught in a fragmentary way; nowhere did one find unifying concepts and educational resources that were capable of generating enthusiasm among a new generation of scientists for the idea of general biology. Schaxel attempted to remedy this problem on several levels. In the first place, he wrote manifestos and reports about the reform of biology courses at universities and medical schools. Among other things, he demanded that experimental biology— which was opposed in many places by tradition-bound professors—be given enhanced status and recognition. He himself had only been able to conduct his experimental work (mainly on axolotl) in Jena with the financial assistance of the Carl Zeiss Foundation. However, Schaxel's activities were not limited to reports and manifestos. During the (short) phase of a communist government in Thuringia, he attempted to put his ideas into action as the undersecretary responsible for universities. As was to be expected, he made many enemies in the process, severely impeding his ability later to act within the professorate. However, Schaxel's campaign to establish general biology as a subject led to a number of changes in academic teaching.

A staunch Marxist, Schaxel attached importance to the dissemination of scientific ideas and insights. It was for this reason that in 1924 he cofounded Urania, a publishing house that, among other things, published a monthly journal and a number of small, inexpensive popular science books. Schaxel wrote a few volumes on biological topics himself (Schaxel 1922, 1924, 1926, 1928, 1931, 1932). They were basically lay presentations of (general) biological knowledge within the framework of a dialectical materialism and thus synthesized the three meanings of "general." They conveyed general (fundamental) biological principles to a broad audience in lay terms, driven by educational/political considerations.

General Biology as the Subject of Textbooks

Nineteenth-century zoology and botany textbooks contained precursors of the idea of general biology. It was common at the time to divide the contents of these books into a general and a special section, with the general section devoted to discussions of the common features and principles of zoology or botany and the special section containing a systematic introduction to the diversity of animal and plant life. The phenomena discussed in the general section included the most recent findings in cell theory, comparative embryology and anatomy, physiology, the theory of evolution, developmental mechanisms, and the theory of heredity. If we compare the general sections of these textbooks over time, we can see how a significant structure of facts gradually emerged from the original *concept* of biology and ultimately led to general biology. Of course, there were major differences between the individual textbooks. For example, botany textbooks were usually more heavily focused on physiology and ecology, while zoology books, though greatly influenced by morphological traditions, normally examined the problems of ontogenesis and evolution in greater detail.

However, alongside botanists and zoologists, physiologists also played an important role in conceptualizing general biology. Experimental physiologists had long been interested in the general and fundamental principles of life and for this reason had close ties to experimental physics and chemistry. This interest is reflected in the many general physiology textbooks that define the field as the theory of life, underscoring the close link to general biology (Verworn 1895). In conceptual terms, though, general biology represented a broadening of general physiology, since the integration of morphology and physiology was one of its aims. The cell played a central role as the basic unit of life and all life processes. As Max Verworn explained: "The cell is the elementary building block of all living matter and the substrate of all the elementary phenomena of life. If physiologists therefore consider it their task to explain the phenomena of life, then it should be obvious that general physiology can only be a cellular physiology" (Verworn 1895, v).

His counterpart in the field of morphology, Oscar Hertwig, changed the title of the second revised edition of his textbook from *Die Zelle und die Gewebe: Grundzüge der allgemeinen Anatomie und Physiologie* (The cell and tissues: Basic characteristics of general anatomy and physiology; Hertwig 1892, 1898) to *Allgemeine Biologie* (General biology; Hertwig 1906). This created a terminological framework for the integration of concepts. Both Verworn and Hertwig interpreted the general principles of life as cell properties; they differed primarily in the way they weighted the methods used to investigate life, with Verworn placing emphasis on physiological experimentation and Hertwig on morphological, histological observations. For both scientists, the idea of general biology was a consistent conceptual extension of their respective research programs, one that also provided a meaningful educational framework for presenting the latest findings in biological research.

General Biology Textbooks

The second edition of Hertwig's work was the first major German-language textbook with "general biology" in the title. The ensuing years saw the publication of additional textbooks, such as the *Lehrbuch der Biologie* by Nussbaum, Karsten, and Weber (1914), *Einführung in die Biologie* by Karl Kraepelin (1912), and *Allgemeine Biologie* by Max Hartmann (1927). These books were instrumental in institutionalizing general biology as a subject in German *Gymnasien* (secondary schools), medical curricula, and, if with some delay, biology or life science departments. Despite all their differences, the textbooks established a canonical version of general biology. Within the framework of teaching, an understanding of biology as the science of life was created that still prevails today. It emerged from (1) Hertwig's idea that general biology was the science treating "the morphology and physiology of the cell and important related questions of life: the elementary structure and basic properties of living matter, the problems of procreation, heredity, development, the nature of species or natural historical groups, etc." (Hertwig 1906, vi), and (2) Max Hartmann's concept of general biology as "a nomothetic science of laws as well as a comparative science of order and an idiographic science of history" (Hartmann 1927, 5).

The common structure that continues to inform all these general biology textbooks today distinguishes (1) the basic building blocks of life: molecules, cells, tissues, organs, organ systems, organisms, and sometimes even populations and ecosystems; (2) the fundamental phenomena of life: metabolism, reproduction, sexuality, development, regulation, behavior, evolution, etc.; and (3) in some texts, even the diversity of life and the interactions between living organisms and their environment. This structure represents the canonical formulation of general biology, which clearly sets the field apart from other branches of science and provides it with an inner cohesion. This structure also privileges experimental research on the basic phenomena of life over the mere systematic cataloguing of the diversity of nature. Furthermore general biology's experimental focus was reflected in the founding of new research facilities.

General Biology and the Establishment of Innovative New Research Institutes

One metric that might be used to determine the impact of new ideas within science is their implementation in the form of scientific disciplines, research institutes, and research programs. This metric lends itself particularly well to the experimental sciences, providing a good yardstick to measure the success of individual branches of research—success that often results from increasing specialization. Although it may seem strange to apply this metric to general biology as well, which is essentially a synthetic and interdisciplinary interpretation of the biological sciences, there is no real contradiction to be found here. The reason is that, especially in the case of theoretical concepts, long-term success depends on whether these concepts can exert a sufficiently large and productive influence on empirical research traditions. For example, in the second half of the nineteenth century, the transformative influence of the Darwinian theory of evolution was a key factor in the reorientation of comparative anatomy and embryology. The related conceptual innovations led to institutional changes at universities, driven by a new generation of researchers (Nyhart 1995). The same can be observed with general biology. The main difference here is that the institutionalization of general biology initially took place for the most part outside the university setting. In the following, I would like to introduce, by means of comparison to the detailed history of the BVA presented in this volume, another of the institutions instrumental in developing general biology and briefly outline its role in the discourse on theoretical and general biology.

After long and intensive discussions on the future of science in the German Empire, the Kaiser Wilhelm Society was founded on January 11, 1911, with the specific mission to advance science. On the occasion of the one-hundredth anniversary of Friedrich Wilhelm University in 1910, Emperor Wilhelm II—citing Wilhelm von Humboldt's comprehensive conception of science and research—proclaimed: "We need institutes that go beyond the university framework and, unhindered by teaching aims but in close contact with academies and universities, serve only research" (vom Brocke and Laitko 1996, 1). One of this initiative's most remarkable innovations was the active involvement of industry and the business community, which, alongside the state and traditional patrons, provided the bulk of funding for the new institutes. In other words, when the Kaiser Wilhelm Society was founded, what was at stake was not noble scientific ideals, but economic and political goals. At the time, reference was often made to the imminent or actual loss of scientific leadership in the life sciences. A major cause of this decline was the inability of many traditional universities to open up to the new specialized disciplines in the biological sciences and above all to devote themselves with the required vigor to experimental research. The reasons for this failure varied, ranging from an adherence to academic tradition to insufficient financial resources at the institutes and for professorships.

In contrast to the traditional focus and self-definition of the life sciences at universities, which resulted, among other things, from the institutional separation of zoology and botany, in philosophical faculties and anatomy and physiology in medical schools, the new experimental research fields, including genetics, biochemistry, and, to a lesser degree, experimental embryology, took an interest in practical and applied research. For this reason, from the outset, "biology"—or whatever one associated with the term—was central to deliberations on the founding of new institutes within the framework of the Kaiser Wilhelm Society. Building on the first manifestos about the future of the biological sciences, including those by the paleontologist Otto Jaeckel and the physiologist Otto Cohnheim, the directors of the Kaiser Wilhelm Society asked leading scientists to write papers on the subject using their own ideas (Sucker 2002). The society received 29 such papers, which served as the basis for planning the new Kaiser Wilhelm Institute for Biology. Despite considerable differences between the individual assessments, four areas emerged and can be seen as representing a general consensus on the future of biology: genetics; experimental embryology and developmental mechanisms; the science of protists; and experimental physiology and biochemistry. All these fields were rooted in an experimental approach. The discussions culminated in the election of Theodor Boveri as the director of the new institute. Boveri completed its conceptual development and was also responsible for its final name—the Kaiser Wilhelm Institute for Biology. With this name he sent a strong signal that the future of the life sciences lay in biology, and particularly in the close link between experimental research and conceptual issues. In the end, on account of poor health, Boveri was unable to take up his post as director of the institute, but his successor, Carl Correns, adopted most of his ideas. Construction began in May 1914, and the first three departments took up their work on April 19, 1916. They consisted of the working groups under Carl Correns (heredity and plant reproduction), Hans Spemann (developmental mechanics and the causal morphology of animals), and Max Hartmann (research on protists). The other two department heads, Richard Goldschmidt (heredity, lower and higher animals) and Otto Warburg (physiology and biochemistry) were prevented from commencing their work by the events of the First World War: Goldschmidt was stranded in the United States and Warburg was serving in the military.

What was the relationship of the Kaiser Wilhelm Institute for Biology to general biology? The answer is complex. The specific history of the founding of the institute—with all the related discussions and debates, and especially the 29 papers by leading zoologists, botanists, physicians and paleontologists—focused primarily on two questions, both of which were closely linked to the development of general biology: What were the important new questions and fields in the life sciences, and in what way were these questions related so that they formed the basis of an independent new branch of science? As has already been seen, these questions dominated the conceptual discussions within the biological disciplines. When the biology institute was founded at the Kaiser Wilhelm Society, these discussions, though often academic, became highly concrete. Much was at stake for all those involved because they now had the chance to make an important contribution to shaping the future of biological research in Germany. Furthermore, it was an opportunity dictated by the availability of financial resources. Boveri and his successor Correns seized the day. One important reason that German biologists were worried about losing their leading role in science was the insufficient resources at their institutions, which made experimental work difficult or even impossible. At the same time, there was a general consensus that further advances in the biological sciences depended on experimental findings. It was only through experimentation that the causal factors underlying the many observations (e.g., in developmental history) could be pinpointed. Additional observations and more detailed descriptions could only exacerbate a data crisis that was already making synthesis work difficult, and undermine the biological disciplines' claim that they were a truly scientific endeavor. The popularity and scientific successes of new institutes devoted to experimental biology—including the Stazione Zoologica in Naples, the Marine Biological Laboratory in Woods Hole, and the BVA in Vienna—pointed to the future of biological research. In this sense, the establishment of the Kaiser Wilhelm Society and its institute of biology marked the culmination of a development that had its roots in the final decades of the nineteenth century. At the same time, the idea behind the institute in Dahlem went beyond those of its predecessors.

In the nineteenth century, experimental research concentrated on the organism and especially on both its life functions (physiology) and development (developmental mechanics and experimental embryology). Marine biological stations lent themselves particularly well to this purpose because the success of such experiments depended on the abundant availability of experimental material. In the early twentieth century the questions explored by physiology and embryology were supplemented by those in genetics, experimental research on evolution, and experimental cell biology and biochemistry. As we have already seen, these questions were linked to the fundamental life processes that went into the formulation of general biology. However, a new type of research institute was needed for the experimental investigation of these processes. Research on genetics and evolution were extremely arduous and time-consuming, especially when the studies did not make use of fruit flies (*Drosophila melanogaster*) (Harwood 1993; Kohler 1994). This kind of long-term research was facilitated by the organizational structure of the Kaiser Wilhelm Institute for Biology, where department heads were able to pursue their interests undisturbed (in accordance with the so-called Harnack principle). However, a central aspect of general biology, one that played a key role in the conceptualization of the Kaiser Wilhelm Institute for Biology, was the integration of different research approaches. Here the directors' personalities and scientific interests were all-important. Boveri's conception of the institute provided something of a blueprint for implementing general biology; of the directors who served after the founding of the institute, it was primarily the geneticist Richard Goldschmidt and the protozoologist Max Hartmann who dealt with these issues. Hartmann's approach to the fundamental problems of biology was shaped to a large degree by his research subject—single-celled organisms—and by his interest in the philosophy of nature. His *Allgemeine Biologie* includes extensive discussions of philosophical questions. He was influenced, on the one hand, by neo-Kantians such as Heinrich Rickert and Wilhelm Windelband and, on the other, by the like-named Nicolai Hartmann. Max Hartmann believed, for example, that his argument for the necessity of general biology was confirmed by Nicolai Hartmann's remark that "what is vital for the desired solution to the problems is an understanding of the problems; what is vital for an understanding of the problems is, in turn, an understanding of the phenomenon as findings" (Hartmann 1927, 10).[7] However, an understanding of the phenomena of biology was possible only within a clearly defined framework, that of general biology, whose development was crucial for advances in experimental and problem-solving research. Consistent with these ideas, Hartmann's synthesis of general biology concentrates on biological laws and on distinguishing biological phenomena from other natural phenomena.

Richard Goldschmidt was one of the most broadly educated zoologists of his time. A student and assistant to Richard Hertwig, he initially explored morphological and physiological questions before turning to genetics. During his time in Munich, he taught extensively and, as he wrote in his autobiography, that experience further broadened his range of interests (Goldschmidt 1960). In Berlin Goldschmidt not only created an integrated research program encompassing genetic, physiological, and evolutionary topics, but also wrote important textbooks and popular works that developed his interpretation of general biology (Goldschmidt 1922, 1927) . Goldschmidt's idea was essentially based on an organismic perspective; he regarded the organism as the site of biological processes. His physiological and developmental-physiological conception of genetics was focused not only on the transmission of abstract factors in line with the Mendelian rules but on explaining the development and evolution of organismic properties. Because of this perspective, Goldschmidt's research was always directed at general questions of biology and he resisted any narrowing of the scope of his research program for the purposes of specialization. As a result, his program always remained one devoted to general biology.

In contrast to the institutions described above, at which a framework of general biology was often used to situate and guide specific empirical research agendas, the founding of the BVA in Vienna was explicitly linked to the primary goal of generalization from the very beginning. "Not specialization but generalization from the experiences we gain is our aim," declared Hans Przibram, the principal founder of the BVA, in a description of the institute's objectives (Przibram 1908/09). This highly programmatic approach, combining experimentation with quantification and mathematization, led to a distinct brand of theoretical biology that went beyond the formulation of generalized laws and provided the kernel of a systems approach to the evolution of life (see chapter 8 in this volume).

General Biology and the Popularization of Science

In the late nineteenth and early twentieth centuries, the popularization of science was an important aspect of middle-class culture in Germany. A large number of associations worked to advance science, organizing lectures and nature excursions and supporting local museums. In some cases, popular magazines and books were published in editions so large that many intellectuals and journalists were able to earn a living from their writing. The desire to learn about the latest scientific findings was not limited to people of specific classes or political affiliations. Engagement with scientific ideas characterized not only the middle classes, which attached great value to a general humanist education, but also the progressive ideologies of social democracy and Marxism (Mocek 2002). As a result, we find popular books and series on scientific topics that, reflecting market logic, were directed at different segments of society. Around 1900 the popularization of science was thus an important aspect of the public discourse. Ideological debates and confrontations were inevitable, especially since certain ideas, above all those relating to the theory of evolution, had significant philosophical consequences. These consequences were successfully highlighted, for example, by Ernst Haeckel in his monistic philosophy.[8] In other words, among the natural sciences, biology held the greatest potential for conflict. It is therefore not surprising that general biology had an important role to play here as well.

Within the context of the popularization of science, we encounter the different meanings of "general" that characterized the discussions as a whole and that reflect different cultural and scientific values. On the one hand, general in the sciences was associated with knowledge that staked a claim to the broadest possible validity for itself. On the other, it was understood to mean that which was essential. It referred to the fundamental principles of a particular scientific field—those things that people absolutely had to know if they wanted to be regarded as educated in their social circles. Finally, popular presentations of science had the goal of being comprehensible in lay terms. This normally meant that they were supposed to be easily accessible to broad circles and did not require a great deal of prior knowledge. These three shades of general are usually difficult to separate, but they allow us to distinguish the different aspects of popular accounts of general biology at the time.

General and Theoretical Biology as Part of Contemporary Culture

The multivolume series Kultur der Gegenwart,[9] edited by Paul Hinneberg, sought to bring together all areas of scientific, humanistic, and technical knowledge within an integrative, authoritative framework.[10] It was based on a specific conception of the unity and order of knowledge that was characteristic of the late nineteenth century. The series aimed to assign all the special disciplines in the theoretical and applied sciences an appropriate place in a comprehensive system of knowledge. This system was determined, on the one hand, by the thematic contexts of the areas of knowledge and, on the other, by their historical development. It was therefore especially important to show the relationship of each scientific field to the overall system of knowledge. Hinneberg clearly describes this objective in his programmatic introduction to the entire work:

Die Kultur der Gegenwart aims to provide a systematically structured, historically grounded overview of contemporary culture by introducing, in broad strokes, the fundamental findings of various fields of culture based on their importance for contemporary culture as a whole and its further development. The work brings together many top names from all scientific and practical fields and offers portrayals of the individual fields written by the most qualified experts in easily comprehensible, well-chosen language. (Hinneberg 1906, vi)

In this introduction we also find a third organizational criterion for contemporary cultural achievement—its importance for the further development of culture. The work as a whole was therefore not a backward-looking summary of past accomplishments. Rather it was conceived as a catalyst for science and culture of the future. Hinneberg was supported in this undertaking by several of Germany's best scientists, and the project also benefited from his contacts as editor of *Deutsche Literaturzeitung*.

The series dedicated to the biological sciences was supervised by Richard von Wettstein, and the individual volumes and chapters in it were written by recognized experts in the various fields, as was also the case for the work as a whole. The four volumes originally planned for biology[11] represented what is probably the most comprehensive summary of all (general) biological knowledge in the early twentieth century. Significantly, the first and most innovative volume is devoted to general biology (Chun and Johannsen 1915). It begins with a brief history of the field, written by the historian of biology Emmanuel Radl. His treatment is followed by two chapters on the research methods of botany and zoology. The following nineteen chapters present the general foundations of biology using concepts central to the field, e.g., homology (Hans Spemann), regeneration in the animal and plant kingdoms (Hans Przibram and Erwin Baur), reproduction (Emil Godlewski), and purposiveness (Otto zur Strassen). Each of these chapters provides a detailed introduction to the respective subjects and discusses at length their implications for the overall understanding of biology. In terms of selection, the focus is on those concepts that influenced the structure of textbooks and figured in discussions of the independence of biology as a science. In that sense, this popular introduction to biology mirrors the conceptual and educational debates on general biology and presents them to a wider audience. In its entirety, the volume provides the most comprehensive introduction to the theoretical problems of general biology from this period. Its audience consisted of the intellectual and economic elite of the German Empire—precisely those circles that actively supported undertakings such as the Kaiser Wilhelm Society. It is no accident that the complete work was dedicated to Wilhelm II. In this sense, the anchoring of general biology in the system of knowledge represented by contemporary culture reflected the importance of the biological sciences for the self-conception of educated circles in the early twentieth century.

These and other, less ambitious book series made the theoretical and conceptual debates on the general principles in the sciences accessible to a wider audience. In this context the various cultural and theoretical meanings of "general" overlapped. For biology, which was just emerging as a scientific discipline in this period, the interaction between meanings and values was particularly important. As a new field of science that claimed to explore and explain life itself and that even called for new institutions and scientific methods as well as a reform of the curricula, it required broad support. It is for this reason that authors of popular works on general biology attempted to generate enthusiasm for biology among their readership. At the same time, they emphasized the special significance that general biology, as the science of life, had for cultural identity. Without knowledge of the fundamental principles of biology, society could not meet the challenges of the (then new) twentieth century—an additional, and prominent, argument in these volumes.

Outlook

The concepts of theoretical and general biology that slowly caught on in the first decades of the twentieth century were complex in terms of their origins and impacts. On the one hand, it was only through the formulation of general biology that biology was able to establish itself as a subject of scientific investigation in research and teaching. In teaching, in particular, this conceptual revolution resulted in a gradual transformation of the curriculum. From science courses in middle schools (*Realschulen*) and high schools (*Gymnasia*) to the preclinical subjects in medical programs, biology, as the science of life, replaced the traditional study of nature and natural history. Schoolbooks and textbooks essentially followed the guidelines of the major synthetic works of general biology. The diverse and complex aspects of life were discussed in textbooks and instructional works from the dual perspectives of the phenomena of life and the conditions of life. The former concentrated on the organism as the basic unit of life and investigated the processes of metabolism, excitability, reproduction, morphogenesis, and development (evolution); the latter examined the relationship between organisms and their environment. On the whole, these reforms were highly successful, especially after their supporters overcame the religiously inspired opposition—particularly strong around 1900—to all forms of natural history and biology classes (Daum 1998). The textbook by Karl Kraepelin, the "father of the movement to elevate biological thought," was a bestseller.[12] The changing titles of Kraepelin's book—from the traditional *Leitfaden für den biologischen Unterricht* (Guide to biology instruction) of the first edition to the more authoritative *Einführung in die Biologie* (Introduction to biology) of later editions—reflect the increasing acceptance of (general) biology as one of the pillars of a modern scientific worldview. A similar development can be observed in the study of medicine. Here, too, the traditional preclinical subjects of zoology, botany, and often also mineralogy were replaced by introductions to general biology (Schaxel 1919; M. Hartmann 1925).

As we have already seen, the popular works on general biology followed a similar pattern, evident also in the popular output of the BVA biologists. On the one hand, they had educational goals with political dimensions, especially within the context of popular and adult education. For example, Julius Schaxel was not only a leading representative of theoretical biology, but also one of the founders of the Marxist publishing house Urania, for which he wrote articles and popular books on general biological topics until he was forced to emigrate. In Vienna, it was among others Hans Przibram and Paul Kammerer who dedicated themselves to popular education. Furthermore, the idea of general biology also helped to integrate the life sciences into a comprehensive scientific humanistic worldview (an outstanding example is the treatment of the biological sciences in the Kultur der Gegenwart series). These efforts were not primarily directed toward educational goals. Rather, they were an expression of a philosophical conception of the unity of knowledge and culture that had its origins in the ideas of Kant, Goethe, and the Humboldt brothers (Clark 2006; Ziche 2006).

From a scientific perspective, the conception of general biology accelerated the reorientation of biological research from traditional descriptive research programs to experimental topics. Natural history, systematics, classification, comparative anatomy, and embryology all focused on the diversity of life, as did phylogeny, which built on knowledge from these other fields. By contrast, experimental biology centered on general problems and principles, including questions related to heredity and sexuality, mechanical explanations of embryogenesis, and the problems of physiological regulation, regeneration, and aging. This research was based on the conceptual claims of the "general." Researchers assumed that they would be able to discover the general principles underlying the observed diversity of life by experimenting with what was special (or individual). In this sense, general biology was the true scientific discipline of biology. However, in addition to theoretical reflections and experimental research programs, the implementation of general biology required new institutional infrastructure, which was reflected in the founding of non-university research institutions such as the BVA and the Kaiser Wilhelm Institute for Biology.

In summary it can be argued that in the early twentieth century, general biology reflected intrinsic scientific as well as cultural factors, which in turn influenced each other in various ways. The conception of life articulated in the systems of general biology was in many ways more comprehensive than those that drew primarily on Darwinian theory and on the philosophical consequences of the theory of evolution propagated by Haeckel and others. A substantial difference lay in the central role played by the concept of the organism in general biology. This organismic focus, also found in theoretical biology (see chapter 8), went hand in hand with the increasing experimentalization of life, which, starting with physiology, came to include psychology, developmental mechanics, and genetics. Nevertheless, as one of the central concepts of general biology, the organism was as culturally determined as the question of the explanation of life (vitalism versus mechanism) and related questions associated with the autonomy of biology. None of these discussions were uniform: they introduced a variety of proposals that covered the full range of political and philosophical positions. In this context, for example, many artists and intellectuals shared the skepticism expressed by a number of biologists about the achievements of modern technology. Furthermore, a widespread sense of the loss of a primal connection to nature motivated a search for alternative views of the natural world. On the other hand, there was also a close link between biology and progressive ideologies and political systems such as Marxism. Here the role played by genetics and eugenics was just as important as the debates on the heritability of acquired properties (neo-Lamarckism).

Around 1900 biology had a significance that extended far beyond its narrow scientific boundaries. That this is still the case today raises the question of just how relevant the history of general biology is to the present. Such questions are not easily answered, but in fact we face a similar constellation of factors today. The life sciences are once again in a crisis brought about by too much data and too little theory. Scientists have responded to this crisis by launching research programs that are both experimental (the "-omics" programs) and theoretical (computational biology or bioinformatics) and that find their reflection in dramatic institutional reforms. Integration, in the sense of inter- and transdisciplinary research, is of great importance and has led to a renaissance of theoretical biology. We are in a phase in which the fundamental principles of education in the life sciences are being intensely debated and these debates are drawing attention to the question of essential or "general" biological principles. The market for popular books on biological topics is booming. At first glance, this all seems familiar. Comparative studies that apply the approach proposed here to the present are surely of interest and can help us better understand the key factors, or so-called agents, of scientific change. In other words, we have the potential to learn from history precisely because history does not repeat itself.

Notes

1. Although for reasons of space this chapter deals mainly with German-language developments, the concept of a general biology was, of course, not restricted to Germany alone. Comparable developments took place in the Anglo-Saxon and French-speaking worlds and include for example Thomas Henry Huxley's introductory course in biology in England (Huxley and Martin 1877) and the biology textbook of Martin's students Sedgwick and Wilson in the United States (Desmond 1997; Geison 1978; Sedgwick and Wilson 1886, 1895).

2. For the popularization of science, see Daum (1998). An outstanding summary of the debates on the role of biology in the medical curriculum can be found in Julius Schaxel, *Über die Darstellung allgemeiner Biologie. Abhandlungen zur Theoretischen Biologie, Band 1* (1919).

3. For an overview see Hagner and Laubichler (2006).

4. Although theoretical biology is currently identified primarily with mathematical biology, its original, more conceptual focus has survived to the present. This integrative tradition is once again receiving increased attention as part of the reorientation of theoretical biology and the philosophy of biology. These developments are reflected in the launch of various scientific journals devoted to specific conceptual issues in biology, including *Biological Theory*, which has been published by the MIT Press since 2006 and by Springer starting in 2012.

5. In this context it should be noted that the conceptual structure of biology, presented for the first time in general biology textbooks, has survived to the present with all its essential elements. Max Hartmann's general biology textbook (Hartmann 1927), which draws on his lectures at the University of Berlin, is a prime example.

6. For the history of the founding of the Kaiser Wilhelm Institute for Biology, see Sucker (2002), as well as vom Brocke and Laitko (1996). For the history of the founding of the Prater Vivarium, see Przibram (1908/09). The Biologische Versuchsanstalt and its founders have also been the subject of a historical novel (Lorenz 1951).

7. Here Hartmann cites Nicolai Hartmann's standard work on epistemology, *Grundzüge einer Metaphysik der Erkenntnis* (1925); quotation on page 30.

8. An overview of the different aspects of Ernst Haeckel's work is presented in the exhibition catalog *Welträtsel und Lebenswunder: Ernst Haeckel—Werk, Wirkung und Folgen*, Stapfia 56 (catalog of the Oberösterreichisches Landesmuseum, Linz, 1998).

9. Paul Hinneberg studied political science and philosophy in Berlin and was the editor of the seventh volume of Ranke's *Weltgeschichte*. In 1892 he was appointed editor of the *Deutsche Literaturzeitung*, a weekly journal of reviews published by the Verband der Deutschen Akademien. His last project, following Kultur der Gegenwart, was Das wissenschaftliche Weltbild, another series of popular books on science. See also Stöltzner (2008).

10. The plan for the entire work included 58 volumes arranged into four groups: group 1: humanities, part 1 (religion, philosophy, literature, music, and art in 14 volumes); group 2: humanities, part 2 (state and society, law and economics, 10 volumes); group 3: mathematics, natural sciences, and medicine in eight parts and 19 volumes; group 4: technology in 15 volumes. Planning for Die Kultur der Gegenwart began around 1900 and the first volumes were published in 1905. Despite the war and economic problems, most of the originally planned 58 volumes were published by 1925.

11. The planned volumes covering biology were: vol. 1, *Allgemeine Biologie*; vol. 2, *Zellen- und Gewebelehre, Morphologie und Embryologie I: Botanischer Teil* and *II: Zoologischer Tei*; vol. 3, *Physiologie and Ökologie, I: Botanischer Teil* and *II: Zoologischer Teil*; vol. 4, *Abstammungslehre, Systematik, Paläontologie, Biogeographie.*.

12. In discussions at the Society of German Scientists and Physicians and in his many textbooks and popular works, Karl Kraepelin, director of the Natural History Museum in Hamburg, was a key advocate for reforming natural history instruction. See, for example, Karl Kraepelin, *Leitfaden für den biologischen Unterricht* (1907). The second edition was entitled *Einführung in die Biologie*.

References

Chun, C., and W. Johannsen. 1915. *Allgemeine Biologie*. Teubner.

Churchill, F. 2015. *August Weismann*. Cambridge, MA: Harvard University Press.

Clark, W. 2006. Die Politik der Onotologie. In *Der Hochsitz des Wissens Das Allgemeine als wissenschaftlicher Wert*, ed. M. Hagner and M. D. Laubichler, 97–128. Zürich, Berlin: Diaphanes.

Daum, A. W. 1998. *Wissenschaftspopularisierung im 19. Jahrhundert: Bürgerliche Kultur, naturwissenschaftliche Bildung und die deutsche Öffentlichkeit, 1848–1914*. München: R. Oldenbourg.

Desmond, A. J. 1997. *Huxley: From Devil's Disciple to Evolution's High Priest*. Reading, MA: Addison-Wesley.

Driesch, H. 1893. *Die Biologie als selbstständige Grundwissenschaft*. Leipzig: Wilhelm Engelmann.

Driesch, H. 1894. *Analytische Theorie der organischen Entwicklung*. Leipzig: Engelmann.

Driesch, H. 1911. *Die Biologie als selbstständige Grundwissenschaft und das System der Biologie*. Leipzig: Wilhelm Engelmann.

Fricke, D. 1964. *Julius Schaxel (1887–1943): Leben und Kampf eines marxistischen deutschen Naturwissenschafters und Hochschullehrers*. Jena: Fischer.

Geison, G. L. 1978. *Michael Foster and the Cambridge School of Physiology: The Scientific Enterprise in Late Victorian Society*. Princeton, NJ: Princeton University Press.

Goldschmidt, R. 1922. *Ascaris. Eine Einführung in die Wissenschaft vom Leben für Jedermann*. Leipzig: Theod. Thomas.

Goldschmidt, R. 1927. *Physiologische Theorie der Vererbung*. Berlin: Julius Springer.

Goldschmidt, R. B. 1960. *In and Out of the Ivory Tower: The Autobiography of Richard B. Goldschmidt*. Seattle: University of Washington Press.

Haeckel, E. 1866. *Generelle Morphologie der Organismen: Allgemeine Grundzüge der organischen Formen-Wissenschaft, mechanisch begründet durch die von Charles Darwin reformirte Descendenz-Theorie.* Berlin: G. Reimer.

Hagner, M., and M. D. Laubichler. 2006. Vorläufige Überlegungen zum Allgemeinen. In *Der Hochsitz des Wissens Das Allgemeine als wissenschaftlicher Wert,* ed. M. Hagner and M. D. Laubichler, 7–21. Zürich, Berlin: Diaphanes

Hartmann, M. 1925. Aufgaben, Ziele und Wege der Allgemeinen Biologie. *Klinische Wochenschrift* 4:2229–2234.

Hartmann, M. 1927. *Allgemeine Biologie. Eine Einführung in die Lehre vom Leben.* Jena: Gustav Fischer.

Hartmann, N. 1925. *Grundzüge einer Metaphysik der Erkenntnis.* Berlin, Leipzig: de Gruyter.

Harwood, J. 1993. *Styles of Scientific Thought: The German Genetics Community, 1900–1933.* Chicago: University of Chicago Press.

Hertwig, O. 1892. *Die Zelle und die Gewebe. Grundzüge der Allgemeinen Anatomie und Physiologie. Erster Theil.* Jena: Fischer.

Hertwig, O. 1898. *Die Zelle und die Gewebe. Grundzüge der Allgemeinen Anatomie und Physiologie. Zeiter Theil.* Jena: Fischer.

Hertwig, O. 1906. *Allgemeine Biologie.* Jena: Fischer.

Hinneberg, P. 1906. *Die Kultur der Gegenwart. Ihre Entwicklung und ihre Ziele.* Berlin: B. G. Teubner.

Hopwood, N. 1997. Biology between university and proletariat: The making of a red professor. *History of Science* 35:367–424.

Huxley, T. H., and H. N. Martin. 1877. *A Course of Practical Instruction in Elementary Biology.* London: Macmillan and Co.

Jahn, I. 1998. *Geschichte der Biologie: Theorien, Methoden, Institutionen, Kurzbiographien.* Jena: G. Fischer.

Kammerer, P. 1915. *Allgemeine Biologie.* Stuttgart: Deutsche Verlagsanstalt.

Kohler, R. E. 1994. *Lords of the Fly: Drosophila Genetics and the Experimental Life.* Chicago: University of Chicago Press.

Kraepelin, K. 1912. *Einführung in die Biologie.* Leipzig, Berlin: Teubner.

Lorenz, F. 1951. *Sieg der Verfemten; Forscherschicksale im Schatten des Riesenrades; Roman.* Wien: Die Buchgemeinde.

Miehe, H. 1915. *Allgemeine biologie; einführung in die hauptprobleme der organischen natur.* Leipzig, Berlin: B. G. Teubner.

Mocek, R. 1974. *Wilhelm RouxHans Driesch.* Jena: VEB Gustav Fischer.

Mocek, R. 1998. *Die werdende Form.* Marburg: Basilisken Presse.

Mocek, R. 2002. *Biologie und soziale Befreiung: Zur Geschichte des Biologismus und der Rassenhygiene in der Arbeiterbewegung.* Frankfurt am Main, New York: Lang.

Nussbaum, M., G. Karsten, and M. Weber. 1914. *Lehrbuch der Biologie für Hochschulen.* Leipzig, Berlin: Wilhelm Engelmann.

Nyhart, L. K. 1995. *Biology Takes Form: Animal Morphology and the German Universities, 1800–1900.* Chicago: University of Chicago Press.

Przibram, H. 1908/09. Die Biologische Versuchsanstalt in Wien: Zweck, Einrichtung und Tätigkeit während der ersten fünf Jahre ihres Bestands (1902–1907), Bericht der zoologischen, botanischen und physikalisch-chemischen Abteilung. *Zeitschrift für biologische Technik und Methodik* 1:234–264, 329–362, 409–433; *Ergänzungsheft*: 231–234.

Reinke, J. 1901. *Einleitung in die Theoretische Biologie*. Berlin: Gebrüder Paetel.

Reiß, C. 2007. No evolution, no heredity, just development: Julius Schaxel and the end of the Evo-Devo agenda in Jena, 1906–1933: A case study. *Theory in Biosciences* 126:155–164.

Roux, W. 1881. *Der Kampf der Theile im Organismus: Ein Beitrag zur Vervollständigung der mechanischen Zweckmässigkeitslehre*. Leipzig: Wilhelm Engelmann.

Sander, K. 1985. *August Weismann (1834–1914) und die theoretische Biologie des 19. Jahrhunderts: Urkunden, Berichte und Analysen*. Freiburg: Verlag Rombach.

Schaxel, J. 1919a. *Grundzüge der Theorienbildung in der Biologie*. Jena: Gustav Fischer.

Schaxel, J. 1919b. *Über die Darstellung allgemeiner Biologie*. Jena: Gustav Fischer.

Schaxel, J. 1922. *Grundzüge der Theorienbildung in der Biologie*. Jena: Gustav Fischer.

Schaxel, J. 1924. *Entwicklung der Wissenschaft vom Leben*. Jena: Urania Verlagsgesellschaft.

Schaxel, J. 1926. *Das Geschlecht bei Tier und Mensch*. Jena: Urania Verlags Gesellschaft.

Schaxel, J. 1928. *Das Leben auf der Erde*. Jena: Urania.

Schaxel, J. 1931. *Vergesellschaftung in der Natur*. Jena: Urania Verlagsgesellschaft.

Schaxel, J. 1932. *Das Weltbild der Gegenwart*. Jena: Urania Verlag.

Sedgwick, W. T., and E. B. Wilson. 1886. *General Biology*. New York: H. Holt and Co.

Sedgwick, W. T., and E. B. Wilson. 1895. An introduction to general biology. New York: H. Holt.

Sucker, U. 2002. *Das Kaiser-Wilhelm-Institut für Biologie: Seine Gründungsgeschichte, seine problemgeschichtlichen und wissenschaftstheoretischen Voraussetzungen (1911–1916)*. Stuttgart: Steiner.

Verworn, M. 1895. *Allgemeine Physiologie: Ein Grundriss der Lehre vom Leben*. Jena: G. Fischer.

vom Brocke, B., and H. Laitko, eds. 1996. *Die Kaiser-Wilhelm-/Max-Planck-Gesellschaft und ihre Institute. Studien zu ihrer Geschichte: Das Harnack-Prinzip*. Berlin, New York: de Gruyter.

Ziche P. 2000. *Monismus um 1900: Wissenschaftskultur und Weltanschauung*. Berlin: VWB, Verlag fr Wissenschaft und Bildung.

Ziche, P. 2006. "Wissen" und "hohe Gedanken." Allgemeinheit und Metareflexion des Wissenschaftssystems im 19. Jahrhundert. In *Der Hochsitz des Wissens Das Allgemeine als wissenschaftlicher Wert*, ed. M. Hagner and M. D. Laubichler, 129–150. Zürich, Berlin: Diaphanes.

7

The Biologische Versuchsanstalt as a Techno-natural Assemblage: Artificial Environments, Animal Husbandry, and the Challenges of Experimental Biology

Christian Reiß

Beyond its particular history as a research space, the Biologische Versuchsanstalt (BVA, or Institute for Experimental Biology) bears broader historiographical significance. It can be understood as both the beginning and the preliminary to the end of the development of certain kinds of research infrastructures in the life sciences, and thus provides a unique focal point in the history of the life sciences around 1900. With its aspiration to synthesis, it can be read as the conscious reflection of the history of the previous 50 years of biological research, set out to lay a new theoretical, methodological, and empirical foundation for the field. At the same time, it was also one of the last and most comprehensive attempts to build a research space comprising all of biology, at least in principle. The already prevailing trend toward specialization, including on the technical level, ultimately superseded this generalist approach at the beginning of the twentieth century. Thus, the ongoing process of disciplinary specialization in a way was mirrored by the high degree of spatial compartmentalization within the BVA.

The history of the BVA can be approached from various perspectives. Conceptually speaking, it was the articulation of a particular vision of scientific and, more specifically, biological research. In a broader cultural sense, the BVA was a spatiotemporal assemblage of concepts, practices, and people, realized in the particular environment of fin-de-siècle Vienna. But it was also a material place in a building with its own history of research in the life sciences. It was a singular research infrastructure, which incorporated a general vision of how biological research should be conducted; the specific conditions of its location; and a highly compartmentalized research space in which to keep, breed, and investigate a broad variety of organisms.

Given the specificity of their intentions, the fact that the founders of the BVA chose a former public aquarium for their enterprise might seem contradictory. Why use an old building designed for an entirely different purpose? But it made perfect sense in the context of both their own practical requirements and one of the historical traditions in which their endeavor was rooted.

In the following, I will analyze the building of the BVA as a research infrastructure. I will contextualize it within the history of research spaces and places in the life sciences of the nineteenth century (Rheinberger et al. 1997; Finnegan 2008; De Bont 2015), especially in the zoological tradition, and will argue that the BVA as a laboratory space was co-produced by the particular history of its building, the history of research spaces in nineteenth-century life sciences, and the synthetic ideas of its founders. Using the concept of *techno-natural assemblages* (Reiß 2012a), I will explore how a former public aquarium was transformed into a state-of-the-art facility for animal husbandry and plant breeding and a vast modular experimental space.

The History of Experimental Research in Nineteenth-Century Zoology and the Question of Spaces and Places

Hans Przibram was not just skilled as an experimenter, but also intimately familiar with the theoretical and methodological developments in the life sciences of his time, which he summarized and commented on in various review papers and books (e.g., Przibram 1907, 1909b, 1910, 1913, 1914, 1929, 1930). As becomes obvious from his presentation of the BVA, the institution was the refined reflection of the preceding decades of experimental research in the life sciences on a theoretical, methodological, and practical level (Przibram 1909a). In creating the BVA as a research space, Przibram analyzed the methodological and infrastructural prerequisites of relevant work of the previous centuries. He conceived of the BVA as a synthesis of those conditions, to provide researchers with the best situation possible without constraining the path of their investigations.

At the turn of the twentieth century, experimental research as a systematic approach in the life sciences already had a considerable history. But biology was still a heterogeneous field with diverse objects, theories, research practices, and disciplinary contexts (Schaxel 1922; Caron 1988; Kanz 2002). Thus, when focusing on the experiment as a method, each branch reflected the development of a particular tradition (Chadarevian 1996).

To understand the ways in which the BVA was set up as an experimental space, one has to place it within a specific tradition of zoological research. The second half of the nineteenth century could be termed the era of field stations in zoology (Kohler 2002; De Bont 2015). In the eighteenth and the early nineteenth century, expeditions set out to collect organisms that were then preserved and sent back home, where they were investigated and kept in the collections of natural history museums. As research practices and instruments became more and more elaborate from the 1850s onward, research facilities were founded in the field to allow for more detailed investigations. This happened especially on the sea coasts, where zoologists set out to investigate the strange life forms of the oceans. But there were also other kinds of field stations, such as freshwater stations and agricultural research stations (Kofoid 1910).

The marine stations in particular quickly became central to zoological research. The most prominent example is the Stazione Zoologica Anton Dohrn in Naples, founded in

1874 by the German zoologist Anton Dohrn, which was one of the models on which the BVA was based. Funded by Dohrn's own money, government grants, and a public aquarium, which also served as a research installation, the Stazione Zoologica was run on the so-called table system (Fantini 2000). Here, governments would rent a table (i.e., a workplace), which they could then give to individual researchers (Kofoid 1910). Very much like Przibram's, Dohrn's vision was of a place for the scientific study of all aspects of (marine) life (De Bont 2009). Nevertheless, the Stazione's fame was based on the descriptive morphology and embryology of marine animals, with experimental practices only gradually emerging (Müller 1975).

From early on, the rise of field stations and especially marine stations as prominent sites of zoological research was paralleled by attempts to break their fixation to specific locations and thus to enable the same kind of investigations in academia's metropolitan centers (Reiß 2012a). Instead of bringing researchers and their methods to the field, the idea was to bring the field to the places of zoological research, i.e., universities, often far away from the various field sites. There were several motivations for this. First, the trips to the stations were expensive and could not be afforded by all zoologists. Second, the crucial part of research was limited to the duration of the residence at the station, which often lasted just a few weeks. Third, it was very difficult to teach the research methods applied at the stations to students at the universities.

While bringing the objects of research to the metropolitan centers of science was common practice in natural history and museum-based zoology from the eighteenth century onward, nineteenth-century zoology called for a different approach. It was not sufficient anymore to collect specimens in the field, preserve them, and send them back to the place of research. For many of the species of interest, especially the aquatic ones, this was not possible without losing crucial traits, like color, certain structures, or even the entire form. This was the central reason why marine stations had been founded in the first place. Furthermore, research practices had developed, and descriptive embryology, in German called *Entwicklungsgeschichte*, required more and more research material to produce more and more detailed investigations of the developmental history of tissues and organ systems. Last but not least, there was a growing interest in organisms that went beyond questions of structure and form. Life processes, and especially the relation between organisms and their environment, started to emerge as a central interest of zoology.[1]

This shift becomes especially obvious in the ways in which new zoological institutes were founded in the German-speaking world (Reiß 2012a). When zoology became an academic discipline from the 1860s onward, it had to demarcate itself from anatomy, which had traditionally been the place for many of those questions zoology was now responsible for (Nyhart 1995). But zoology also had to deal with its links to natural history, a discipline that was often replaced by the new zoology professorships and institutes. A central part of this process consisted in getting rid of the natural history collections in one way or another to free up space for what was considered to be the proper occupation of zoologists.

At the new institutes, such as in Würzburg, Freiburg, and Leipzig, scientists were struggling for light, space, and water both, with their colleagues from other disciplines and with the conditions of the mostly ancient buildings. All these elements were needed for the microscopical studies of the development of tissues and organs, for observations of animal life, for experiments, and, most important, for keeping the necessary animals as research material.

When Przibram looked back at this period at the beginning of the twentieth century, it was exactly this last point that he still stressed. He pointed out that in contrast to the situation in physics, chemistry, and physiology—the exemplary experimental sciences of the nineteenth and early twentieth centuries—biological experiments were not based on "the design of complicated apparatus, but in developing the kind of husbandry necessary for keeping organisms over a longer period," Although his statement extended to all kinds of organisms, he emphasized that while botany already had a long tradition of scientifically growing a broad variety of plants in botanical gardens, zoology was a long way from developing the same sort of practical knowledge (Przibram 1909a, 235).

This implicit hint at the institutional differences between botanical and zoological gardens, where the former had traditionally been closely linked to scientific research while the latter had a much shorter history and a less distinct institutional profile—resonated with similar statements made a few decades earlier by the preceding generation of what could be termed breeder-zoologists (Reiß 2012a). Even though scientific research had more often than not been part of the agenda when zoological gardens and public aquaria were founded in the second half of the nineteenth century, a productive connection similar to that of botany and botanical gardens could almost never be established.[2]

It was the zoologists of the generation before Przibram who made the first attempts to develop such practices, which he still found capable of further development. This fact and the problems those earlier scientists had faced show what a long way zoology had to go from collecting dead specimens to keeping living ones (Reiß 2012a, 2012b). But it was the experience and the practical knowledge gathered in that earlier context on which the concept of the BVA rested. The practical foundation of these earlier experiments in zoology and of the concept of the BVA was the creation of techno-natural assemblages (*künstliche Naturräume*), technologically stabilized spaces in which organisms could be kept and bred under quasi-natural conditions. On the one hand, such conditions were essential to provide the necessary taxonomic variety of research material, which zoology relied on in contrast to the more pragmatic approach of physiology (Przibram 1909a, 235). On the other hand, the quasi-natural conditions provided by the techno-natural assemblages were the basis of the nascent experimental investigation of the relation between organisms and their environment, a key problem of the life sciences after Darwin and part of the research program of the BVA (Przibram 1903, 153; Bowler 1992).

The Aquarium and the Role of Techno-natural Assemblages in Nineteenth-Century Zoology

The most illustrative example of a techno-natural assemblage is the aquarium, a model that also played a key role in the concept of the BVA. The aquarium, in the narrow sense of the word, was invented in Great Britain in the 1850s by the London chemist Robert Warington (Vennen 2013). Keeping aquatic animals in containers filled with water was a much older practice, but the term *aquarium* specifically referred to the combination of water, animals, plants, and other items in a way that would resemble nature not just in appearance but also in function. The idea was to re-create nature based on scientific principles, with the ultimate goal of establishing a self-sustaining system that would stay in balance without further human intervention (Reiß and Vennen 2014). Even though that goal was hardly ever reached, the aquarium as an aesthetic and functional model of aquatic nature became popular in Europe, situated between science, entertainment, and economic interests (Reiß 2012a). To be able to keep animals and plants under what were considered to be conditions as close to nature as possible, elaborate technologies and practices were developed by the aquarium-fancying community. The water was ventilated and the temperature regulated with special devices and practices, and knowledge was developed about the husbandry of a broad range of aquatic animals (Reiß 2012b).

From the beginning, the aquarium as a techno-natural assemblage emerged as a "boundary object" (Star and Griesemer 1989) between actors with different interests. On the one hand, zoologists relied on aquaria as a source to supply growing demand for research material and as an instrument for experimental research. On the other hand, fanciers outside academic zoology combined pet keeping and an enthusiasm for nature and natural history. But often a clear-cut distinction between the two groups was not possible. For example, the German zoology professors Carl Gottfried Semper (Würzburg) and Karl August Möbius (Kiel) could be considered enthusiasts as well, publishing in fanciers' journals and developing technical solutions while at the same time using the aquarium for their research, both to keep animals and as an experimental tool (Reiß 2012a; Reiß and Vennen 2014). In contrast, the German practical naturalist and aquarium fancier Marie von Chauvin became famous for her experimental studies on amphibian reproduction, development, and evolution, which she conducted together with August Weismann in Freiburg as well as on her own (Weismann 1875; Chauvin 1876; Reiß 2012a, 2014).

The case of Paul Kammerer connects the aquarium as a techno-natural assemblage and the field of aquarium fancying with the history of the BVA, illustrating the fluid boundaries between science and enthusiasm (see also chapter 3). As a teenager, Kammerer started out as an aquarium enthusiast, developing elaborate skills in husbandry, especially of reptiles and amphibians. He was a member of the local aquarium club Verein für Aquarien Lotus of his hometown, Vienna, and a corresponding member of the club Triton—Verein für Aquarien- und Terrarienkunde zu Berlin in Berlin, which was modeled after academic societies and had an international outreach (Gliboff 2006); and also published in fanciers' journals (Kammerer 1901). While Kammerer was still a student, Przibram hired him as an assistant for animal husbandry at the BVA (Przibram 1903). Kammerer made a career as an experimental zoologist, tragically ended by his suicide. The skills in the husbandry and experimental treatment of reptiles and amphibians that Kammerer was famous for were directly linked to the knowledge and practices of aquarium enthusiasts. In his research he explicitly followed up on work such as that of Marie von Chauvin (Kammerer 1904, 1907). At the beginning of the twentieth century, knowledge and practices of this type were still not yet part of zoological methodology but had to be transferred from other areas by figures like Kammerer.

The kind of experiments for which this practical knowledge was needed had their origin in the zoological research of the second half of the nineteenth century. The question of the interrelatedness of reproduction, ontogenesis, and evolution as phenomena of animal life has a long history. With the nineteenth-century emergence of development as a central concept in the life sciences and the 1859 publication of Charles Darwin's *On the Origin of Species*, investigation of the mechanisms of these phenomena was intensified. Techno-natural assemblages in general, and the aquarium in particular, were crucial instruments in this endeavor. From the 1860s onward, almost all zoological institutes were equipped with aquaria and other devices to keep and study animals with respect to these questions (Reiß 2012a). The importance of techno-natural assemblages became particularly visible in the newly built institutes at the universities of Leipzig, Würzburg, Freiburg, and Kiel. There, the devices were integral parts of the structures of the new buildings. And it was Carl Gottfried Semper's new institute in Würzburg that Przibram explicitly referred to as an early example of the kind of research and the kind of building he thought of when planning the BVA (Przibram 1909a).

Semper became full professor of zoology and comparative anatomy in Würzburg in 1871. After extensive studies in morphology and an expedition to Southeast Asia between 1858 and 1865, he explicitly extended his research program for his newly founded zoological institute to "experimental biological studies" (Schuberg 1891a, x). Based on his interest in Darwin's theory of evolution, Semper intended to investigate the influence of the environment on the organism. As it turned out, one of his main challenges would be the transformation of the zoological collection, which he was appointed to head, into an appropriate space for his research. Even in Semper's first experimental study on the great pond snail (*Lymnaeus stagnalis*, today *Lymnaea stagnalis*; Semper 1874a), problems occurred because of conditions in the rooms of the collection, then situated in the old main building of the university. Initially, his intention was to find out whether the species was capable of self-fertilization or parthenogenesis. But it turned out that the situation in what used to be an ancient monastery was unfavorable for Semper's investigation. He complained about the lack of light and the dusty air, which caused all sorts of unwanted parasites to grow in the water and ultimately led to the death of the snails (Semper 1874a).

Although his efforts to solve the problem led to a new experiment on the influence of environmental factors on the growth of the snails, Semper began to campaign to the university administration for a new building. An appropriate site had already been bought in 1875, but it took until 1889 for the new zoology center to be opened. In the intermediate years, Semper and his academic assistants Johann Wilhelm Spengel, Maximilian Braun, and August Schuberg had to improvise and tinker to adapt the given situation at least rudimentarily to their needs. Thus, Semper and Spengel became active participants in the discourse of aquarium fanciers. Due to the insufficient water supply in the institute's rooms, they developed their own technical solutions for ventilating the water and wrote about them in the journal *Der Zoologische Garten*, which was the most important forum for animal keeping and breeding (Semper 1874b; Spengel 1875). They also had personal contacts with nonacademic aquarium breeders, who visited their institute and commented on their inventions (Buck 1874; Dorner 1874).

Toward the end of the 1870s, Semper successfully negotiated for two rooms, which were turned into sophisticated aquarium rooms (Braun 1878a). Although located in the basement of the old university building, they provided better light conditions, a stable temperature, and a connection to the municipal water supply system. But there were also facilities to keep reptiles, birds, and mammals for research and teaching purposes (Braun 1878b). Thus, despite the rather provisional conditions, Semper and his assistants managed to transform the rooms of the old university building into spaces for animal breeding and keeping. With their own efforts in tinkering and invention and with the help of non-academic aquarium fanciers, they developed the basis for Semper's program of "experimental biological studies" (Schuberg 1891a, x).

When the new institute building was finally opened in 1889, its design reflected Semper's research program or, as Schuberg put it, "the building in a way embodies the dominating ideas." (Schuberg 1891b, 1) But it also reflected the specific knowledge and practices collected and the technologies developed during the time in the provisional rooms. The building and the surrounding garden were built as highly compartmentalized spaces with a variety of techno-natural assemblages. The institute had an integrated greenhouse with separate heating, a large number of aquaria and concrete basins, a beehive, and a pigeonry. The large garden, which Schuberg saw as the most important facility of the institute, had a pond, more concrete basins, a rivulet, and an artificial grotto. All the spillover water from the internal and external basins was collected in the rivulet and flowed through the grotto to supply the pond. Thus, the new institute provided discrete but interconnected spaces to provide for a variety of different environments to keep and breed animals.

While Semper's agenda and the design of the new facilities might suggest a focus on experimental studies, the institute's actual publication output does not confirm that assumption (Schuberg 1891a; Nyhart 1995).[3] Besides his early work with snails, Semper was mostly interested in morphology, evolutionary theory, developing his own theory of the origin of vertebrates, and the relation between organisms and their environment (Semper 1875), about which he published in his influential work *Die natürlichen Existenzbedingungen der Thiere* (Semper 1880). Similarly, his students conducted their research in an evolutionary framework, in which more or less experimental approaches only rarely were applied (Nyhart 1995).

The situation in Würzburg is exemplary for many of the new zoological institutes and professorships that were founded in Germany in the 1860s and 1870s (Reiß 2012a). The new research agenda of zoology (Nyhart 1995) called for new facilities and thus negotiations for space; tinkering and improvisation, as well as contacts with nonacademic fanciers, were commonplace. Natural history collections had to be replaced by, or at least complemented with, living animals kept and bred in techno-natural assemblages. The research conducted with these new objects shows that experimentation as a methodology was not the central incentive to build them. Even though zoologists like August Weismann (Freiburg), Semper (Würzburg), and Möbius (Kiel) did conduct investigations that had an experimental character from early on, the experiment became an explicit and central epistemological device only over the course of time (Reiß 2012a; Reiß 2014; Reiß and Vennen 2014).

Much more important for the integration of the techno-natural assemblages was that they enabled the keeping and breeding of animals for descriptive studies and to teach the growing number of students. Even more than in other disciplines in the German sciences, the laboratory emerged not just as the "placeless place" (Kohler 2002) in the epistemological sense, but as a place where research and teaching could be practiced in close connection (Nyhart 1995; Dierig 2006). It was then that the presence of living animals in zoological research spaces and the possibility of controlling environmental conditions in the techno-natural assemblages quickly led to the emergence of experimental research. The progression from living animals for descriptive studies to experimental research also happened at many of the field stations, such as the Stazione in Naples (Müller 1975).

This progression demonstrates one of the central differences between zoology and physiology, the emblematic experimental life science of the nineteenth century. While physiology followed a functionalist approach and aimed toward identifying and explaining general properties of life, zoology operated under a morphological agenda, with an interest in particularities on various taxonomic levels (Nyhart 1995). Besides the fact that this focus on form did not call for experimental investigations as strongly as did physiology's interest in function (Canguilhem 2008), zoology had different demands when it came to the choice of research animals and the ways in which they had to be kept and treated in the laboratory. Zoology could not focus on a small number of species to investigate a particular function; rather, at least in principle, it had to keep the entire animal kingdom in view (Hopwood 2011).[4] Furthermore, zoologists had to investigate their research animals over a much longer period of time, sometimes even breeding them under controlled conditions, a requirement that Przibram explicitly points out with respect to the initial conception of the BVA (Przibram 1909a). The approach of developmental physiology (*Entwicklungsphysiologie*) or developmental mechanics (*Entwicklungsmechanik*), which was thought to overcome some of these problems, was only in the process of formation, with Przibram as one of its main protagonists and the BVA as one of its central places. The aim was to fuse zoology's interest in form and development, as studied in (descriptive) embryology or *Entwicklungsgeschichte*, with physiology's focus on general principles, causal explanations, and experimental methodology.

All this helps to explain Przibram's claims regarding the status of experimental research in zoology in general (Przibram 1909a). And it helps to provide a more detailed understanding of the idea behind the BVA. The fact that experimentation in zoology and other branches of what would later become biology was such a dispersed approach was due to the fact that it was neither part of a methodological turn—as in physiology in the 1840s and 1850s (Dierig 2006)—nor inscribed into the respective knowledge production of those fields. The institutional prevalence of a tradition rooted in natural history with an emphasis on taxonomy and systematics can be seen in the zoological professorships at the big, prestigious universities such as Berlin and Munich (Nyhart 1995). While Semper in Würzburg, Weismann in Freiburg, Leukart in Leipzig, and Möbius in Kiel made successful efforts to portray themselves as pioneers of a new zoology and to distance themselves from that tradition, the research agenda in places like Berlin and Munich followed this shift only after a delay of about 20 years. Rather, the experiment as the prevailing method in some branches of zoology developed on a personal and local level, and often under contingent circumstances. Przibram's own work and the foundation of the BVA were integral to the process of changing this situation.

The BVA as a Techno-natural Assemblage

With this background, the fact that the Prater Vivarium was chosen as the location for the BVA can by no means be regarded as a coincidence. While the close connection between zoology and aquarium fancying provided the context, the particular configuration and equipment of the building as a techno-natural assemblage offered a favorable basis for Przibram's plans. Specifically built and designed to keep, breed, and publicly exhibit aquatic animals, the place, though already rather outdated and decayed, had many of the requirements necessary to fulfill the agenda of the BVA.[5]

Nevertheless, the building had to be extensively adapted to its new function (figure 7.1): a public aquarium had to be transformed into a biological laboratory. The aim was to create conditions under which "small animals and plants from terrestrial, fresh- and saltwater habitats from different regions of the world" (Przibram 1903, 154) could be kept, bred, and investigated. Similar to the first wave of institutes of German zoology, the distance between lab and field should be erased to obviate the need for travels (Wettstein 1902) and thus to enable methodologically more solid kinds of biological research. The underlying themes of control and precision were among the major influences that made up the BVA's unique research profile (Coen 2007). Przibram himself had experienced the benefits and problems of research at marine stations (e.g., Przibram 1901a, 1901b). While the advantages in terms of organismal variety and availability, as well as the exclusive focus on research, were certainly appreciated, a trip to these places was nevertheless difficult to arrange in conjunction with academic and private duties at the home institution. In addition, the visits were expensive and thus not suitable for the growing number of students, whose teaching was largely based on practical courses.

The keeping and breeding of organisms was dependent upon the controllability of the conditions of the artificial environments of the BVA. As many environmental factors (*äußere Factoren*) as possible should be controllable, to study their influence on the

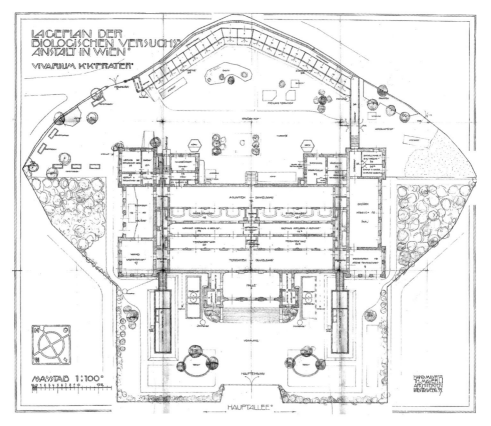

Figure 7.1
Original site plan from ca. 1908 showing the layout of the BVA plot and how the BVA building is divided up by function. The map was published in an addendum to Przibram's report on the first five years of the institute (Przibram 1909a).

research organisms (Przibram 1903). Especially with respect to aquatic animals and thus to aquaria, the problem of the controllability of conditions for experiments was closely tied to the conditions of their husbandry, and so some of the scientific questions had their origin in practical problems, as the example of Semper and the snails had shown.

While the idea behind the foundation of the BVA was to allow for all kinds of biological research, with an emphasis on experimental and quantitative approaches, the initial focus nevertheless lay on experimental morphology, heredity, biochemistry, and biophysics (Przibram 1909a; Kofoid 1910). Accordingly, the building and the laboratories were equipped with a broad range of facilities to keep and breed as many different species as possible. While the chemical and physical laboratory (figure 7.1, no. 29) followed the standards already established in those disciplines (Meinel 2000; Hoffmann 2001), the biological laboratories, which were at the same time animal-keeping facilities, still posed a major challenge, since there was no directly applicable model and thus no such specialized equipment as existed in physics and chemistry departments (Eggert et al. 1905). Przibram therefore relied on a broad variety of relevant experiences from zoological institutes like Semper's and marine stations like the Stazione Zoologica in Naples, but also from public aquaria and fanciers.

The Central Supply Systems

The BVA had three central systems, which supplied the entire building with freshwater, saltwater, and (compressed) air. The most important but also the most challenging aspect was the water supply system. Thanks to the building's history, two aquarium systems, a salt- and a freshwater one, were already available. In the course of the renovations, these systems were extended beyond the aquaria and the whole building was equipped with salt- and freshwater circulation. In combination with a system for compressed air, which was installed throughout the entire building, most of the rooms provided for basic laboratory conditions.

The saltwater system in particular needed special technical solutions due to the corrosive effect of saltwater, a topic on which Przibram elaborated extensively in his description of the BVA. The corrosive effects of saltwater not only affected the circulation system, but also the water itself, with eroded metals contaminating the water and altering or even killing its inhabitants (Przibram 1909a). These problems were solved with the help of suitable materials like rubber, wood, and appropriate metals; the installations used were by no means standard technologies. Przibram gives detailed descriptions of these solutions, which were specifically designed for the purposes of the BVA. Highlighting the innovative character of the infrastructure both drew attention to the state-of-the-art research conditions and made the technologies available for other researchers, as illustrated by the publication of supply sources. The operation of saltwater aquaria, especially, was still a difficult endeavor in the early twentieth century, which was the reason marine stations were of such importance and the project of the BVA was viewed with such interest. The problems started with the supply of saltwater, which the BVA faced as did the stations (Przibram 1909a). There were basically two solutions. Saltwater was either obtained from the ocean or produced artificially.[6] But neither method worked satisfactorily. The BVA received its saltwater from the Austrian marine station in Trieste, located on the Adriatic coast on the other side of the Alps. The water was stored in barrels and shipped by train, which meant additional costs and logistical challenges. An additional problem was the then still operative state monopoly on salt, a problem already reported about in fanciers' journals in the nineteenth century (Reiß 2012a). The BVA needed special permission from the Austrian tax authorities to be allowed to import large quantities of saltwater, and had to guarantee that they would not use it to make salt and break the monopoly. In order for the saltwater from the Adriatic to be useful for as long as possible, it had to be filtered and ventilated continuously to keep it fresh. As Przibram points out with considerable pride, the saltwater obtained in Trieste could thus be used for several years. Both filtration and ventilation systems had again to be purpose-built, based on the ones developed and used at the marine station in Trieste by the station's director, the Austrian zoologist Carl Cori.

Compared to the saltwater system, the freshwater system was less of a challenge. Freshwater could be easily obtained, and the materials needed for the system were the same ones used in public and private water supply systems. The BVA used water from its own well, from the public water supply, and rainwater, which was collected on the estate. Furthermore, the staff produced its own distilled water. The freshwater system supplying most rooms of the building as well as the aquaria used water from the public water supply, which came from the Alps. The different kinds of water available (salt, well, alpine, rain, and distilled) nicely illustrate the technological challenges of establishing a research facility like the BVA and the level of effort made by its protagonists. And they foreshadow the institution's research program and the complexity of the considerations that had to be taken into account prior to the BVA's opening. Water, of course, was a basic requirement for all research at the BVA, not just the biological.

The ventilation apparatus was a third system that ran through the entire building. In contrast to the water systems, it was much simpler; different kinds of air did not need to be provided. An air pump with an electric motor filled big tanks with compressed air, which was accessible through outlets in most of the rooms.

Facilities for Animal Husbandry

The BVA provided for as great a variety of environments as possible. Aquaria were central, and not only on account of the building's history. The ways in which the aquaria were integrated into the overall system shows how much thought had to be put into not just the well-being of the organisms but also the working conditions of the researchers. Eight main aquaria were located in the central hall of the BVA (figure 7.1, nos. 18 and 18b). These could be filled with either salt- or freshwater. Additional basins (six on the northern end of the estate and two in the east, close to the stables) and two ponds (on the forecourt of the estate, next to the main entrance) existed in the garden surrounding the building. Animals for research were kept and bred there. Smaller, mobile aquaria were used to separate animals for various investigations, to bring them to the laboratories for experiments, and to keep them during and after the experiments. Thus, a wide variety of aquatic environments were provided. With additional chemical or physical modifications, the aquaria could potentially be filled with five different kinds of water. The outdoor basins and ponds offered additional possibilities. The combination of large, static and small, mobile aquaria allowed for an easy circulation of organisms throughout the building.

Next to the main aquaria was the section for terraria (figure 7.1, nos. 20 and 20b), in which reptiles, insects, and other such animals were kept. Again, various forms of terraria were specifically designed by Przibram and Kammerer for a variety of purposes (Przibram 1909a), turning them from mere devices to keep animals into scientific instruments central to the planned research. Between the aquarium and the terrarium section was another section with a hothouse and a coldhouse (nos. 19 and 19b), which were used mostly to breed plants. Furthermore, there was a warm-temperature room, which also housed an aviary. Beneath this main part was a cellar that was used to simulate cave-like conditions when needed for studies, for example of the olm (Kammerer 1907; Berz 2009).

As with the outdoor basins and ponds, the techno-natural assemblages (and thus the laboratory facilities) were not restricted to the main building. The garden had additional large terraria (two at each side of the main entrance of the building and more at the eastern end of the estate, near the stables). In addition to the indoor ones, there were two outdoor greenhouses with high and low temperatures (figure 7.1, nos. 33 and 34). Furthermore, there were stables (no. 35) and cages on the outside terrain for birds and mammals. All of these spaces, and the garden, were also used to grow plants for botanical studies.

The BVA was designed to house, at least in principle, almost any kind of organism, no matter what environmental conditions its keeping demanded. As the list of animals kept at the BVA provided by Przibram shows, they took any organisms they could get hold of (Przibram 1909a). The list totals 700 different animal species (including microorganisms), 100 of which were also very successfully bred.[7] In the context of the research program of the BVA and the tradition of zoological research in which it was situated, keeping and breeding were not merely a prerequisite for the actual investigations, but already a vital part thereof. For all these organisms, specific practices of husbandry and feeding had to be practiced or newly developed; such tasks further emphasized the breeder aspect of experimental zoologists (Przibram 1909a; Reiß 2012b). These environmental conditions, in turn, were not just requirements for successful experiments and valid results. They also gave hints to promising starting points for experimental manipulation.

The animal husbandry and plant breeding facilities themselves already pointed to a number of possibilities for altering the conditions external to the organisms. Temperatures, as well as humidity in some areas, could be minutely regulated in the aquaria, terraria, and greenhouses. Exposure to light could be controlled (figure 7.1, nos. 30 and 30a). Furthermore, the specific composition of salt- and freshwater could be altered. As we have seen, the ability to control these factors was both necessary to adjust the techno-natural assemblages to meet the broad range of conditions needed for the well-being of the different plants and animals, and essential as a means of conducting experiments concerning the influence of environmental conditions on organisms. Investigations were done either in the main work space (figure 7.1, no. 10) or, for more specific experiments, in two laboratories, one for warm temperatures (no. 17) and one for moderate ones (no. 9). These facilities were complemented with more standard laboratory instruments and practices, including chemicals, electricity, machines to simulate variations in gravity (e.g., rotating discs), and the techniques of developmental physiology. Thus, the BVA building, in its entire elaborate technical configuration, must itself be viewed not just as a huge techno-natural assemblage but as a massive experimental apparatus. It was designed as a highly compartmentalized and fundamentally modular space for experimental biology, to allow environmental conditions for the organisms as well as laboratory conditions for the scientists to be stabilized and thus to be simultaneously adapted to the requirements of both humans and nonhumans.

Conclusion

The BVA as an experimental space must be understood as the peak of a development in experimental biology, and especially experimental zoology, that had its beginning half a century earlier in the European coastal marine stations and in the new zoological institutes in Germany. It was conceived as a synthesis of the dispersed experimental approaches that had emerged in those contexts. Central to this endeavor was the building of the BVA—the former Prater Vivarium, taken over and adapted throughout for its new purpose and thus turned into a huge techno-natural assemblage—in which a broad range of animal and plant species could be kept, bred, and researched. It functioned as a modular experimental space, designed to study the influence of external or environmental conditions on organisms. The building's former history as a public aquarium facilitated the modification and highlighted the historical connection between aquarium fancying, marine stations, and breeder zoology. Thus not only did design of the BVA reflect the methodological developments of experimental research in biology; it also incorporated knowledge on environmental conditions and influences, as well as the practices and technologies that had earlier been developed to stabilize and control them. Remarkably, the work of the BVA in this area shows that there was still a crucial dependence on the expertise of nonacademic fanciers.

While in the years before the BVA existed, zoological research spaces were yet to emerge as techno-natural assemblages and full-fledged experimental spaces, the period during and after the BVA's term saw an increasing specialization of what had now become the biological laboratory proper as a highly specialized experimental space. In its comprehensive and general approach, the BVA was a unique exception in the history of the life sciences, uniting the history, aspirations, and challenges of experimental biology at the turn of the twentieth century in a focal point. August Schuberg's characterization of the zoological institute in Würzburg fits the BVA even more accurately: "Thus the way in which a new building is equipped and furnished shows the mode and direction of research that is supposed to be practiced there: in a way, the building becomes the embodiment of the prevailing ideas" (Schuberg 1891b, 1).

While Schuberg meant to point out the novelty of the Würzburg institute, it is rather the history of the BVA that makes visible the diverse and complexly layered traditions out of which experimental biology emerged.

Notes

1. The study of the life of organisms in their milieu was the field of study to which the term "biological" was initially applied, and the way it was used throughout the nineteenth century (Nyhart 2009; Toepfer 2011).

2. The Stazione in Naples with its public aquarium and the cooperation between the experimental physiologist Emil DuBois-Reymond and the Berlin public aquarium to study the electric eel (Dierig 2006) must be regarded as exceptions.

3. See also the institute's own in-house journal, *Arbeiten aus dem Zoologisch-Zootomischen Institut in Würzburg*, published between 1875 and 1891.

4. The variety of research animals used in nineteenth-century physiology was still rather broad and only became limited at the turn of the century (Logan 2002).

5. A similar example was Johann Jakob von Uexküll's Institut für Umweltforschung in Hamburg (Mildenberger 2007). Coming from a similar disciplinary background as Przibram, Uexküll used the decayed aquaria of the Hamburg Zoo, which were built and run by Karl August Möbius, for his research.

6. Recipes for artificial saltwater already circulated in aquarium fancier journals in the nineteenth century (Reiß 2012a).

7. The numbers were provided by Klaus Taschwer, who went to the effort of analyzing the list in Przibram's report.

References

Berz, P. 2009. The eyes of the olms. *History and Philosophy of the Life Sciences* 31 (2): 215–240.

Bowler, P. J. 1992. *The Eclipse of Darwinism: Anti-Darwinian Evolution Theories in the Decades around 1900.* Baltimore, MD: Johns Hopkins University Press.

Braun, M. 1878a. Die Einrichtung des zoologischen Instituts Würzburg zur Zucht der Thiere. I. Aquarien. *Zoologischer Anzeiger* 1:34–36.

Braun, M. 1878b. 1. Die Einrichtung des zoologischen Instituts Würzburg zur Zucht niederer Wirbelthiere. *Zoologischer Anzeiger* 1:128–130.

Buck, E. 1874. Die Stromerzeugungsmaschine für das Süsswasser-Aquarium. *Der Zoologische Garten* 15:148–150.

Canguilhem, G. 2008. *Knowledge of Life*. New York: Fordham University Press.

Caron, J. A. 1988. "Biology" in the life sciences: A historiographical contribution. *History of Science* 26:223–268.

Chadarevian, S. d. 1996. Laboratory science versus country-house experiments: The controversy between Julius Sachs and Charles Darwin. *British Journal for the History of Science* 29 (1): 17–41.

Chauvin, M. v. 1876. Ueber die Verwandlung der mexicanischen Axolotl in Amblystoma. *Zeitschrift für Wissenschartliche Zoologie* 27 (4): 522–535.

Coen, D. R. 2007. *Vienna in the Age of Uncertainty: Science, Liberalism, and Private Life*. Chicago, London: University of Chicago Press.

De Bont, R. 2009. Between the laboratory and the deep blue sea: Space issues in the marine stations of Naples and Wimereux. *Social Studies of Science* 39 (2): 199–227.

De Bont, R. 2015. Stations in the Field: A History of Place-Based Animal Research, 1870–1930. Chicago, London: University of Chicago Press. </bok

Dierig, S. 2006. *Wissenschaft in der Maschinenstadt. Emil Du Bois-Reymond und seine Laboratorien in Berlin*. Göttingen: Wallstein.

Dorner, H. 1874. Eine neue Verbesserung der Zimmeraquarien. *Der Zoologische Garten* 15 (2): 41–45.

Eggert, H., C. Junk, C. Körner, and E. Schmitt. 1905. Hochschulen im allgemeinen. Universitäten und Technische Hochschulen. Naturwissenschaftliche Institute. In *Handbuch der Architektur. 4. Teil: Entwerfen, Anlage und Einrichtung der Gebäude. 6. Halbband: Gebäude für Erziehung, Wissenschaft und Kunst. Heft 2: Hochschulen, zugehörige und verwandte wissenschaftliche Institute*, ed. E. Schmitt. Stuttgart: Alfred Kröner Verlag.

Fantini, B. 2000. The "Stazione Zoologica Anton Dohrn" and the history of embryology. *International Journal of Developmental Biology* 44:523–535.

Finnegan, D. A. 2008. The spatial turn: Geographical approaches in the history of science. *Journal of the History of Biology* 41 (2): 369–388.

Gliboff, S. 2006. The case of Paul Kammerer: Evolution and experimentation in the early 20th century. *Journal of the History of Biology* 39 (3): 525–563.

Hoffmann, C. 2001. The design of disturbance: Physics institutes and physics research in Germany, 1870–1910. *Perspectives on Science* 9 (2): 173–195.

Hopwood, N. 2011. Approaches and species in the history of vertebrate embryology. In *Vertebrate Embryogenesis: Embryological, Cellular and Genetic Methods*, ed. F. J. Pelegrini, 1–20. New York: Humana Press.

Kammerer, P. 1901. Die Querzahnmolche (Amblystoma). *Blätter für Aquarien- und Terrarien-Freunde* 12:166–69, 177–80.

Kammerer, P. 1904. Beitrag zur Erkenntnis der Verwandtschaftsverhältnisse von Salamandra atra und maculosa. *Archiv für Entwicklungsmechanik der Organismen* 17:165–264.

Kammerer, P. 1907. Die Fortpflanzung des Grottenolms (*Proteus anguineus Laur*). *Verhandlungen der kaiserlich-königlichen zoologisch-botanischen Gesellschaft in Wien* 57: 277–92.

Kanz, K. T. 2002. Von der "biologia" zur Biologie. Begriffsentwicklung und Disziplingenese vom 17. bis 20. Jahrhundert. In *Die Entstehung biologischer Disziplinen II. Beiträge zur 10. Jahrestagung der DGGTB in Berlin 2001*, ed. U. Hoßfeld and T. Junker, 9–30. Berlin: VWB—Verlag für Wissenschaft und Bildung.

Kofoid, C. A. 1910. *The Biological Stations of Europe*. Washington, DC: Government Printing Office.

,Kohler, R. E. 2002. *Landscapes and Labscapes: Exploring the Lab-Field Border in Biology*. Chicago, London: University of Chicago Press.

Logan, C. A. 2002. Before there were standards: The role of test animals in the production of empirical generality in physiology. *Journal of the History of Biology* 35 (2): 329–363.

Meinel, C. 2000. Chemische Laboratorien: Funktion und Disposition. *Berichte zur Wissenschaftsgeschichte* 23 (3): 287–302.

Mildenberger, F. 2007. *Umwelt als Vision. Leben und Werk Jakob von Uexkülls (1864–1944)*. Stuttgart: Steiner.

Müller, I. 1975. Die Wandlung embryologischer Forschung von der deskriptiven zur experimentellen Phase unter dem Einfluss der Zoologischen Station in Neapel. *Medizinhistorisches Journal* 10:191–218.

Nyhart, L. K. 1995. *Biology Takes Form: Animal Morphology and the German Universities, 1800–1900*. Chicago, London: University of Chicago Press.

Nyhart, L. K. 2009. *Modern Nature: The Rise of the Biological Perspective in Germany*. Chicago, London: University of Chicago Press.

Przibram, H. 1901a. Experimentelle Studien über Regeneration. *Archiv für Entwicklungsmechanik der Organismen* 11 (2): 321–345.

Przibram, H. 1901b. Experimentelle Studien über Regeneration. Zweite Mittheilung. *Archiv für Entwicklungsmechanik der Organismen* 11 (2): 507–527.

Przibram, H. 1903. Die neue Anstalt für experimentelle Biologie in Wien. *Verhandlungen der Gesellschaft Deutscher Naturforscher und Ärzte* 74: 152–155.

Przibram, H. 1907. *Experimental-Zoologie. 1. Embryogenese. Eine Zusammenfassung der durch Versuche ermittelten Gesetzmäszigkeiten tierischer Ei-Entwicklung (Befruchtung, Furchung, Organbildung)*. Leipzig, Wien: Franz Deuticke.

Przibram, H. 1909a. Die Biologische Versuchsanstalt in Wien. Zweck, Einrichtung und Tätigkeit während der ersten fünf Jahre ihres Bestandes (1902–1907), Bericht der zoologischen, botanischen und physikalisch-chemischen Abteilung. *Zeitschrift für biologische Technik und Methodik* 1: 234–64, 329–62, 409–33, S1–S34.

Przibram, H. 1909b. *Experimental-Zoologie. 2. Regeneration. Eine Zusammenfassung der durch Versuche ermittelten Gesetzmässigkeiten tierischer Wieder-Erzeugung (Nachwachsen, Umformung, Missbildung)*. Leipzig, Wien: Franz Deuticke.

Przibram, H. 1910. *Experimental-Zoologie. 3. Phylogenese. Eine Zusammenfassung der durch Versuche ermittelten Gesetzmäszigkeiten tierischer Art-Bildung (Arteigenheit, Artübertragung, Artwandlung)*. Leipzig, Wien: Franz Deuticke.

Przibram, H. 1913. *Experimental-Zoologie. 4. Vitalität. Eine Zusammenfassung der durch Versuche ermittelten Gesetzmässigkeiten tierischer Lebenszustände (Kolloidform, Wachstum, Bewegung)*. Leipzig, Wien: Franz Deuticke.

Przibram, H. 1914. *Experimental-Zoologie. 5. Funktion. Eine Zusammenfassung der durch Versuche ermittelten Gesetzmässigkeiten tierischer Verrichtung (Ausübung, Wechselwirkung, Anpassung)*. Leipzig, Wien: Franz Deuticke.

Przibram, H. 1929. *Experimental-Zoologie. 6. Zoonomie. Eine Zusammenfassung der durch Versuche ermittelten Gesetzmässigkeiten tierischer Formbildung (experimentelle, theoretische und literarische Übersicht bis einschliesslich 1928)*. Leipzig, Wien: Franz Deuticke.

Przibram, H. 1930. *Experimental-Zoologie. 7. Zootechniken. Eine Zusammenfassung der für Versuche mit Tieren verfügbaren Forschungsweisen (Fragestellung, Versuchsführung, Bearbeitung)*. Leipzig, Wien: Franz Deuticke.

Rheinberger, H.-J., M. Hagner, and B. Wahrig-Schmidt, eds. 1997. *Räume des Wissens: Repräsentation, Codierung, Spur*. Berlin: Akademie-Verlag.

Reiß, C. 2012a. Gateway, instrument, environment: The aquarium as a hybrid space between animal fancying and experimental zoology. *NTM Zeitschrift für Geschichte der Wissenschaften, Technik und Medizin* 24 (4): 309–336.

Reiß, C. 2012b. Wie die Zoologie das Füttern lernte. Die Ernährung von Tieren in der Zoologie im 19. Jahrhundert. *Berichte zur Wissenschaftsgeschichte* 35 (4): 286–299.

Reiß, C. 2014. August Weismanns frühe Evolutionsforschung: Experiment und Theorie im künstlichen Naturraum. *Rudolstädter naturhistorische Schriften* 20:11–29.

Reiß, C., and M. Vennen. 2014. Muddy waters. Das Aquarium als Experimentalraum (proto)ökologischen Wissens, 1850–1877. In *Stoffe in Bewegung. Beiträge zu einer Wissenschaftsgeschichte der materiellen Welt*, ed. K. Espahangizi and B. Orland, 121–142. Berlin: Diaphanes.

Schaxel, J. 1922. *Grundzüge der Theoriebildung in der Biologie*. Jena: Gustav Fischer.

Schuberg, A. 1891a. Carl Semper. *Arbeiten aus dem Zoologisch-Zootomischen Institut in Würzburg* 10 (1): i–xxii.

Schuberg, A. 1891b. Das neue zoologisch-zootomische Institut der Königl. Julius-Maximilians-Universität zu Würzburg. *Arbeiten aus dem Zoologisch-Zootomischen Institut in Würzburg* 10 (1): 1–12.

Semper, C. G. 1874a. Ueber die Wachsthums-Bedingungen des Lymnaeus stagnalis. *Arbeiten aus dem Zoologisch-Zootomischen Institut in Würzburg* 1:137–167.

Semper, C. G. 1874b. Bemerkungen über den Apparat zum Halten von niederen Seethieren. *Der Zoologische Garten* 15 (2): 46–48.

Semper, C. G. 1875. *Die Verwandtschaftsbeziehungen der gegliederten Thiere*. Würzburg: Stahel'sche Buch- und Kunsthandlung.

Semper, C. G. 1880. *Die natürlichen Existenzbedingungen der Thiere*. Leipzig: Brockhaus.

Spengel, J. W. 1875. Der Durchlüftungsapparat für Zimmeraquarien. *Der Zoologische Garten* 16:451–453.

Star, S. L., and J. R. Griesemer. 1989. Institutional ecology, "translations" and boundary objects: Amateurs and professionals in Berkeley's Museum of Vertebrate Zoology, 1907–39. *Social Studies of Science* 19 (3): 387–420.

Toepfer, G. 2011. *Analogie–Ganzheit*. Volume 1, Historisches Wörterbuch der Biologie. Geschichte und Theorie der biologischen Grundbegriffe. Stuttgart, Weimar: Metzler.

Vennen, M. 2013. Die Hygiene der Stadtfische und das wilde Leben in der Wasserleitung. Zum Verhältnis von Aquarium und Stadt im 19. Jahrhundert. *Berichte zur Wissenschaftsgeschichte* 36 (2): 148–171.

Weismann, A. 1875. Über die Verwandlung des mexikanischen Axolotl in ein Ambylstoma. *Zeitschrift für Wissenschartliche Zoologie* 25 (Supplement): 297–334.

Wettstein, R. 1902. Oesterreichische biologische Stationen. *Neue Freie Presse*, 24. August 1902: 14–15.

IV RESEARCH AGENDA

8 The Substance of Form: Hans Przibram's Quest for Biological Experiment, Quantification, and Theory

Gerd B. Müller

When Hans Leo Przibram gathered his friends and academic allies Wilhelm Figdor and Leopold von Portheim to found the Biologische Versuchsanstalt (BVA) in Vienna in 1902, he had something very special in mind. As he envisaged it, the research facility should be able to experimentally address a suite of open questions in biology that were at the core of the scientific debate of the period, and at the same time it should advance the transformation of biology into an exact science. Since the universities were not equipped for large-scale experimental work that required the prolonged rearing of live organisms, nothing less than an independent institution had to be created—not in opposition to the university but as an expansion of its scientific capacities. To Przibram (1908/09a) the enterprise was not an extravagance but a "requisite of modern biology."

The chapters in the present book demonstrate how a suite of specific conditions—scientific, technological, societal, and financial—played together in the founding and operation of the BVA. Here I will highlight the work at its zoological department, which was headed by Hans Przibram personally (figure 8.1). Whereas the botanical, the physicochemical, and the physiological departments were predominantly guided by research questions germane to those respective fields, the zoological department most fully instantiated the new programmatic approach Przibram had conceived. That program foresaw the systematic experimental study of "all major questions of biology," with special attention given to evolutionary change, embryological development, external conditions affecting organismal processes, and functional adaptation. The ultimate goal was the "generalization of the experiences thus obtained" (Przibram 1908/09a), leading toward (what we would call today) a causal theory of organismal form. How did Hans Przibram come to choose this particular and—at the time—rather unusual approach?

Mentors and Motives

At the turn from the nineteenth to the twentieth century the exact composition of the newly emerging scientific discipline of biology was controversial (see chapter 6). Darwin's

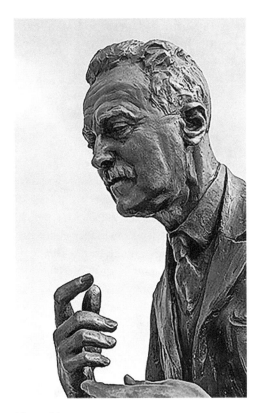

Figure 8.1
Sculpture of Hans Przibram reinstated in the entrance hall of the Austrian Academy of Sciences in 2014. Courtesy
of the author.

"dangerous idea" had fully impregnated the natural sciences, but was the evolutionary
process a characteristic that united all living systems? Could the predictions of evolutionary theory be tested empirically? Was natural selection all there was to evolutionary
change? Was the future of humanity a mere struggle for existence, or could thoughtful
planning based on scientific evidence ameliorate the human condition? Were there specific
methodologies by which biological knowledge could be established? And what was the
position of the young discipline of biology within the overall advancement of the natural
sciences?

These were some of the issues under vivid discussion when Hans Przibram took up
his studies of zoology at the University of Vienna in 1894. His principal advisor would
be Berthold Hatschek, then head of the Second Institute of Zoology at the University of
Vienna. Hatschek was a student and close friend of Ernst Haeckel and a representative
of a new kind of zoology that explicitly went beyond the traditional description and

classification of organismal phenomena. An expert in the study of larval development, Hatschek placed special emphasis on the possibility of testing evolutionary and systematic propositions. At the University of Vienna, he was a main instigator of the change from typological description to mechanistic analysis in zoological research (Müller and Nemeschkal 2015).

Hatschek's role in the academic orientation of Przibram and in the foundation of the BVA cannot be overestimated. His ambition to forge a connection between morphology, ontogenetic development, and phylogenetic descent was in line with the Haeckelian spirit (e.g., Haeckel 1866), but he gained personal prominence for his studies of larval organization and his model-oriented approach to the study of organismal development. In particular, his meticulous work on the embryogenesis of amphioxus (*Branchiostoma lanceolatum*; Hatschek and Tuckey 1893) established its position as the prototypic model for chordate and vertebrate development, and the book is still a classic. At the same time Hatschek was broadly acquainted with the emerging causal-analytical and theoretical biological thought of the time. Early on he paid attention to chromosomal inheritance, Weismann's germ plasm theory, and Mendel's rules, and he devised an independent "hypothesis of organic inheritance" (Hatschek 1905). He was equally interested in physiological and developmental processes of growth, organization, and regulation, and he formulated a series of ideas concerning the direct influences of environmental conditions and functional properties of the organism on the cells of the germ line—much in disagreement with Weismann's concept of the strict independence of the germ plasm.

While Hatschek was a firm evolutionist and a follower of Darwin, he was also critical of certain aspects of Darwinism, a sentiment shared by numerous biologists in Vienna and elsewhere. In his lecture before the Philosophical Society on the occasion of Darwin's 100th anniversary (in which he predicted millions of human deaths if racial ideals should come to guide societal progression) he called the principle of "individual selection" insufficient to account for the "multitudinous simultaneous phylogenetic modifications of organisms" (Hatschek 1909). Against natural selection he pitted the principle of "functional self-adaptation" within organisms and argued (in a neo-Lamarckian vein) its reciprocal influence on the germ line. This idea was rooted in a well-developed conception of "organism" that we would call "systemic" today. The "heritable effects of functional adaptation" and "the harmonic cooperation of its parts … [are] the properties that make the organism a real Organism." Thus, organisms represent "dispositions of organization … that are interconnected by thousands of internal interdependencies" (Hatschek 1902). He emphasized that these interrelations consisted, on the one hand, of heritable (phylogenetic) components, but were also "permanently renewed and individually re-acquired" through continued physiological activity. By this approach Hatschek intended to supplement mere variation of (typological) basic forms with emergent individuality. That fundamental criticism of Darwinism was to become one of the strongest motivations for Hans Przibram.

Another important influence on Przibram was the general experimental turn in biological research (see chapters 4 and 5). Hatschek was not an experimentalist himself, even though he clearly recognized the importance of this approach. But another former disciple of Haeckel (and an acquaintance of Hatschek from his student days), Wilhelm Roux, became a leader in the field of *Entwicklungsmechanik* and the founder of the most important publication series for experimental biology, the *Archiv für Entwickelungsmechanik der Organismen* (a journal still in existence today as *Development, Genes, and Evolution*). It would become the preferred publication venue for the experimental results obtained at the zoological department of the BVA. Przibram credits Roux for having provided the decisive impulse to follow an experimental direction of research (Przibram 1922a).

Besides these conceptual and methodological incentives for the founding of an experimental institution in Vienna, another, more pragmatic consideration was likely at play. During Przibram's student days, Berthold Hatschek had been the director of a marine biology research station of the University of Vienna in the northern Adriatic, the k. k. Zoologische Station Triest, founded in 1875. This facility provided easy access to a plethora of marine organisms, for which even Haeckel envied Hatschek (Michler 1999). Przibram had visited the station, as well as those in Naples and Roscoff, and surely he had recognized their enormous value for experimental work. But also, to Hatschek the idea of having a (partly) marine station in close proximity, within Vienna, must have been compelling. Actually, Hatschek seems to have been the one who suggested that Przibram buy the former Vivarium building (Przibram 1903, Loisel 1907), and subsequently he served on the BVA's board of directors for many years. When Trieste was no longer a part of Austria due to the territorial consequences of World War I, a national facility gained even greater significance.

Many elements of Hatschek's universe can be found in Przibram's program for the BVA, but the complete devotion to experimentation was of his own making. And he added another, even more unusual component to it, one not present in the programs of any other biological institution of the time: quantification and mathematization. From the very beginning he intended to make it "a task of the Biologische Versuchsanstalt ... to enable a quantitative treatment of biological problems" (Przibram 1913). The goal was nothing less than "a mathematical theory of organic life, based on quantitative measurement and linked with geometrics, physics, and chemistry" (Przibram 1923c). This way, he frequently pronounced, it would be possible to "make biology an exact science"—the second explicit objective for the BVA, likely derived from the close interaction with his brother Karl, a physicist. This chapter will examine to what extent these goals were fulfilled.

Thus, we find Przibram's program for the BVA governed by a suite of theoretical considerations: a critical stance regarding the power of Darwinian selection, a mechanistic (in the sense of Roux) and organismal conception of biology, an environment-controlled view of development and inheritance, and an experimental approach that permitted quantification and mathematical formalization. Przibram saw his program in deliberate contrast to the prevailing selectionist paradigm derived from Darwin's theory, and he chose to take an alternative perspective in which it was not natural selection acting from the outside on limitless chance variation, but the internal form-generating processes would act in dependence of environmental and functional factors, which would represent the primary determinants of organismal change. Though influenced by contemporary discourse and several creative minds in his academic environment—with a conspicuous restraint of reference to Haeckelian thought—Przibram's program was a highly genuine one.

Experimental Zoology of Qualities

Rarely in biology has a theoretically motivated program been followed so strictly in its practical realization. As we have seen elsewhere (e.g., in chapter 7), the entire BVA was specifically designed and equipped to promote experimentation involving whole organisms, providing facilities to keep animals throughout their entire life cycles, and over several generations, under definable environmental conditions. This attitude followed Przibram's conviction that "the fundament for success in biological experimentation lies not in contriving complicated apparatuses, but in the development of maintenance techniques for keeping organisms over extended periods of time." This was necessary because "whereas it may suffice the physiologist to keep his experimental objects alive for as long as he cares to study a certain function and then, for further observation, take fresh specimens, for the biologist [!], the important thing is to follow changes in form continuously, throughout a longer experimental duration" (Przibram 1908/09a). Accordingly, the techniques of successful animal keeping were elaborated in the most sophisticated ways (e.g., Przibram 1908/09b), and the BVA established facilities for keeping a plethora of organisms from very different habitats as well as means and devices to control for a wide range of environmental parameters, such as temperature, humidity, light intensity, background color, salinity, gravitation, and others (e.g., Przibram 1913). The zoological department made extensive use of these facilities.

Equally diverse were the research questions, the kinds of experimentation, and the methodologies used at the zoological department. With regard to the guiding themes, we may distinguish between an early phase and a later, post-1912 phase of research. The principal issues addressed in the first phase pertained to embryological development, regeneration, transplantation, and homeosis. In addition, the influence of external parameters on organismal development and the inheritance of induced modifications were target issues for experimentation. Whereas other topics were also studied to a lesser degree or as individual exceptions, the themes listed above stand out because they yielded extensive amounts of empirical results and publications. Many of these are collected in Przibram's magnum opus, *Experimental-Zoologie*, which appeared in seven volumes from 1907 to 1929. Some of these works will be discussed below.

Regeneration was by far the most prominent issue during the early years of experimentation at the zoological department. This comes as no surprise, because Hans Przibram's own principal expertise was in regeneration studies. Already his thesis with Hatschek had focused on regeneration in crustaceans (Przibram 1899), and regeneration remained the preferred topic throughout his entire career up until his very last publications. Altogether Przibram's work contains more than 30 publications on regeneration, including several comprehensive treatises and books. He was an international authority on the subject, which at the time was pursued in zoological and botanical research worldwide, yet hardly anywhere nearly as comprehensively and systematically as at the BVA.

The primary ambition of the regeneration studies at the zoological department was to determine whether all classes of organisms were capable of regeneration, since a number of exceptions had been claimed to exist among nonvertebrates and vertebrates alike. To this end coelenterates, worms, insects, crustaceans, myriapods, spiders, mollusks, and other invertebrates were analyzed with respect to their regenerative potentials; amphioxus, amphibians, and birds were also studied in that regard by various members of the department. The studies were carried out preferably on body appendages, such as limbs, wings, antennae, eyes, or whole body sections, to such extent that one is tempted to say that "anything that could be cut off was cut off." Przibram's own interest continued to concern crustaceans, in which he studied, for instance, the capacity of reversed heterochely (*Scherenumkehr*; Przibram 1905a, 1907a), i.e., the tendency to regenerate a claw of the contralateral identity. But his most preferred organism for regeneration studies (and other investigations) became the praying mantis, of which he studied both the North African (*Sphodromantis bioculata*) and the European species (*Mantis religiosa*). He devoted a large number of papers to the partial or total deletion of their antennae, legs, eyes, and other body parts, describing and measuring the respective regenerative processes (e.g., Przibram 1906a, 1917a) (figure 8.2).

The BVA scientists came to the conclusion that no true exceptions existed for the capacity to regenerate, and that a number of general rules of regeneration could be determined. These rules concerned the relationship of regeneration with growth and with organismal age, the speed of regenerative processes, the functional role of the affected tissues, the quantitative and qualitative features of regeneration, the evolutionary modification of regenerative potential, the origins of regenerative cells and tissues, and the relationships of regeneration with teratological, hypo- or hypermorphotic, heteromorphotic, and atavistic structures (for an overview see Przibram 1909a). During the early period of the BVA, we see a large number of workers occupied with topics of regeneration, also at the botanical department (see chapter 9 in this volume). Among them was Paul Weiss, who held the adjunct assistant position in the zoological department from 1925 to 1927 and would later become a leading developmental biologist in the United States (Brauckmann 2013). Weiss had focused on amphibian limbs (e.g., Weiss 1923, 1925a), and his and Przibram's (1924a) own work on that topic led to an extensive

Figure 8.2
Three stages of regeneration of a grasping limb in *Sphodromantis bioculata*. (a) Larva stage 10. (b) Nymph stage 11. (c) Imago stage 12. Drawings by Hans Przibram (1906a).

controversy with the Yale embryologist Ross G. Harrison concerning the axes and polarities in transplanted salamander limbs. Whereas Harrison saw the inversion of symmetries in transplanted and rotated limb buds as determined by the flank of the embryo, Przibram interpreted the results as a consequence of the regeneration properties of the transplant. Their dispute is probably unique in the history of biology, because it was conducted via the extensive use of models. Przibram (1924a) criticized and reinterpreted Harrison's conclusions using sketched and two-dimensional cardboard models of salamander limbs (figure 8.3a) and eventually responded to Harrison (1925) even with the help of three-dimensional models (figure 8.3b) (Przibram 1927a).

In combination with regeneration studies, and independently from them, transplantation experiments were another preferred topic at the zoological department. The questions to be answered included whether it was possible to transmit properties from one organism to another by the transfer of body parts, whether these transplanted parts would follow the developmental "program" of the donor organisms or whether "instructive" influences from

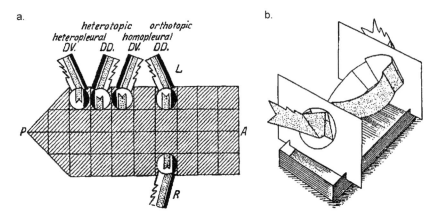

Figure 8.3
Przibram's cardboard models of salamander limb development. (a) 2-D model from Przibram 1924a. (b) 3-D model from Przibram 1927a.

the host tissues would prevail, and how the host would be able to integrate foreign structures. Transplantations involved preferentially the eyes and limbs of vertebrates and invertebrates, and even entire heads of insects. Przibram and his coworkers developed the method of "autophoric transplantation," the term designating the retention of the transplant by the own forces of the recipient (Przibram 1922a, 1923a). The transplanted eyes of insects and vertebrates regained functionality to a surprising degree (see overviews in Przibram 1926) and were even tested with behavioral experiments such as those of Auguste Jellinek (1923). An extensive number of transplantation studies involving different organ systems were carried out by other members of the zoological department, including Finkler, Koppányi, Uhlenhuth, Wiesner, and Weiss. In particular the head transplants by Walter Finkler (1923a, 1923b, 1923c) raised much attention and were controversially discussed in *Nature* (Calman 1924; Przibram 1924b). But they were also picked up by the public media and added to the image of sensational science at the BVA. The consequences of these experiments with regard to the potential for future transplantations in humans was clearly recognized (Przibram 1926).

In hindsight, one of the most important topics to be addressed experimentally at the zoological department—again, mostly by Hans Przibram—was *homoeosis* (spelled "homeosis" in English). The term had been introduced by Bateson (1894) to denote the phenomenon that a body part sometimes is found to have changed into "the likeness" of a different body part. Homeosis was in fact the phenomenon that eventually led to the notion of major genetic switches that could turn on or off entire suites of coordinated genetic events in the development of a body part, and the subsequent discovery of homeotic genes was awarded with the Nobel Prize in Physiology or Medicine in 1995. Between 1910 and 1919 Przibram alone published eight papers on homeosis, clearly recognizing the importance of this particular kind of variation. Characteristically, he first attempted to define and classify the phenomenon by distinguishing three kinds of homeosis, which he would come to call substitutive, additive, and translative (Przibram 1910a). In subsequent papers he described induced homeosis in regenerating antennae of the praying mantis (Przibram 1917a, 1919a) as well as in crabs Przibram (1917b), hymenoptera (Przibram 1918), and beetles (Przibram 1919b). Other members of the department to mention induced homeosis were Leonore Brecher (1924b) and Franz Friza (Friza and Przibram 1930).

Whereas isolated cases of regenerative homeosis had been reported previously (e.g., Herbst 1896), the systematic experiments and their interpretation at the BVA were significant insofar as they not only showed that homeotic events could reliably result from regeneration, but also demonstrated the developmental conditions that give rise to homeotic regeneration, a process later shown in vertebrates to depend also on the chemical exposure of the regenerate (Müller et al. 1996). In his English summary of regeneration and transplantation studies, Przibram (1925d) points out that one of the most consistently reproduced cases of experimental homeosis was the regeneration of feet instead of amputated antennae in the Indian stick insect *Carausius* (*Dixippus*) *morosus*, a model case extensively studied at the BVA by Leonore Brecher (1924b). Original photographic plates of these "antennapedia" regenerates exist in the Zoological Collection at the Department of Theoretical Biology at the University of Vienna (figure 8.4).

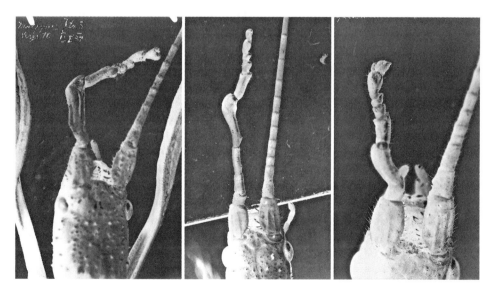

Figure 8.4
Three homeotic regenerates resulting from amputations of the left-side antenna in *Carausius* (*Dixippus*) *morosus*. Forelimbs that exhibit different degrees of differentiation have formed instead of the antennae. Unpublished originals of photographic glass plates from Przibram's slide collection dated 1931, most likely obtained from an experimental series conducted by Leonore Brecher. Courtesy of the Zoological Collection of the University of Vienna.

Aside from direct intervention with development and growth, the field of experimentation most closely related to the evolutionary dimension of the BVA's program was the influence of environmental parameters on all organismal processes. The parameters considered and manipulated included temperature, chemical factors, light and other forms of radiation, background coloration, electrical and magnetic influences, humidity, mechanical factors, gravity, and more. As mentioned above, the entire facility had been equipped for the technical control of environmental conditions, including numerous automated devices. In many cases these self-regulating appliances proved essential, as Przibram (1923b) notes in the case of the temperature chambers: "Only due to these devices were we able to perform quantitative temperature experiments on living organisms with sufficient exactitude, as to demonstrate unequivocally the relationship of temperature with the processes of life." A vast literature resulted from these studies, and only a few selected themes can be discussed here.

For instance, the determinants of and influences on animal coloration were a favored topic at the zoological department. More than 50 papers on animal coloration were published between 1910 and 1935, of which the principal authors were Abolin, Brecher, Eisler, Finkler, Kammerer, Koppányi, Kunio, and Przibram. Again, the variety of organisms used in these studies was extensive, including fish, amphibians, reptiles, insects, crustaceans, and others, and the perspectives and techniques of analysis were multifold, ranging from the study of molecular composition to cellular and physiological regulation to various kinds of experimental interventions. At the same time, the topic of coloration was probably the most integrative one at the BVA, involving members of all departments, since it included histological, physical, chemical, hormonal, and behavioral aspects.

Besides Przibram, Kammerer, and others, the most productive worker on animal coloration was Leonore Brecher (figure 8.5a) who produced more than 20 papers on butterfly pupae but also on representatives of other groups, such as the stick insect *Dixippus* (Przibram and Brecher 1922). Using careful variations of the lighting and color conditions in which butterfly caterpillars (predominantly *Pieris brassicae* and several species of *Vanessa*) underwent pupation, Brecher investigated the effects on the coloration of the pupae. Her analysis of the resulting color polymorphisms and the possible pathways of transmission of color information to the caterpillars formed one of the most extensive studies of environmental influences on phenotypic traits at the time (e.g., Brecher 1917, 1919, 1922a, 1924a). In addition, Brecher showed the effects on the production of different pigments in the pupae and the requirement of visual stimulation of the caterpillars for the color modifications to take place. Apart from the experimental testing of the influence of environmental conditions, the physiological and chemical factors underlying animal coloration were also investigated (Przibram and Brecher 1919) and discussed with Wolfgang Pauli from the Physicochemical Department (see chapter 10 in this volume).

a.   b.

Figure 8.5
(a) Leonore Brecher in 1923. Courtesy of Privatarchiv Eisert. (b) Paul Kammerer in 1923. Public domain.

The kind of experimentation most widely associated with the BVA is that in which one of the approaches discussed so far was combined with breeding. As pointed out above, the reason for this emphasis was the deep conviction that "inheritance represents only one aspect of the problem of descendance. The second lies in change through external factors" (Przibram 1908/09a). Hence, testing the potentials for the inheritance of acquired characteristics was a major goal of experimentation, albeit a complicated one that required not only experimental skills but also and especially those of animal breeding. Several members of the zoological department undertook experiments of that kind (including Brecher, Megusar, Uhlenhuth, and Werber), but the master of the trade clearly was Paul Kammerer (figure 8.5b). Kammerer had worked at the zoological department since its founding and was its adjunct assistant from 1914 to 1923. Much has been written about his dazzling character, his sensational experiments, the alleged faking of results, his involvement with politics, society, music, and more (for a new biography see Taschwer 2016). But essentially Kammerer was a very gifted terrarian—one of the very reasons why Przibram hired him early on in the establishment of the institute. He was ideally suited, indeed, to develop the "maintenance techniques for keeping organisms over extended periods of time."

Together with Przibram, Kammerer conceived of several multigenerational studies that were based on the variation of one particular experimental design: Start with the selection of a suitable life history trait that can be shown to depend on environmental influence. Then choose at least two related species that represent opposite extremes with regard to that trait, and gradually compel the organisms to live and reproduce under the environmental condition of the opposite form. Breed (without selection) for several generations under the artificial condition, thus modifying the conditions for development and the trait's phenotype. Then allow the experimental offspring to again reproduce under the original conditions. If the induced phenotype persisted through further generations in the absence of the artificial environmental stimulus, the heritability of the modified trait was supposed to have been demonstrated.

Let us look at an example. Kammerer (1904, 1908) chose two species of salamander that differ with regard to their reproduction mode. One was *Salamandra atra*, a viviparous species without aquatic larvae, in which females usually give birth to only two individuals, which have metamorphosed already before birth. The second was *Salamandra maculosa*, an oviparous/ovoviviparous species whose females give rise to as many as 70 aquatic larvae that will metamorphose after several months into terrestrial salamanders. In opposing modifications of the reproduction mode initiated through subtle alterations of temperature and water availability, *Salamandra atra* was coerced to lay eggs or premetamorphic larvae into water, and *Salamandra maculosa* was forced to live and reproduce terrestrially, delaying birthing until the larvae had metamorphosed. Kammerer reports that, as a consequence of the experimental regimes, the terrestrial species gave rise to aquatic larvae with the number of offspring increased, and the aquatic species gave birth to small juveniles with numbers of offspring decreased, showing that the reproductive behaviors and life histories of both species could be environmentally altered. Furthermore, he observed that subsequent generations derived from the experimental offspring of both species continued the induced behavior for some time, even after the original environmental conditions were reestablished. *Salamandra atra* from forced water-breeding deposited some of its larvae into water, and *Salamandra maculosa* from forced live-bearing gave birth to metamorphosed juveniles. With prolonged exposure of the experimental parents to the original conditions the effect gradually disappeared, and eventually the offspring of both species returned to their original reproduction modes. The descriptions of methodological detail, pros and cons, failures and successes, as well as the discussion of the results in light of Weismann's concepts are meticulous (Kammerer 1908).

Although we cannot be sure how strictly controlled these experiments were and whether (probably inadvertent) selection took place, Kammerer was certainly not alone in reporting such results. To the contrary, both within and outside the BVA similar kinds of experiments rendered extensive amounts of data (e.g., Jollos 1934) on what was known at the time by the term "enduring modifications" (Jablonka and Lamb 1995). At the BVA, Przibram obtained comparable results with his temperature-treated rats (see below) and Kammerer

(1913) with color modification experiments in salamanders, which were also continued by Przibram (1922b) and by Przibram and Dembowski (1922), sometimes raising forceful objections (e.g., Herbst 1919, 1924). Likewise, Leonore Brecher showed in her comprehensive work on butterflies that induced modifications of pupal coloration could be transmitted to subsequent generations (Brecher 1922b).

A similar setup as the one described above was used in the case of the infamous midwife toads. Again opposing modifications of reproduction mode stood at the beginning of the experiment, this time using a toad (*Alytes obstetricans*) and a frog (*Hyla arborea*). The toads mate on land and lay their eggs on land, and the males carry the strings of developing eggs on their hind legs (hence the term midwife toad) until the tadpoles hatch into water. The frogs mate in water and deposit their eggs in the water, and the tadpoles remain aquatic until metamorphosis. Again Kammerer coerced both species to assume the opposite reproduction mode. The frogs were forced to mate on land (or in small patches of water) and lay their eggs on land (where they were kept humid), and the toads were forced to mate and deposit their eggs in water. The resulting "water toads" were continuously bred in the aquatic environment over several generations, using different strains, some of which lasted more than ten years.

Leaving aside the frogs, the results obtained from the toads are quite different from those usually reported in truncated descriptions of alleged fraud. First and foremost, the contentious "nuptial pads" were *not* a goal of experimentation or even at the center of attention. Kammerer (1906, 1909, 1919) published three major papers altogether on his experiments with the midwife toads, which, in combination with additional information in Kammerer 1911, make up over 200 pages. The "real" results he reported were significant effects on the timing and rates of larval development (water larvae sped up metamorphosis, which usually takes very long in *Alytes*, and land larvae slowed down), a prolongation of the larval state for much longer than usual (by removing the embryos from the egg while the gills were still present), and a modification of several life history traits (indicating interesting relations between temperature, egg size, egg number, time of metamorphosis, and resulting animal size). He describes large water eggs, delayed metamorphosis, neotenic larvae (that could even reach sexual maturity), and extremely large animals. Thus, actually, Kammerer's work on *Alytes* demonstrates an exquisite case of experimentally induced heterochrony and an enduring behavioral modification, i.e., descendants of water-raised larvae maintain the water-breeding behavior. Recent interpretations raise the intriguing possibility that such outcomes can be attributed to epigenetic effects (Vargas et al. 2016).

The contentious nuptial pads are not mentioned at all in the first paper on the midwife toads and tree frogs (Kammerer 1906) and are merely noted in passing in the second paper (Kammerer 1909), listed among other morphological changes seen in the modified water toads. Only the third publication (Kammerer 1919) concentrated on the nuptial pads, mostly because William Bateson, the contemporary British authority on genetic inheritance and evolution, had qualified them as potentially crucial acquired effects, and

Kammerer had set up a new breeding line especially for that purpose in 1909. However, he repeatedly pointed out that Bateson's interpretation was erroneous: "In his tendency to regard the generation and maintenance of the nuptial pads as experimentum crucis, Bateson will be shattered when he remembers that *Alytes* ... must necessarily be derived from anurans that already possessed the pads" (Kammerer 1909). Kammerer correctly interpreted the appearance of nuptial pads as a case of induced atavism and not as the acquisition of a new characteristic (as Przibram also noted in his 1930 commentary in Nature). Still he regarded them as a useful demonstration of the reactivation and inheritance of the responsive capacity for pad generation. The atavistic aspect of this kind of experiment was intensively discussed by Kammerer and other scientists, including August Weismann (see Kammerer 1919, 1923).

Here is not the place to delve into the muddle of the midwife toad "case" that continues to haunt the BVA science, with its vast literature that either condemns or vindicates Kammerer (but see the more differentiated accounts by Gliboff 2006, Taschwer 2016, and Vargas et al. 2016). Little attention seems to have been paid to the fact that the alleged fake concerns a character that was not crucial for Kammerer's contentions regarding the inheritance of acquired traits. To the contrary, he regarded the occurrence of the nuptial pads as "by no means a conclusive proof of the inheritance of acquired characteristics" (Kammerer 1923). None of the allegations and counter allegations can be conclusively assessed today, and the growing disrepute of Kammerer also had an interesting international aspect (see chapter 13). But it can be said with confidence that given Kammerer's meticulous reports on positive and negative results from several breeding lines of experimental midwife toads (including photographs of the pads and of histological sections, taken by four different photographers), the singularity of one tampered-with museum specimen, examined more than a decade after the original study had taken place (Noble 1926), holds no scientific significance whatsoever. No one will be able to exclude inadvertent or intentional manipulation at the time of the original experimentation, but the incriminated specimen provides no proof that this was the case.

Independently from the nuptial pads, based on the BVA experiments and other accumulated evidence, by 1912 Przibram considered the problem of the inheritance of acquired characteristics essentially solved. He stated that "it is no longer a question whether characters that were brought to appearance through environmental changes in the bodies of the parents can reappear in the offspring that was returned to the original conditions. It has been shown experimentally in nearly all major groups of animals and plants that this is the case. Now the question is open by which *mechanisms* the modification of the offspring was realized" (Przibram 1912).

In what amounts to a significant turn in the experimental goals of the zoological department (and the BVA altogether), Przibram proposed a new program, based on the observation that "few efforts were made to study the physicochemical influences on the gonads, although far-reaching conclusions on their independence from the soma (Weismann) or the impossibility of external factors that modify the soma to also reach the germ cells (Semon) were forcefully pronounced" (Przibram 1912). Whereas in Weismann's (1892) concept each differentiated cell type in development was dependent on the action of a single determinant, provided by "biophores" that were generated by components (ids) of the idants (units of hereditary information) of the germ plasm (chromosomes in the nuclei of the germ cells), Przibram and his colleagues from the BVA argued that the hereditary information also needed an environment to become properly expressed or disturbed in its expression. Hence, contrasting with Weismann's germ plasm theory in a similar fashion as he had contrasted with Darwin's selectionism in the previous program, Przibram (1912) called the new program "Die Umwelt des Keimplasmas"—the environment of the germ plasm.

The factors considered to affect the germ cells and the germ plasm were chemical agents, humidity, osmotic differences, mechanical forces, gravitation, electricity, light and other forms of radiation, temperature, and more. Even though the new thematic orientation again led to successful experimentation, the outbreak of World War I and the tightening of the financial situation of the BVA did not permit a similar blossoming as took place with the work of the previous phase. But by 1925 around 20 publications had appeared under the heading "Die Umwelt des Keimplasmas." The effects of temperature, for instance, were studied most intensely during the period following 1912. Temperature was taken as a parameter that could represent the energy balance between the organism's metabolism and the external world. Przibram and others focused on the relationship between temperature and tail length in rats and found a number of interesting correlations (see below).

In 1932, Hans Przibram envisaged yet another change of emphasis in the experimental agenda of the BVA. Most likely based on his intensive interaction with his brother Karl, a physicist, and in an attempt to secure the future of the BVA, together with his co-director Leopold von Portheim he urged the Austrian Academy of Sciences "to establish a laboratory that is devoted to the examination of the effects of Radium on animals and plants." Those hopes were dashed by the political events and their horrible consequences for the BVA and its staff. Yet, despite the hindrances to the work of the BVA throughout its existence, the overall output of experimental results and publications by the zoological department is stunning. The above overview demonstrates that amid all the attention given to perceived fraud, oversimplified Lamarckism, and sensational experimentation at the BVA, the immense amount of serious research has been severely neglected. But the BVA's other chief objective, the quantification of biology, was even more thoroughly overlooked.

Experimental Zoology of Quantities

Besides the rigorous establishment of the experimental method, Przibram's second major goal with the BVA was to overcome the traditional descriptive treatment of biological phenomena—which prevailed even in experimental studies—by adopting the quantitative and mathematical approach underlying other "exact" sciences. He notes that "physics and chemistry are brilliant examples of what may be achieved … by quantitative experiment and mathematical formulation based thereon," and argues his "firm conviction that we must proceed on the same lines if we wish Biology to develop into an exact science." Przibram (1931) adds: "This has been my aim in all my work."

The quantitative approach Przibram envisaged for biology owed much to physics. Conceivably, Przibram's brother Karl, the vice chair of the Vienna Institut für Radiumforschung had been a major source of inspiration. Coen (2006) points out how statistical methods were exchanged between physics and biology and detects a telling similarity between Karl Przibram's statistical treatment of irregular normal distributions of electrical charges and Hans Przibram's analyses of arthropod growth. But in order to do statistics, quantitative data were required, and, hence, Przibram placed major emphasis on furnishing the BVA with equipment that would enable such quantification. By this he not only meant the measurement apparatuses to be used for recording the effects of experimentation but also the devices for the quantitative control of environmental parameters used in performing experiments or in the keeping of the experimental organisms. Furthermore, his quantitative approach included methods of labeling and enumeration of the experimental specimens (Przibram 1913). In the same systematic attitude that pervaded all of Przibram's undertakings, he intended to progress from the acquisition of quantitative data to statistical analysis and to mathematization.

Przibram was keen to familiarize himself and the audience of experimental zoology with other approaches to biological quantification, such as those of the mathematical biologist D'Arcy Thompson (1917), the botanist Julius MacLeod (1919), or the physiologist Jacques Loeb (1922). Even though he favorably commented on these authors' views regarding the general necessity of quantification in biology, Przibram (1922c) found their approaches wanting, due to the lack of experimental confirmation. His own quantitative studies, and most of those performed at the BVA, were based on experimentation and predominantly concerned subjects such as growth of organismal shape and size, speed of regeneration, temperature relations, and aspects of metabolism and aging.

The study of temperature relations provided one of the most straightforward ways by which the quantitative approach could be implemented. Different measurement and recording devices could be used in combination with automated temperature chambers, which were built as walk-in rooms, to enable experiments without the need to remove the animals from the desired temperature conditions. A large number of different organisms and experimental setups were used in the study of temperature effects (see Przibram 1923b for an overview), but one of the most consistent and long-lasting studies was the one on "heat rats" (e.g., Przibram 1917c, 1924c, 1925a, 1925b, 1925c). Based on the work of Sumner (1909), Przibram developed his own intricate experimental system specifically designed for the study of temperature-dependent processes. Using different strains of rats, such as

Mus decumanus (today *Rattus norvegicus*) and *Mus rattus* (today *Rattus rattus*), he combined controlled variation of temperature regimes with breeding.

In an impressive amount of publications (mostly belonging to the series "Environment of the germ plasm"), Przibram (sometimes with coauthors) systematically analyzed the relationship between external temperature, body temperature, and the tail length of rats in both genders and at different ages. Breeding the rats over several generations in different temperature regimes, ranging from +5 to +40 degrees Celsius, and followed by the eventual reinstatement of newborns into either maintained, colder, or warmer temperature conditions compared to those of the parent generations, Przibram (1925b) showed what he called an acquired heritable "reaction potency" to temperature conditions in the descendant rats, expressed (and measured) as tail length. Extensive data sets were obtained on the relations between external temperature, body temperature, and tail length in different species, age classes, and genders. Besides the quantitative growth estimates for different temperature conditions (e.g., Przibram 1925a, 1925c), a number of generalizations could be drawn (Przibram 1923b), of which the most interesting were those on the heritability of "temperature primed" metabolic states (*Wärmestimmung*) (Przibram 1925b). The "heat rats" model was also used to investigate the relationships between body temperature and hormonal activity, another major field of research at the BVA (see Logan 2013).

The application of quantitative methods to biological research seems to have been more important to Przibram than a sophisticated mathematization. In fact, he opposed the application of advanced forms of mathematics, which he regarded as "heavy baggage." Nevertheless he proceeded to write several articles and three books (figure 8.6) on mathematical biology, some of the first anywhere in the world. Already in 1908 he published a small volume on the "Application of elementary mathematics to problems of biology," based on lectures held at the University of Vienna in 1907. There he argued that mathematization should be used not only in the obvious case of physiology but also in dealing with problems of organismal growth, form generation, inheritance, speciation, and psychology (behavior). Drawing strongly on the examples of chemistry and physics, he described, in the simple and didactic manner of a lecture, several examples of the mathematical treatment of biological phenomena. Later, in the 1922 volume "Form und Formel im Tierreiche" (Form and formula in the animal kingdom) and in several other individual publications (some coauthored with other BVA scientists), he presented applied cases of selected problems of biological quantification. These concerned the developmental progression of biological forms (e.g., surface-to-volume relations, division planes); the absolute and relative speeds of growth, regeneration, movement, and aging; and various aspects of weight, mass, and geometry of biological traits.

By contrast, Przibram's famous 1923 volume *Aufbau Mathematischer Biologie* (Structure of mathematical biology) contains no formulae, but rather represents a comprehensive research program defining the goals of a mathematical approach to biology. Among other issues, he called for a "quantitative systematics" of the kind we later find elaborated in

Figure 8.6
Hans Przibram's three books on mathematical biology, published in 1908, 1922, and 1923. Courtesy of the author.

approaches such as numerical taxonomy (Sneath and Sokal 1973). Other themes were biophysics, biometrics, heritability, periodicity, bionomics, and a kind of energetics. Przibram did not share the idea proposed by others that biology required the introduction of new principles pertaining to biological processes alone; rather, he concluded that it would be sufficient to derive a mathematical theory of organic life from quantitative measurement, the application of geometric, physical, and chemical principles, and the precise definition of biological terms. In the manner of chemistry and physics, he attributed major significance to the understanding of measurement error.

The mathematical approach probably reached its greatest elaboration in the study of growth. Przibram's (1905b, 1908) quantitative theory of growth drew on the temperature and regeneration experiments as well as on analogies to crystal growth that he had examined (e.g., Przibram 1906b). Comprehensive measurements of size, mass, weight, and timing of organismal features at different stages of development allowed the establishment of rules of absolute and relative growth. The approach was not so much geometrical but was rather directed toward correlations between temporal and spatial processes of development and regeneration. For instance, based primarily on detailed studies of mantid and crustacean growth, it was proposed that arthropods increase at each molting by a doubling of the body volume (Przibram and Megusar 1912). The generalization of this rule has become known as Przibram's law and is still cited as such.

Przibram's pioneering work in this domain can be seen as a direct forerunner of the growth model developed by Ludwig von Bertalanffy (1938), who elevated the mathematization of the subject to the domain of differential equations, and the Bertalanffy growth function is still in use today. It is not by coincidence that Bertalanffy should be the one who continued the work on growth and theoretical biology. As a student of the Vienna Circle philosopher Moritz Schlick, Bertalanffy had close ties with the BVA scientists, especially with Paul Kammerer and Paul Weiss. Bertalanffy's views on homeostasis and the behavior of open biological systems is based on the same thermodynamic principles that had already inspired Hatschek and Przibram. Thus the development of organismal system theories, both by Weiss (1925b, 1973) and by Bertalanffy (1933, 1950, 1952), is closely related to the BVA enterprise and to Przibram's quest for the mathematization of biology. Indeed, Bertalanffy's reputation as one of the inspirations for a physics of open systems (see Prigogine and Wiame 1946) goes back to the roots of his work in that of Przibram and Hatschek.

A Theory of Form

Whereas Hans Przibram's declared intentions with the establishment of the BVA were the promotion of experimentation and quantification, the principal motivation underlying the entire enterprise was the desire to understand the laws that govern animal form. In this regard, Przibram was much dissatisfied with the explanations provided by contemporary Darwinian theory. In several places he reiterated his critique of natural selection, arguing that it could not account for variation beyond the established reaction potential of a lineage and neither for the origin of new characters (e.g., Przibram 1910b). He deplored that Darwinian theory "distracts research from tedious experimentation" and instead promotes the "heaping of speculation on speculation" (Przibram 1909b). In his critique of Darwinism, he shared the views of many local and international biologists, such as D'Arcy Thompson, whose entire book *On Growth and Form* (1917) deemphasized evolution as the fundamental determinant of the form and structure of living organisms and instead favored the roles of physical laws and mechanics. Przibram (1922d) expressed the view that Thompson had been much inspired by his own and his brother's work and mentions a letter in that regard by Thompson to his brother Karl. A reference to Hans Przibram at the beginning of chapter 2 in the 1917 edition of *On Growth and Form* is no longer present in later editions, probably because of Thompson's growing dissatisfaction with Przibram's zealousness.[1] Several other Viennese biologists and medical scientists, such as Breuer, Kassowitz, von Wettstein, and Tandler—including Przibram's mentor Hatschek—had been equally outspoken against the notion that Darwinian theory contained a sufficient explanation of form and structure of living organisms (see Hatschek 1909).

There is no single publication in which Przibram laid out a complete theory of biological form, but the remarks he made in multiple places make clear that to him the dominant causal factors of form generation lay in the physics of biological processes and the environmental influences on those processes. These relationships he cast in a series of general laws that we find best summarized in *Form und Formel im Tierreiche* (Przibram 1922d), in *Zoonomie* (Przibram 1929), and in *Connecting Laws in Animal Morphology* (Przibram 1931). There Przibram formulated, among others, a law of differentiation (as a function of spatial division and reduction of developmental potentialities), a law of growth (as a function of chemical metabolism), a law of induction (as a function of instructions emanating from developmentally more advanced regions), a law of competition (as a function of developmental rivalry for material by independent processes), a law of morphological modification (as a function of environmental influences on the temporal and topological relations of genetically determined and parallel-acting processes), and a law of disposition in physiological and morphological reactions (as a function of aftereffects of previous environmental stimuli).

Przibram conceded that these law-like generalizations were still scant and, for the most part, could not yet be cast in fully chemical and physical terms or satisfactorily based on the comprehensive quantification and mathematization he desired. But he regarded these generalized consequences of experimentation as the first steps toward a systems-analytical and organism-based theory of form. In all these considerations Przibram did not negate the existence of heritable factors, but he took them not to be the causally decisive determinants of organismal form. He was well aware of the genotype-phenotype distinction (Przibram 1929) and extended it to what he called "geno-structure" and "pheno-structure." Whereas geno-structure represented a set of potentialities for organismal features, pheno-structure designated the actually realized structures resulting from the interaction of external and internal determinants. Here we can discern kernels of the concepts of "morphospace" and "developmental constraint" underlying some current notions of phenotypic evolution (McGhee 1999). Przibram (1929) acknowledged that his program concentrated on the physical pheno-structure and suggested that also "the nature of the geno-structure of the living substance needs to be examined by further biological and biochemical research."

In his theoretical apprehension of biological form as generated from the "living substance," Przibram took a distinctly different perspective from contemporary mathematical approaches, such as D'Arcy Thompson's (1917). Whereas Thompson concentrated on mathematical and geometric kinds of biological lawfulness, Przibram pursued causal lawfulness. Or, as Dorothy Wrinch (who—like Joseph Woodger[2] and other students of Thompson—had visited Przibram in Vienna and, together with Bernal, Needham, Waddington, and Woodger, became one of the founders of the Theoretical Biology Club in Cambridge) had put it: "Whereas Thompson calculated form, the Przibram brothers did experiments" (Senechal 2012, see also chapter 1). Przibram saw morphological structure essentially governed by equilibrium processes, both in a developmental and an energetic sense. This view was influenced by the thermodynamical physics developed in Vienna by Ludwig Boltzmann shortly before the flourishing of quantitative zoology at the BVA. In much the same way as the thermodynamically variable approach replaced the static—and typological—conception of classical physics, quantitative probabilistic biology was meant to replace classical biology. Whereas Thompson's approach perpetuated a typological notion, in which irregularities represent mere distractions from a general principle, Przibram felt that it was precisely the irregularities and the deviations from mathematical regularity that were essential in realistic form generation, because they were an outcome of stochastic generative dynamics. The deviations held the key to an explanation of regular form!

From this theoretical perspective, organismal form was seen as a contingent consequence of individually variable development, which meant that no sharp borders could be defined for biological entities—a decisive step away from typological thought. To Przibram, the development of individual form was decidedly non-preformistic, but rather resulted from the continuous reciprocal balancing among all interacting factors in a way that was not fully epigenetic either. That mode of form generation he termed "apogenetic" (Przibram 1930b). Whereas in epigenetic development the potentialities would be fewer at the beginning and would increase through the growing number of interactions that lead to differentiation, apogenetic development would start from a complex yet rather uniform initial condition, from which the differentiated states would be achieved not by the addition of new potentialities but by their successive abatement toward the periphery of the developing organism. Akin to later views on the initial multiplicity and subsequent "weeding out" of basic body plans (Gould 2002), the concept of decreasing potentialities in development was extended by Przibram to the evolutionary generation of organismal diversity, which he felt could just as well have started from a multiplicity of spontaneous forms instead of the gradual accruement of complexity out of a singular or few ancestral states.

The substance of Przibram's theory of form is the outline of a dynamical morphology based on physical equilibrium conditions (Przibram 1907b, 1908). Phenotypes were viewed as energetically variable "open systems" that should potentially be characterizable by the same principles as thermodynamical systems. Akin to statistical mechanics, stable macroscopic phenomena would result from the permanent turnover at the microscopic level in an emergent way. One of the overlooked consequences of this concept is the fact that biological equilibria are often reached in discontinuous ways, for instance when the reaction potential of a system is governed by bimodality. Discontinuities in development could give rise to discontinuous variation in evolution (Przibram 1927b). Przibram saw this principle supported by non-Gaussian distributions of phenotypic variations, and his brother's statistical approach was able to handle such kinds of distributions (Coen 2006). These notions ran counter to one of the fundamental properties of Darwinian theory, namely the (assumed) continuous, incremental, and gradual nature of organismal variation, but they were in agreement with Bateson's (1894) findings of an abundance of discontinuous variation in natural populations (Lange and Müller 2017).

Conclusions

At the beginning of the twentieth century, the fashionable themes of evolutionary debate were Darwinian selection, Mendelian inheritance, and Pearsonian variation—paving the way to a population-theoretical model of evolutionary variation. By contrast, the research program at the BVA focused on the relations between external physical conditions, internal rules of generation, and nongenetic inheritance, aiming for a theory of organismal form and phenotypic change. The work of Hans Przibram and his zoological department challenged the prevailing theory of evolution on all four of its fundamental pillars: Selectionism was countered by environmental induction; genetic inheritance was confronted with nongenetic transmission; gradual variation was opposed by discontinuous change; and genetic determination was contrasted with rules of development. The resulting theory, elaborated only in fragmented ways, proposed physical and environmental principles as the predominant determinants in the evolution of biological form.

It is hardly surprising that such a massive attack on all fronts of the prevailing orthodoxy in evolutionary theory caused equally strong counterreactions. The institute became a perceived hub of neo-Lamarckian belief and a prime target for attacks against that position, which culminated in the notorious Kammerer affair. The overemphasis on this scientifically irrelevant scandal, the tremendous successes of the population-dynamical account of evolutionary theory, and the complete destruction of the BVA at the end of World War II have hampered the recognition of its unique historical role in the causal analysis of organismal form.

Many of the topics raised and studied by Przibram and the zoological department have gained renewed attention today. Environmental induction has been reevaluated in the context of developmental plasticity and has been assigned a significant role in the evolutionary process (West-Eberhard 2003). Multiple forms of transgenerational inheritance, including epigenetic inheritance, are now recognized (Jablonka and Lamb 2006; Danchin et al. 2011), even though their respective importance in evolutionary change has not yet been fully determined. Also the occurrence of discontinuous forms of variation has recently been reemphasized (e.g., Goldbeter et al. 2007; Jaeger et al. 2012; Lange et al. 2014) and is found particularly relevant in the evolution of phenotypic novelty (Peterson & Müller 2016). And the influence of physics in the origin of primary body plans and morphological novelty is now thought to have great significance for phenotypic evolution (Newman et al. 2006; Newman and Bhat 2008). The recent recognition by mainstream science of quantitative development (Oates et al. 2009), often without any mention of earlier approaches, is another reflection of the renewed attention to the BVA's themes.

In light of the recent advances in developmental and evolutionary theory, we may see Hans Przibram as a forerunner of two significant contemporary fields: evo-devo and theoretical biology. The BVA and its scientists had established a new experimental and quantitative approach to the study of the relationship between environment, development, and evolution, an approach that played a major part in the theoretical discourse regarding the causality of biological evolution. That discourse highlighted the interconnections of experimental approaches, mathematical models, historical and comparative analyses, and conceptual systems (see chapter 6). In many ways the BVA also embodied the quest to root the theoretical foundations of biology in an organicist conception of life (Nicholson and Gawne 2015). Although sometimes cast in unfamiliar wording, Przibram's approach establishes him as a conceptual predecessor of a systems notion of biology, the traces of which can already be seen in his mentor Hatschek's work and which came to full fruition with Paul Weiss and Ludwig von Bertalanffy (see Drack et al. 2007), both of whom had been strongly influenced by the BVA. It provided a first step in the systematic investigation of the relations between external conditions, internal rules of development, and different forms of transgenerational inheritance in the evolutionary generation of organismal diversity. Whereas the discussion of the relative importance of these factors in governing organismal evolution is still ongoing (see, e.g., Laland et al. 2014; Noble et al. 2014), the general validity of many of the key concepts developed by Hans Przibram and his colleagues at the BVA has been vindicated by the biological sciences of today.

Acknowledgments

I am much obliged to Heidemarie Pollack from the Department of Theoretical Biology at the University of Vienna, who tirelessly searched for original literature and discovered Przibram's photographic plates of Brecher's experiments. The chapter benefited greatly from discussions with Hans Nemeschkal and Sabine Brauckmann as well as from comments by Tim Peterson and Dan Nicholson.

Notes

1. Letter by Thompson to Woodger on November 29, 1925; Woodger Archives, University College London.

2. For the decisive influence his visit in 1926 to Przibram and the BVA had on Woodger's intellectual development see Nicholson and Gawne 2013.

References

Bateson, W. 1894. *Materials for the Study of Variation, Treated with Especial Regard to Discontinuity in the Origin of Species*. London: Macmillan.

Bertalanffy, L. v. 1933. *Modern Theories of Development: An Introduction to Theoretical Biology*. Trans. J. H. Woodger. Oxford: Oxford University Press.

Bertalanffy, L. v. 1938. A quantitative theory of organic growth. *Human Biology* 10:181–213.

Bertalanffy, L. v. 1950. The theory of open systems in physics and biology. *Science* 111:23–29.

Bertalanffy, L. v. 1952. *Problems of Life: An Evaluation of Modern Biological and Scientific Thought.* New York: Harper & Brothers.

Brauckmann, S. 2013. Weiss, Paul Alfred. In *eLS Encyclopedia of Life Sciences.* Chichester: John Wiley & Sons. Ltd. doi: 10.1002/9780470015902.

Brecher, L. 1917. Die Puppenfärbung des Kohlweisslings, *Pieris brassicae* L. I.–III. Teil: Wirkung unsichtbarer und sichtbarer Strahlen. *Archiv für Entwickelungsmechanik* 40:88.

Brecher, L. 1919. Die Puppenfärbung des Kohlweisslings, *Pieris brassicae* L. IV. Teil: Wirkung unsichtbarer und sichtbarer Strahlen. *Archiv für Entwickelungsmechanik* 45:273–319.

Brecher, L. 1922a. Die Puppenfärbung der Vanessiden (*Vanessa io, V. urticae, Pyrameis cardui, P. atalanta*). *Archiv für Entwickelungsmechanik* 50:209–308.

Brecher, L. 1922b. Nachwirkung von Lichtmodifikationen in Finsternis. Puppenfärbungen des Kohlweisslings *Pieris brassicae* IX u. Puppenfärbungen der Vanessiden III. *Akademischer Anzeiger* 24/25.

Brecher, L. 1924a. Die Puppenfärbung des Kohlweisslings, *Pieris brassicae* L. VIII. Teil: Die Farbanpassung der Puppen durch das Raupenauge. *Archiv für mikroskopische Anatomie und Entwicklungsmechanik* 102:501–516.

Brecher, L. 1924b. Die Bedingungen für Fühlerfüsse bei *Dixippus* (*Carausius*) *morosus.* Homöosis bei Arthropoden, VII. Mitteilung. *Archiv für mikroskopische Anatomie und Entwicklungsmechanik* 102:549–572.

Calman, W. T. 1924. Chimaeras Dire. *Nature* 114:11–12.

Coen, D. R. 2006. Living precisely in fin-de-siècle Vienna. *Journal of the History of Biology* 39:493–523.

Danchin, E., A. Charmantier, F. A. Champagne, A. Mesoudi, B. Pujol, and S. Blanchet. 2011. Beyond DNA: Integrating inclusive inheritance into an extended theory of evolution. *Nature Reviews Genetics* 12:475–486.

Drack, M., W. Apfalter, and D. Pouvreau. 2007. On the making of a system theory of life: Paul A. Weiss and Ludwig von Bertalanffy's conceptual connection. *Quarterly Review of Biology* 82:349–373.

Finkler, W. 1923a. Kopftransplantation an Insekten. I. Funktionsfähigkeit replantierter Köpfe. *Archiv für Entwickelungsmechanik* 99:104–118.

Finkler, W. 1923b. Kopftransplantation an Insekten. II. Austausch von Hydrophilusköpfen zwischen Männchen und Weibchen. *Archiv für Entwickelungsmechanik* 99:119–125.

Finkler, W. 1923c. Kopftransplantation an Insekten. IIII. Einfluss des replantierten Kopfes auf andere Körperteile. *Archiv für Entwickelungsmechanik* 99:126–133.

Friza, F., and H. Przibram 1930. Johnston'sche Sinnesorgane in den Fühlerfüssen (Aristopoden) der *Sphodromantis* und *Drosophila.* Zugleich Homoeosis VIII. *Akademischer Anzeiger* 17.

Gliboff, S. 2006. The case of Paul Kammerer. Evolution and experimentation in the early 20th century. *Journal of the History of Biology* 39:525–563.

Goldbeter, A., D. Gonze, and O. Pourquié. 2007. Sharp developmental thresholds defined through bistability by antagonistic gradients of retinoic acid and FGF signaling. *Developmental Dynamics* 236:1495–1508.

Gould, S. J. 2002. *The Structure of Evolutionary Theory.* Cambridge, MA: Harvard University Press.

Haeckel, E. 1866. *Prinzipien der generellen Morphologie der Organismen.* Berlin: Reimer.

Harrison, R. G. 1925. The effect of reversing the medio-lateral or transverse axis of the fore-limb bud in the salamander embryo (*Amblystoma punctatum* Linn.) *Wilhelm Roux' Archiv für Entwicklungsmechanik* 106:469–502.

Hatschek, B. 1902. Entgegnungen von Dr. Berthold Hatschek. In *Vorträge und Besprechungen über die Krisis des Darwinismus. Wissenschaftliche Beilage zum 15. Jahresbericht der Philosophischen Gesellschaft an der Universität Wien*, ed. M. Kassowitz, R. v. Wettstein, B. Hatschek, C. Ehrenfels, and J. Breuer. Leipzig: Barth.

Hatschek, B. 1905. *Hypothese der organischen Vererbung*. Leipzig: Wilhelm Engelmann.

Hatschek, B. 1909. *Darwin's 100. Geburtstag. Wissenschaftliche Beilage zum 22. Jahresbericht der Philosophischen Gesellschaft an der Universität Wien*. Leipzig: Barth.

Hatschek, B., and J. Tuckey. 1893. *The Amphioxus and Its Development*. New York: MacMillan & Co.

Herbst, C. 1896. Über Regeneration von antennenähnlichen Organen an Stelle von Augen. *Archiv für Entwickelungsmechanik* 2:554.

Herbst, C. 1919. Beiträge zur Entwicklungsphysiologie der Färbung und Zeichnung der Tiere. Der Einfluß gelber, weißer und schwarzer Umgebung auf die Zeichnung von *Salamandra maculosa*. *Abhandlungen der Heidelberger Akademie der Wissenschaften. 7. Abh. d. Math.-nat. Klasse*. Heidelberg: Winter.

Herbst, C. 1924. Beiträge zur Entwicklungsphysiologie der Färbung und Zeichnung der Tiere. 2. Die Weiterzucht der Tiere in gelber und schwarzer Umgebung. *Archiv für mikroskopische Anatomie und Entwicklungsmechanik* 102:130–167.

Jablonka, E., and M. J. Lamb. 1995. *Epigenetic Inheritance and Evolution*. Oxford: Oxford University Press.

Jablonka, E., and M. J. Lamb. 2006. *Evolution in Four Dimensions*. Cambridge: MIT Press.

Jaeger, J., D. Irons, and N. Monk. 2012. The inheritance of process: A dynamical systems approach. *Journal of Experimental Zoology B* 318:591–612.

Jellinek, A. 1923. Die Replantation von Augen VII. Dressurversuche an Ratten mit optisch verschiedenen Dressurgefäßen. *Archiv für mikroskopische Anatomie und Entwicklungsmechanik* 99:82–104.

Jollos, V. 1934. Inherited changes produced by heat-treatment in *Drosophila melanogaster*. *Genetica* 16:476–494.

Kammerer, P. 1904. Beitrag zur Erkenntnis der Verwandtschaftsverhältnisse von *Salamandra atra* und *Salamandra maculosa*. Experimentelle und statistische Studie. *Archiv für Entwickelungsmechanik* 17:165–264.

Kammerer, P. 1906. Experimentelle Veränderung der Fortpflanzungstäigkeit bei Geburtshelferkröte (*Alytes obstetricans*) und Laubfrosch (*Hyla arborea*). *Archiv für Entwickelungsmechanik* 22:48–140.

Kammerer, P. 1908. Vererbung erzwungener Fortpflanzungsanpassungen. I. und II. Mitteilung: Die Nachkommen der spätgeborenen *Salamandra maculosa* und der frühgeborenen *Salamandra atra*. *Archiv für Entwickelungsmechanik* 25:7–51.

Kammerer, P. 1909. Vererbung erzwungener Fortpflanzungsanpassungen. III. Mitteilung: Die Nachkommen der nicht brutpflegenden *Alytes obstetricans*. *Archiv für Entwickelungsmechanik* 28:447–545.

Kammerer, P. 1911. Mendelsche Regeln und Vererbung erworbener Eigenschaften. *Verhandlungen des naturforschenden Vereines in Brünn* 49:72–110.

Kammerer, P. 1913. Vererbung erzwungener Farbveränderungen. IV. Mitteilung: Das Farbkleid des Feuersalamanders (*Salamandra maculosa* Laurenti) in seiner Abhängigkeit von der Umwelt. *Archiv für Entwickelungsmechanik* 36:41–93.

Kammerer, P. 1919. Vererbung erzwungener Formveränderungen. I. Mitteilung: Die Brunftschwiele des Alytes-Männchens aus Wassereiern. *Archiv für Entwickelungsmechanik* 40:323–370.

Kammerer, P. 1923. Breeding experiments on the inheritance of acquired characters. *Nature* 111:637–640.

Laland, K., T. Uller, M. Feldman, K. Sterelny, G. B. Müller, A. Moczek, E. Jablonka, et al. 2014. Does evolutionary theory need a rethink? *Nature* 514:161–164.

Lange, A., and G. B. Müller. 2017. Polydactyly in development, inheritance, and evolution. *Quarterly Review of Biology* 92:1–38.

Lange, A., H. L. Nemeschkal, and G. B. Müller. 2014. Biased polyphenism in polydactylous cats carrying a single point mutation: The Hemingway model for digit novelty. *Evolutionary Biology* 41:262–275.

Loeb, J. 1922. *Proteins and the Theory of Colloidal Behavior*. New York: McGraw-Hill.

Loisel, G. 1907. Rapport sur une mission scientifique dans les jardins et établissements zoologiques publiques et privés de l'Allemagne, de L'Autriche-Hongrie, de la Suisse et du Danemarck. In *Nouvelles Archives des Missions Scientifiques et Littéraires, XV*, pp. 251–256. Paris: Imprimerie Nationale.

Logan, C. A. 2013. *Hormones, Heredity, and Race*. New Brunswick, NJ: Rutgers University Press.

MacLeod, J. 1919. *The Quantitative Method in Biology*. Manchester: Manchester University Press.

McGhee, G. R. 1999. *Theoretical Morphology*. New York: Columbia University Press.

Michler, W. 1999. *Darwinismus und Literatur*. Wien: Böhlau.

Müller, G. B., and H. L. Nemeschkal. 2015 . Zoologie im Hauch der Moderne: Vom Typus zum offenen System. In *Reflexive Innensichten aus der Universität Wien: Wiener Disziplingeschichten zwischen Wissenschaft, Gesellschaft und Politik*, ed. K. A. Fröschl, G. B. Müller, T. Olechowksi, and B. Schmidt-Lauber, pp. 355–369. Wien: V&R Vienna University Press.

Müller, G. B., J. Streicher, and R. J. Müller. 1996. Homeotic duplication of the pelvic body segment in regenerating tadpole tails induced by retinoic acid. *Development Genes and Evolution* 206:344–348.

Newman, S. A., and R. Bhat. 2008. Dynamical patterning modules: Physico-genetic determinants of morphological development and evolution. *Physical Biology* 5:2–14.

Newman, S. A., G. Forgacs, and G. B. Müller. 2006. Before programs: The physical origination of multicellular forms. *International Journal of Developmental Biology* 50 (2–3): 289–299.

Nicholson, D. J., and R. Gawne. 2013. Rethinking Woodger's legacy in the philosophy of biology. *Journal of the History of Biology* 47 (2): 243–292.

Nicholson, D. J., and R. Gawne. 2015. Neither logical empiricism nor vitalism, but organicism: What the philosophy of biology was. *History and Philosophy of the Life Sciences* 37 (4): 345–381.

Noble, D., E. Jablonka, M. Joyner, G. B. Müller, and S. Omholt. 2014. The integration of evolutionary biology with physiological science. *Special Issue of the Journal of Physiology* 592 (11).

Noble, G. K. 1926. Kammerer's Alytes. *Nature* 118:209–210.

Oates, A. C., N. Gorfinkiel, M. González-Gaitán, and C.-P. Heisenberg. 2009. Quantitative approaches in developmental biology. *Nature Reviews Genetics* 10:517–530.

Peterson, T., and G. B. Müller. 2016. Phenotypic novelty in EvoDevo: The distinction between continuous and discontinuous variation and its importance in evolutionary theory. *Evolutionary Biology* 43:314–335.

Prigogine, I., and J. M. Wiame. 1946. Biologie et thermodynamique des phénomènes irréversibles. *Experientia* 2 (11): 451–453.

Przibram, H. 1899. *Die Regeneration bei den Crustaceen*. Dissertation. Wien: Arbeiten des Zoologischen Instituts.

Przibram, H. 1903. Die neue Anstalt für experimentelle Biologie in Wien. *Verhandlungen der Gesellschaft deutscher Naturforscher und Ärzte* 74:152–155.

Przibram, H. 1905a. Die "Heterochelie" bei dekapoden Crustaceen. Zugleich: Experimentelle Studien über Regeneration. Dritte Mitteilung. *Archiv für Entwickelungsmechanik* 19:181–247.

Przibram, H. 1905b. Quantitative Wachstumstheorie der Regeneration. *Zentralblatt für Physiologie* 19:682–684.

Przibram, H. 1906a. Aufzucht, Farbwechsel und Regeneration einer ägyptischen Gottesanbeterin (*Sphodromantis bioculata* Burm.). *Archiv für Entwickelungsmechanik* 22:149–206.

Przibram, H. 1906b. Kristall-Analogien zur Entwickelungsmechanik der Organismen. *Archiv für Entwickelungs-mechanik* 22:207–287.

Przibram, H. 1907a. Die "Scherenumkehr" bei decapoden Crustaceen. Zugleich: Experimentelle Studien über Regeneration. Vierte Mitteilung. *Archiv für Entwickelungsmechanik* 25:266–343.

Przibram, H. 1907b. Equilibrium of animal form. *Journal of Experimental Zoology* 5:259–263.

Przibram, H. 1908. *Anwendung elementarer Mathematik auf biologische Probleme. Nach Vorlesungen gehalten an der Wiener Universität im Sommersemester 1907. Roux' Vorträge und Aufsätze über Entwicklungsmechanik III.* Leipzig: Engelmann.

Przibram, Hans. 1908/09a. Die Biologische Versuchsanstalt in Wien. Zweck, Einrichtung und Tätigkeit während der ersten fünf Jahre ihres Bestandes (1902–1907). Bericht der zoologischen, botanischen und physikalisch-chemischen Abteilung (Teil 1 und 2). *Zeitschrift für biologische Technik und Methodik* 1:234–264.

Przibram, H. 1908/09b. Die Biologische Versuchsanstalt in Wien. Zweck, Einrichtung und Tätigkeit während der ersten fünf Jahre ihres Bestandes (1902–1907). Bericht der zoologischen, botanischen und physikalisch-chemischen Abteilung. (2. Fortsetzung). *Zeitschrift für biologische Technik und Methodik* 1:409–433.

Przibram, H. 1909a. *Experimental-Zoologie. 2. Regeneration. Eine Zusammenfassung der durch Versuche ermittelten Gesetzmässigkeiten tierischer Wieder-Erzeugung (Nachwachsen, Umformung, Missbildung).* Leipzig, Wien: Franz Deuticke.

Przibram, H. 1909b. Anstalten für experimentelle Abstammungslehre. *Dokumente des Fortschritts* 2:443–445.

Przibram, H. 1910a. Die Homoeosis bei Arthropoden. *Archiv für Entwickelungsmechanik* 29:587–615.

Przibram, H. 1910b. *Experimental-Zoologie. 3. Phylogenese. Eine Zusammenfassung der durch Versuche ermittelten Gesetzmässigkeiten tierischer Art-Bildung (Arteigenheit, Artübertragung, Artwandlung).* Leipzig, Wien: Franz Deuticke.

Przibram, H. 1912. Die Umwelt des Keimplasmas. I. Das Arbeitsprogramm. *Archiv für Entwickelungsmechanik* 33:666–681.

Przibram H. 1913. Die Biologische Versuchsanstalt in Wien. Ausgestaltung und Tätigkeit während des zweiten Quinquenniums (1908–1912). Bericht der zoologischen, botanischen und physikalisch-chemischen Abteilung. *Zeitschrift für biologische Technik und Methodik* 3:163–245.

Przibram, H. 1917a. Fühlerregeneration halberwachsener Sphodromantis-Larven. Zugleich: Aufzucht der Gottesanbeterinnen, IX. Mitteilung, und: Homoeosis bei Arthropoden, III. Mitteilung. *Archiv für Entwickelungs-mechanik* 43:63–87.

Przibram, H. 1917b. Transitäre Scherenformen der Winkerkrabbe, *Gelasimus pugnax* Smith. Zugleich: Experimentelle Studien über Regeneration, V. Mitteilung, und: Homoeosis bei Arthropoden, II. Mitteilung. *Archiv für Entwickelungsmechanik* 43:47–62.

Przibram, H. 1917c. Direkte Temperaturabhängigkeit der Körperwärme bei Ratten (*Mus decumanus* und *M. rattus*). Die Umwelt des Keimplasmas VI. *Archiv für Entwickelungsmechanik* 43:37–46.

Przibram, H. 1918. Regeneration beim Hautflügler *Cimbex axillaris* Panz. Zugleich: Homoeosis bei Arthropoden. VI Mitteilung. *Akademischer Anzeiger* 17.

Przibram, H. 1919a. Fangbeine als Regenerate. Zugleich: Aufzucht der Gottesanbeterinnen, IX. Mitteilung, und Homoeosis bei Arthropoden. IV. Mitteilung. *Archiv für Entwickelungsmechanik* 45:39–51.

Przibram, H. 1919b. Fussglieder an Käferfühlern. Zugleich: Homoeosis bei Arthropoden. V. Mitteilung. *Archiv für Entwickelungsmechanik* 45:52–68.

Przibram, H. 1922a. Autophoric transplantation: Its theory and practise. *American Naturalist* 56:548–559.

Przibram, H. 1922b. Der Einfluss gelber und schwarzer Umgebung der Larve auf die Fleckenzeichnung des Vollmolches von *Salamandra maculosa* Laur. forma typica. *Archiv für Entwickelungsmechanik* 50:108–146.

Przibram, H. 1922c. Veröffentlichungen über quantitative Biologie 1916–1919 in englischer Sprache. Referate. *Archiv für Entwickelungsmechanik* 50:326–337.

Przibram, H. 1922d. *Form und Formel im Tierreiche. Beiträge zu einer quantitativen Biologie.* Wien: Deuticke.

Przibram, H. 1923a. Die Methode autophorer Transplantation. Zugleich: Die Replantation von Augen. I. *Archiv für mikroskopische Anatomie und Entwicklungsmechanik* 99:1–14.

Przibram, H. 1923b. *Temperatur und Temperatoren im Tierreiche. Beiträge zu einer quantitativen Biologie.* Wien: Deuticke.

Przibram, H. 1923c. *Aufbau mathematischer Biologie. Abhandlungen zur theoretischen Biologie.* Berlin: Borntraeger.

Przibram, H. 1924a. Achsenverhältnisse und Entwicklungspotenzen der Urodelenextremitäten an Modellen zu Harrisons Transplantationsversuchen. *Archiv für mikroskopische Anatomie und Entwicklungsmechanik* 102:604–623.

Przibram, H. 1924b. Chimaeras dire: Transplantation of heads of insects. *Nature* 114:347.

Przibram, H. 1924c. Erhöhung der Körpertemperatur junger Wanderratten (*Mus decumanus*) über den Normalwert und ihr Einfluß auf die Schwanzlänge. *Archiv für Entwickelungsmechanik* 102:731–740.

Przibram, H. 1925a. Direkte Temperaturabhängigkeit der Schwanzlänge bei Ratten, *Mus* (*Epimys*) *decumanus* Pall. und *M.* (*E.*) *rattus* L. Die Umwelt des Keimplasmas. XI. *Wilhelm Roux' Archiv für Entwicklungsmechanik* 104:434–496.

Przibram, H. 1925b. Die Schwanzlänge der Nachkommen temperaturmodifizierter Ratten, *Mus* (*Epimys*) *decumanus* Pall. und *M.* (*E.*) *rattus* L. Die Umwelt des Keimplasmas. XIII. *Wilhelm Roux' Archiv für Entwicklungsmechanik* 104:548–610.

Przibram, H. 1925c. Das Anwachsen der relativen Schwanzlänge und dessen Temperaturquotient bei den Ratten *Mus* (*Epimys*) *decumanus* Pall. und *M.* (*E.*) *rattus* L. Die Umwelt des Keimplasmas. XIV. *Wilhelm Roux' Archiv für Entwicklungsmechanik* 104:611–648.

Przibram, H. 1925d. Transplantation and Regeneration: Their bearing on developmental mechanics. A review of the experiments and conclusions of the last ten years (1915–1924). *British Journal of Experimental Biology* 3:313–330.

Przibram, H. 1926. *Tierpfropfung. Die Transplantation der Körperabschnitte, Organe und Keime.* Braunschweig: Vieweg.

Przibram, H. 1927a. Deutungen spiegelbildlicher Lurcharme. (Zur Verständigung mit R. G. Harrison u. a.) *Wilhelm Roux' Archiv für Entwicklungsmechanik* 109:411–448.

Przibram, H. 1927b. Diskontinuität des Wachstums als eine Ursache diskontinuierlicher Variation bei *Forficula*. Eine theoretische Erörterung. *Wilhelm Roux' Archiv für Entwicklungsmechanik* 112:142–148.

Przibram, H. 1929. *Experimental-Zoologie. 6. Zoonomie. Eine Zusammenfassung der durch Versuche ermittelten Gesetzmässigkeiten tierischer Formbildung.* Wien: Franz Deuticke.

Przibram, H. 1930a. The acquired characters of *Alytes. Nature* 125:856–857.

Przibram, H. 1930b. Apogenetische Theorie der Entwicklung der Organismen. *Forschungen und Fortschritte* 6:68.

Przibram, H. 1931. *Connecting Laws in Animal Morphology: Four Lectures held at the University of London.* London: University of London Press.

Przibram, H., and L. Brecher. 1919. Ursachen tierischer Farbkleidung. I. Vorversuche an Extrakten. *Archiv für Entwickelungsmechanik* 45:83–198.

Przibram, H., and L. Brecher. 1922. Die Farbmodifikationen der Stabheuschrecke *Dixippus morosus* Br. et Redt. Zugleich: Ursachen tierischer Farbkleidung. VI. *Archiv für Entwickelungsmechanik* 50:147–185.

Przibram, H., and J. Dembowski. 1922. Der Einfluss gelber und schwarzer Umgebung der Larve auf die Fleckenzeichnung des Vollmolches von *Salamandra maculosa* Laur. forma typica. Zugleich: Ursachen tierischer Farbkleidung. V. *Archiv für Entwickelungsmechanik* 50:108–146.

Przibram, H., and F. Megusar. 1912. Wachstumsmessungen an *Sphodromantis bioculata* Burm 1. Länge und Masse. Zugleich: Aufzucht der Gottesanbeterinnen. IV. Mitteilung. *Archiv für Entwickelungsmechanik* 34:680–741.

Senechal, M. 2012. *I Died for Beauty: Dorothy Wrinch and the Cultures of Science*. Oxford: Oxford University Press.

Sneath, P. H. A., and R. R. Sokal. 1973. *Numerical Taxonomy*. San Francisco: Freeman.

Sumner, F. B. 1909. Some effects of external conditions upon the white mouse. *Journal of Experimental Zoology* 7:97–155.

Taschwer, K. 2016. *Der Fall Paul Kammerer. Das abenteuerliche Leben des umstrittensten Biologen seiner Zeit.* München: Carl Hanser Verlag.

Thompson, D. W. 1917. *On Growth and Form*. Cambridge: Cambridge University Press.

Vargas, A. O., Q. Krabichler, and C. Guerrero-Bosagna. 2016. An epigenetic perspective on the midwife toad experiments of Paul Kammerer (1880–1926). *Journal of Experimental Zoology B* 00B:1–14.

Weismann, A. 1892. *Das Keimplasma. Eine Theorie der Vererbung*. Jena: Fischer.

Weiss, P. A. 1923. Die Regeneration der Urodelenextremität als Selbstdifferenzierung des Organrestes. *Naturwissenschaften* 11:669–677.

Weiss, P. A. 1925a. Die seitliche Regeneration der Urodelenextremität. *Wilhelm Roux' Archiv für Entwicklungsmechanik* 104:395–408.

Weiss, P. A. 1925b. Tierisches Verhalten als "Systemreaktion": Die Orientierung der Ruhestellungen von Schmetterlingen (*Vanessa*) gegen Licht und Schwerkraft. *Biologia Generalis* 1:165–248.

Weiss, P. A. 1973. *The Science of Life: The Living System—A System for Living*. Mount Kisco: Futura Publishing.

West-Eberhard, M. J. 2003. *Developmental Plasticity and Evolution*. Oxford: Oxford University Press.

9 Growth, Development, and Regeneration: Plant Biology in Vienna around 1900

Kärin Nickelsen

The study of plants at the Biologische Versuchsanstalt (BVA) has so far received only little attention. There was no botanical Paul Kammerer or Eugen Steinach; there was no spectacle and no scandal. And there was no visionary theorist who would have come close to Hans Przibram's ambition and achievements. It is thus no wonder that the literature on this remarkable research institution has until now concentrated on the zoological and physiological departments of the BVA.[1] Yet, two of the three founders, Wilhelm Figdor and Leopold von Portheim, were plant scientists; and, as I argue in this chapter, one cannot fully appreciate the BVA's work and functioning without taking the botanical section into account. For pragmatic reasons, the chapter will concentrate on the years up to 1918, only glimpsing upon the interwar period.

I will, first, briefly introduce the botanical protagonists and give an idea of how the field was conceived of in Vienna at the time; and then, second, examine how plant sciences were organized and implemented at the BVA. The main part of this chapter is devoted to the analysis of two examples: the investigation of anisophylly and the study of regeneration processes, both of which were subjects of long-term projects of the institute. Both examples reveal that the study of plants formed an integral part of the BVA's activity. Within their field of expertise, Figdor and Portheim, together with their associates, picked up some of the most contested research themes of the time, while their findings also contributed to Przibram's program of comparative analysis across biological realms.

The Subject and the Actors

Neither Figdor nor Portheim were botanists in the traditional sense of the word. They were "plant physiologists," that is, experts in an academic field that, at the time of the BVA's foundation, was still very young.[2] Its emergence began in the 1850s, when plant physiology became a subject at some universities, such as Vienna, Prague, Breslau (today's Wrocław), Würzburg, and Leipzig, and at schools of agriculture and forestry, such as Tharandt and Bonn. In 1857, Julius Sachs was the first to receive a habilitation in plant physiology at the University of Prague. This award was by no means a matter of course, and resistance was high among his colleagues. In particular, the chemist of the Prague philosophical faculty, Friedrich Rochleder, who was devoted to plant chemistry in the tradition of Justus von Liebig, strongly advised against the denomination. Sachs ought to choose a different subject for his venia legendi, Rochleder argued, otherwise his class "would be finished after two to three hours of lecturing."[3] Sachs insisted, successfully—but his positions would be professorships in botany, including his chair at the University of Würzburg, where he spent most of his academic career.

The first chair to be explicitly devoted to the anatomy and physiology of plants was established in 1873 at the University of Vienna; and after the two favorite candidates had declined, the appointing committee in charge agreed on Julius Wiesner to be given the position.[4] The choice proved extremely fortunate. Wiesner held the Vienna chair of plant physiology for almost four decades and turned it into a lively institution.[5] He became an eminent authority in the educated Vienna circles, including the university, the Imperial Academy of Sciences, and the Herrenhaus, the upper house of the parliament. Wiesner's influence on the field of plant sciences was effective far beyond this local context. In several ways he maintained intimate connections with the BVA. He became the first chair of the Kuratorium, which was the BVA's academic board; furthermore, Wiesner continuously held a research table in the BVA; that is, he was entitled to free use of working space and material.[6] But the strongest influence of Wiesner on the BVA came, no doubt, through the academic training of both its botanical founders, Figdor and Portheim, whose research at the BVA displayed a strong continuity with that of Wiesner's institute in terms of research focus and experimental practices.[7]

Wilhelm Figdor (figure 9.1) began his studies in Bonn, but he soon moved to Vienna, where he continued to read plant sciences. He completed his doctoral dissertation in 1891, and afterward held a position as Hochschulassistent at Wiesner's chair.[8] In this function, Figdor also accompanied Wiesner on some of the latter's expeditions, such as trips to Java and Ceylon, and he received his habilitation in 1899. Figdor continued to teach classes, and ten years later, in 1909, the year of Wiesner's retirement, Figdor became Außerordentlicher Professor of plant physiology, although without remuneration. He was promoted to an actual Extraordinariat in 1923, when Hans Molisch, who was Wiesner's successor, went on an extended trip to Japan and was in need of someone to replace him.[9] In the three years of Molisch's absence, Figdor (not least through Wiesner's support) was thus heading the university's Institute of Plant Physiology.

Leopold von Portheim (figure 9.2) also started his academic education elsewhere (at Prague) and then moved to Wiesner's institute in Vienna. Like Figdor, Portheim became one of Wiesner's assistants and accompanied the professor on research trips, for instance, into Yellowstone National Park in the United States.[10] Portheim's own scientific profile was less prominent than Figdor's. He preferred to pursue research projects in cooperation with others, and he was uninterested in keeping institutional ties with the university.

Figure 9.1
Wilhelm Figdor in 1925. Public domain.

However, Portheim's proximity to Wiesner is documented by the fact that he was one of the three editors of the widely received Festschrift published in 1903, the year of Wiesner's 65th birthday (which was also the 30th anniversary of the founding of Wiesner's institute).[11]

It seems worthwhile, then, to glance briefly at Wiesner's concept of plant sciences. In the broadest sense of the term, Wiesner argued in a public lecture, plant physiology ought to encompass "the whole investigation of the structure, the development and the life of the plant."[12] He particularly recommended four areas for further investigation: plant physiology in its narrow sense, that is, the attempt to trace life processes back to their foundations in physics and chemistry; organography, which not only described the shape and structure of plant organs but also their development and variability; a thoroughly reformed systematics that took into account the functional and chemical properties of the plants; and, in particular, the biology (*Biologie*) of plants[13]—and the latter is what not only Figdor and Portheim but also many others within the BVA became mostly devoted to. The term is open to misunderstanding, because today it is used as a generic. Wiesner, in contrast, gave a concise definition of the field as the study of plants' reproduction, ontogeny,

Figure 9.2
Leopold von Portheim in the greenhouse (ca. 1910). Courtesy of the ÖAW Archive.

phylogeny, adaptation, and evolution; and he devoted a full monograph to it, in order to explore this hitherto almost uncharted territory.[14]

This usage of the term *Biologie* helps to explain why Przibram chose to call his institution Biologische Versuchsanstalt. In the "draft of the institute's program," Przibram underlined that, so far, there had been no institution "that was specifically devoted to experimental biology"—whereas many other areas had found their places at universities and elsewhere, Przibram wrote (implying that for him those other areas did not fall under "biology"). The "big problems" that Przibram identified in the field of biology were, in line with Wiesner, embryogenesis, regeneration, evolution, and the external influences on vitality and form, all of which Przibram thought to be in intimate connection with each other. This perspective also illuminates Przibram's comment that while physiologists only needed to study an organism until the functions had become clear, biologists were in need of long-term experimentation.[15]

Installing Plant Biology at the BVA

When it was founded, in 1903, there was no plant physiology at the BVA—at least, as far as terminology goes. Next to the Zoological Department, there was, in line with traditional biological disciplines, a Botanical Department, although the research programs of both these divisions strongly deviated from classical zoology and botany.[16] The Botanical Department was jointly headed by Figdor and Portheim; originally both were also, together with Przibram, heading the whole institution. However, when the BVA became part of the Imperial Academy of Sciences in 1911, Figdor withdrew from the institution's directorate, for reasons unclear, while he continued to act as head of the Botanical Department.[17] Financial considerations may partly explain Figdor's withdrawal; but a personal asymmetry in the triad may have been more important. Several observations indicate that Przibram and Portheim were by far closer to each other than to Figdor: the Przibram and Portheim families were on excellent terms; Przibram and Portheim traveled together to Egypt and the Sudan; and they were both active members of the adult education center of Vienna, the Volksheim Ottakring.[18] Perhaps unsurprisingly, given this situation, the cooperation of the three founders was not without tension. Already at the first meeting of the newly founded Kuratorium of the BVA, in 1911, Przibram and Portheim presented the request to enlarge the BVA by a fifth department, namely Plant Physiology. The latter would be headed by Figdor on his own. The request was granted, and the Kuratorium even supported a one-time supplementary allowance in order to carry out the necessary architectural changes.[19] The latter was due to Figdor's desire to have his new department set up as separately and independently as possible of the Botanical Department, which would henceforth be headed by Portheim.[20]

There is no reason to assume that the founding of the new department was motivated by the need to broaden or redirect the institute's research program: the Botanical Department had already been concerned with questions of plant physiology in Wiesner's sense. In the first year of the BVA's existence, projects had been initiated that included studying the influence of oxalic acid on the germination of geranium plants; the regeneration of split root tips; the bipolarity of the bean's hypocotyl; and the influence of the air's electricity on plants.[21] Processes of adaptation, development, growth, and regeneration continued to dominate in the following decades—that is, in the Vienna terminology, issues of the plants' biology.[22]

These questions were high on the research agenda of plant scientists elsewhere as well. The remarkable—and unique—feature of the BVA, however, was the comparative treatment of these questions, in both the botanical and the zoological departments. From the very beginning, Przibram had envisaged the BVA as institution that not only was devoted to an underexplored area of study—experimental biology—but also tried to overcome disciplinary limitations. He planned it to become an institution that "would enable, through a combination of the two branches [of experimental biology] in one room, to study the interrelation of plants and animals."[23] Przibram was convinced that a comparative perspective would reveal remarkable similarities in the general patterns of growth, development, and adaptation. Symptomatic of this general attitude is the report that Przibram presented after the first five years of the BVA's existence, in which he listed the BVA's research on a range of systematic topics without separating the study of plants from the study of animals. With regard to questions of embryogenesis, for example, Przibram described, first, Paul Kammerer's experiments on induced parthenogenesis in fish and, second, the "very promising" observations made by Portheim and Loewi on the artificially induced appearance of membranes in pollen tubes. There is little evidence of joint projects in the BVA, with explicitly shared responsibilities between, for example, botany and zoology,[24] but there is ample indication that the members closely followed each other's work.

The foregoing may suffice as a brief introduction to the BVA's general program in plant biology and its institutional setting. In the following I shall more closely analyze two of the themes that were studied. Each was pursued with substantial effort, that is, resulted in more than a passing paper; and both exemplify the program of the BVA in general. They are, first, the phenomenon of anisophylly, which provides an illustrative case of how the plant scientists studied the interaction of internal and external factors in the shaping of plants; and second, questions of regeneration and development, whose study at the BVA serves as an example of how botanical work featured in Przibram's vision of research across the realms of nature.

Anisophylly; or, The Interplay of Internal and External Factors

In a number of plant species, leaves growing from one and the same shoot, even from the same node, are very different in shape or size. This phenomenon, called anisophylly, that is, unequal leafage, is particularly striking in species with branches that grow in horizontal orientation. Typically, leaves on the lower side of the shoot are larger than those on the upper side of the shoot—the paradigmatic example is the spike-moss (e.g., *Selaginella selaginoides*). In a slightly different form, anisophyllous leaves also appear at the lateral branches of trees and shrubs with decussate leafage, such as in elder and maple species (e.g., *Acer platanoides*). These differences in the leaves' morphology started to attract the botanists' interest in the second half of the nineteenth century.[25] It was completely obscure (and controversially debated), how these differences were brought about. Was anisophylly part of the inherently fixed constitution of a plant, or was it the effect of external influences, such as light and gravitation? Both options found their critics and defenders, and soon the phenomenon of anisophylly was perceived as a crucial exemplar of general questions at the heart of the discipline: How did internal and external factors interact in shaping the organism? And was it possible that an organism's local adaptations became part of its hereditary constitution?

Wiesner had been one of the pioneers in the study of anisophylly. He suggested that Figdor pursued the subject further, and in the first years of the BVA's existence, these questions were still Figdor's main research focus. Przibram became keenly interested in this project. In his report of the BVA's research in 1908, Przibram proudly stated—in more general terms than Figdor ever used—that Figdor had confirmed that the influence of external factors on the plant's leafage was substantial. Even in cases of (apparently) hereditarily fixed anisophylly, Przibram reported, Figdor had succeeded in experimental induction of morphological changes during development.[26] From Przibram's perspective, Figdor's work hence contributed in important ways to his own favorite field of research: to "experimental morphology." This was how Przibram chose to label the investigation of the causes of organic form.[27] Questions related to this field were at the heart of many of the BVA's research projects, and there was a general attitude of keeping an open mind to the range of relevant factors. Although Przibram principally accepted the neo-Darwinian thesis that there was no inheritance of acquired properties, he found it deeply unsatisfactory to leave the issue at that: "For [the neo-Darwinian], natural selection is the only true answer, while the actual, real causes are no longer under debate."[28] These "real causes" of an organism's morphology might include the inherited constitution; but they might also include the processes of development and the effect of external factors, Przibram claimed. Which of these factors was of more influence than others in the shaping of specific structures was an empirical question, to which no a priori answer was available. Przibram was thoroughly convinced that similar patterns prevailed in animals and plants, and these patterns had to be found through experimental work. Yet, while he was ready to investigate the zoological part, he needed his colleagues to clarify the botanical issues. Figdor's work, therefore, was more than welcome.

In 1909, Figdor summarized his investigations of anisophylly in a monograph.[29] The book was dedicated to Wiesner, his former mentor and teacher; and throughout the study, Figdor duly reported (and supported) Wiesner's earlier work on the theme. The first three parts of the book comprehensively presented the different forms of anisophylly in a wealth of genera and species of plants, using data taken both from existing literature and from Figdor's own research. The fourth part was the most interesting (and challenging): it was here that Figdor eventually addressed the causes of anisophylly. When the phenomenon was first discovered, it had been explained, by Wiesner among others, as a result of the effects of gravitation or light, Figdor recalled. However, in 1880 the Munich-based botanist Karl Goebel had drawn attention to the fact that in some cases (such as in *Goldfussia anisophylla* and *Centradenia grandifolia*) neither of these external factors provided a satisfactory explanation. Goebel thought that in these instances the anisophyllous appearance was due to a factor which he called the "internal symmetry" of the plant's architecture. After some debate, Wiesner had agreed, and so did Figdor many years thereafter.[30]

However, acknowledging the efficacy of a factor that had an unmistakably metaphysical tint was prone to raise suspicious associations, which Figdor wanted to avoid at all cost. He therefore hastened to emphasize the purely rational and empirical nature of his own research. The problem was addressed in one of the first sections of the chapter, when Figdor exposed how he was going to analyze the causes of anisophylly: "I will constrict myself, and I declare this once and for all, to reporting only those matters of fact that came to the fore in the course of experimentation or that are based on observed phenomena in natural environments for which, as far as we know, there is an unequivocal mode of interpretation."[31]

Figdor underlined that it was a purely empirical question, and far from easy to determine, whether a plant species' anisophylly was inherently fixed or externally prompted. Even long-standing forms of anisophylly might be due to the influence of one or more external factors, Figdor explained, that had exerted their influence on the plant for several generations. This would be revealed only if the phenomenon were investigated and turned out to be reversible under altered circumstances. However, in order to be meaningful, Figdor claimed, these investigations had unconditionally to comply with the rules of experimental methodology:

Regarding the experimental setup, I would just like to add briefly that impeccable results are only to be expected, if the influence of no more than one single factor is brought to bear on the test object in question, while all other external conditions ought to remain as stable as possible … One might think that all of this is very clear, and the same holds true for the fact that in research of this kind, the age of all organs, the circumstances of induction and correlation and many other factors have to be taken into account. However, it may not be completely unnecessary to point out these principles, as it is precisely in the area of interest here, namely morphology, that so many sins in this respect have been committed in the past.[32]

The character of these "sins"—a drastic term that, no doubt, was chosen with care—was not spelled out in detail, but the idealistic tradition of morphology immediately comes to mind, which invoked causes for organic form that were declared impossible to single out in experiments. The German Naturphilosophie was still a popular, thoroughly negative—dystopian!—*lieu de memoire* in the sciences around 1900.[33]

Figdor himself was not in doubt about how to proceed. There was, he claimed, a rather straightforward procedure to identify and describe the causes of anisophylly. If external factors were of influence, Figdor argued, one ought to be able to make this influence discernible and plain. There were two ways: either one could try and force an anisophyllous plant to lose that property and grow isophyllous leaves; or, vice versa, try to induce anisophylly in a plant with ordinary leafage. In short, one ought to test whether external factors were able to prompt the plant out of its usual way of growing. The tricky part was to alter only one factor at a time, a requirement if valid causal inferences were to be drawn.

Figdor carried out experiments along both lines of reasoning and studied extensively the influences of light and gravitation on anisophylly. For these experiments, the BVA's excellent instrumentation was indispensable. To begin with, Figdor was in need of a premium-quality clinostat: a device, originally invented by the plant physiologist Sachs in Würzburg, that leveled out the effect of gravitation on the plants' growth by slowly rotating them in horizontal orientation. Figdor and Portheim had a motor-driven variant constructed for their botanical laboratory (figure 9.3). The exclusion of gravitational

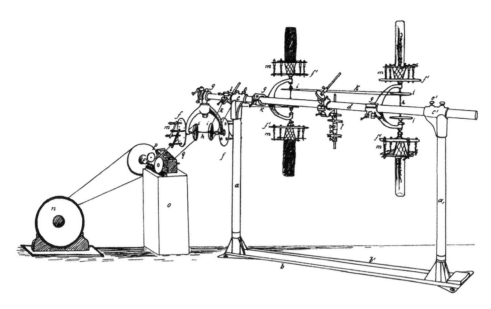

Figure 9.3
Motor-driven clinostat constructed by Figdor and Portheim for their research at the BVA. Plants were to be mounted on the shaft and slowly turned around, so that gravitational effects were leveled out. Reproduced from Przibram (1908), p. 260.

effects was crucial, if, for example, Figdor wanted to find out whether light was the decisive cause of anisophylly in one species or another. Equally important was the exclusion of light, if gravitation was investigated: this is why the BVA's dark chambers were so crucial, as well as the BVA's specific phainometer—a high-precision apparatus to measure light intensities—and the plant beds that allowed for illumination only from above or from below. Making generous use of this rich equipment and instrumentation, Figdor produced a wealth of experimental findings. He examined species from the genera *Abies*, *Acer*, *Aesculus*, *Anagallis*, *Centradenia*, *Monophyllea*, *Selaginella*, *Streptocarpus*, and many more.[34]

By 1909, a number of these experiments had been running for six years already, in order to observe the effect of external factors either on succeeding generations of annual plants or on the long-term development of perennial plants. Thus, Figdor's project was exactly the kind of investigation that Przibram had in mind when he claimed that long-term studies were indispensable for biologists.

However, it is striking that hardly any general conclusions were drawn by Figdor, such as would have been in line with Przibram's report of 1908. Figdor crafted his findings very carefully, providing individual analyses and conclusions for each genus or even species. The main factors of influence varied greatly, although some recurrent phenomena emerged, as Figdor conceded. The influence of light was frequently found to be of relevance. In the Nordmann fir (*Abies nordmannia*), for example, even the slightest difference of illumination between the lower and upper side of the branches was shown to be sufficient to induce anisophyllous effects. Gravitation was also regularly conformed to be relevant—but sometimes it was only the specific combination of light and gravitation that proved effective (as in the case of the blue pimpernel, *Anagallis grandiflora*, today's *A. morelli*). In some species, the course and intensity of life processes, such as photosynthesis and transpiration, also seemed to play a role, although the exact nature of that role remained unclear. Other species, finally, firmly resisted all external influences they were exposed to. A conspicuous example of this behavior was the tropical plant *Monophyllaea horsfieldii*, in which the cotyledons are extremely different in size—and would remain that way no matter how the environment changed. Figdor suspected, in line with Wiesner, that in such cases a long-term adaptation prevailed, which had become part of the plant's fixed constitution.[35] This is also one of the very few cases, in which Figdor explicitly commented on the contested question of the heritability of acquired properties: an issue that he usually avoided.

Finally, Figdor underlined that in nature all these factors (and possibly more) were in constant interaction with each other, leading to mutual reinforcement or alleviation of their influences. This he was unable to imitate and analyze in his laboratory. Even more difficult was the analysis of the interaction of these external factors with the most important factor that prompted anisophylly (according to Wiesner), namely "the original position of the leaf primordia in the vegetative organism."[36] After all, it was this position that determined the extent to which the shoots were exposed to light, gravitation or other factors, Figdor stated.[37] However, he did not address the obvious question, by which factors these primordial positions were determined. "Internal symmetry" would have been one possible answer. Figdor could also have referred to Wiesner's theory of "plasomes," which were conceived of as the constitutive elements of the organism, which assembled according to their inherent principle of *Enharmonie*.[38] This idea grossly violated Figdor's confinement to observable entities; and notwithstanding all loyalty to his mentor, he made no mention of it.

Figdor's monograph on anisophylly was the only book-length study that the BVA's botanical departments produced on the interplay of internal and external factors of morphology. But there were many small projects, both by Figdor and by Portheim, along similar lines—that is, in pursuit of the "real causes," in Przibram's sense, of the organism's shape. However, Figdor's study stands out, not only in its depth and extension but also in its outspoken treatment of methodology. Not even Przibram dealt with that subject with the same clarity. The requirements of experimentation that Figdor so strongly endorsed had already become standard at the time in physics and chemistry, but were still contested in organismal botany and zoology. Figdor pointedly described the need to design two situations that, at best, differed only in one single factor, so that, if the effect differed, causal inferences regarding the efficacy of the test factor were justified.[39] It was obvious that this ideal was hard to attain in the study of living beings, but, from Figdor's perspective, that was no excuse for not even trying; rather, it made methodological awareness all the more important—particularly in fields like morphology, which only recently had become the subject of experimentation.

His methodological reflections, furthermore, reveal Figdor's more fundamental notion of how to investigate natural phenomena. Although Figdor conceded that in nature plants are subject to a network of factors, which he would never be able to reconstruct in its full complexity, he clearly conceived of the situation as determined and, thus, open to classic experimental analysis. Figdor's aim was to identify individual, causally relevant factors for the growth of leaves, without losing sight of the fact that plant species are highly diverse in their reaction to the environment; and he was convinced that the approach chosen for the anisophylly project was the only way to do so. Yet, there was no room in this methodological framework for formulating more sweeping hypotheses of any kind. Figdor was not looking for general patterns or regularities; and he was not interested in probabilities, quantification, or mathematization, which elsewhere have been identified as characteristic of the BVA's zoology.[40] Given this difference in interest, it is not surprising that, as far as we know, Figdor never cooperated directly with the Zoological Department of the BVA. This did not prevent Przibram, however, from integrating Figdor's results into his own cross-disciplinary program, as will be demonstrated in the next section.

The Principles of Regeneration

The term "regeneration" refers to the capability of an organism to restitute the bodily constitution after lesion, or to form new organs after a full loss. Around 1900, it was the subject of intense study, closely related to the study of ordinary development, in search of regularities and essential factors in the growing and shaping of organic form.[41] As is well known, regeneration and transplantation of animal organs, or parts of organs, became central research themes for Przibram and Kammerer. But the BVA's plant scientists, Figdor and Portheim, also published on this topic; the latter even wrote the chapter on regeneration in plants in one of the common handbooks of the time.[42]

The fact that plants were able to grow new leaves, roots, and shoots in case of loss and lesion was hardly new around 1900. Unlike animals, plants retain spots of embryonic tissue throughout their existence—the so-called vegetative points or meristems—which can be activated to grow new organs. Gardeners had long exploited this phenomenon, for instance, in their trimming of bushes and trees. The underlying mechanism, however, was a subject of controversial debate. It was generally assumed at the time that, in contrast to animals, there was no "true" regeneration in plants, that is, no direct restitution of lost organs. The opinion was that plants rather replace these losses with the help of adventive shoots that might grow at some distance from the site of lesion. This was the position defended by eminent figures such as the aforementioned Karl Goebel. However, the same Goebel drew attention to a spectacular exception to this rule: the odd case of a fern's frond (of the species *Polypodium heracleum*) that, after it had been split lengthwise at the tip, continued to grow in cleavage.[43] Goebel strongly encouraged further investigation of similar phenomena, as the findings were likely to shed light on the complicated issues of cell development and differentiation.[44]

It was in these years, and very likely inspired by Goebel's findings, that Figdor also turned to the regenerative capacities of plants; and as an object of study he chose the heart's-tongue fern (*Scolopendrium scolopendrium*).[45] Figdor replicated Goebel's approach—a tiny cut, fractions of a millimeter, at the very tip of the frond—and also in this case, a cleavage resulted so that, as in the *Polypodium*, the frond seemed to have doubled. Figdor was unsure how to explain this phenomenon. Significantly, however, he referred for further information to a zoological paper that described similar cleavages in the regeneration of a catfish's whiskers. The latter was published in a somewhat obscure journal for aquarium keepers. In the cited issue, however, Kammerer also had a paper, so that one may suspect that he had given Figdor a hint after having learned about Figdor's dealing with cleavage.[46]

Figdor became interested in this phenomenon; and in parallel to his studies of anisophylly, he started to investigate regeneration in species of the Gesneriaceae family. In some of these, the foliage is reduced to the two cotyledons, one of which develops into a leaf-like organ (while the other almost disappears). Owing to its nature as a cotyledon, the leaf retains meristematic properties in its tip and at the base. In *Monophyllaea horsfieldii*, Figdor observed in young seedlings the formation of adventive leaves in response to injuries.[47] He then studied species of the genus Streptocarpus. No regeneration was observed if lateral parts of the lamina or the leaf's tip were removed. However, if the leaf was cut in halves from the tip, either both halves or at least one of them displayed regenerative reactions that were similar to the cleavage in the fern's frond. Figdor claimed this proved that leaves of higher plants were also capable of "true regeneration"—that is, restitution at the site of lesion—against Goebel and others.[48] A similar reaction was observed by Figdor's associate Georg Stingl, after the latter had carefully cut the tip of very young roots (of various plant genera, including *Zea*, *Vicea*, *Picea*, *Larix*, etc.). Within 48 hours, cleavage occurred in reaction to the injury: the one vegetative point at the tip of the root was replaced by two newly formed vegetative points at either branch of the cleavage.[49] Later, Figdor also reported evidence of restitution in the marine (unicellular but macroscopic) alga *Dasycladus clavaeformis*.[50]

These papers are interesting in themselves, since they contributed important findings to a highly contested issue. But they gain even further import and relevance in view of a talk given by Hans Przibram in 1906, which was presented to the Versammlung Deutscher Naturforscher und Ärzte in Stuttgart, Germany, and entitled "Regeneration as a General Phenomenon in the Three Realms of Nature."[51] In this talk, Przibram compared regenerative processes in crystals, plants, and animals. His goal was ambitious: Przibram claimed that regeneration was a "primordial, general phenomenon," that could not possibly be explained by referring to its acquisition in the course of organismal evolution. This was in direct opposition to the neo-Darwinian August Weismann, who had argued that the capacity of regeneration is a derived property, produced by evolutionary forces in those species that are particularly prone to lose limbs or organs—such as crabs or starfish.[52] Przibram, in contrast, attempted to show that all organisms, including all vertebrates, even mammals, have this capacity, and that the same capacity was present in crystals.[53] In order to argue this point, Przibram stated, he would have to demonstrate the following: "(1) the general existence of this property in all forms capable of growing; (2) the explanation of apparent or actual exceptions; and, finally, (3) a common cause for this phenomenon in the three realms [of nature]."[54]

To fulfill the first requirement, Przibram demonstrated the similarity in the patterns of regeneration in all three realms of nature. A series of tables with small sketches was presented as visual evidence of this claim. In a first table, phenomena in crystals and plants were juxtaposed and recurrent patterns were identified (figure 9.4). Then, in a second table (not reproduced here), very similar phenomena and patterns were presented as having been observed in animals. It was in particular the "doubling through cleavage" that Przibram drew attention to, as a pattern of regeneration that clearly occurred in all three realms. If the cleavage went too deep, Przibram argued, a break of the structure may occur, occasionally with a subsequent tripling of the organ. While the examples from the animal and mineral realms mostly came from Przibram's own work (and that of his assistants, including Kammerer), the findings in plants as well as the pertinent sketches were taken from the work of Figdor (doubling in leaves and fronds), of Stingl (doubling in roots), and of a frequent guest at the BVA, the Prague plant physiologist Bohumil Nemec (break and tripling).[55] As in the report of 1908, in which Przibram had boldly restated Figdor's findings on anisophylly (cited above), Przibram did not hesitate to generalize the botanists' results: cleavage in a juvenile state, Przibram stated, "allows demonstration of the capability of regeneration *in all organs of plants*."[56]

Przibram then explained the apparent exceptions that might be adduced to reject the thesis that regeneration was a primordial, general phenomenon. In most cases, he argued, the experiments these authors referred to had been conducted under inappropriate conditions or the organism had already been too old. Przibram emphasized that researchers at the BVA were able to prompt regeneration in a number of species that up to then had been considered incapable of it, including olms of the genus *Proteus*.[57] (Przibram had to concede, however, that "higher animals traded their capacity of regeneration for an optimized economy and the potential of a long-term memory"[58]).

Finally, Przibram suggested the same common cause for regeneration in all cases: locally accelerated growth. (It was very important to him that crystals also "grow," although they are not alive, so that neither growth nor regeneration could be conceived of

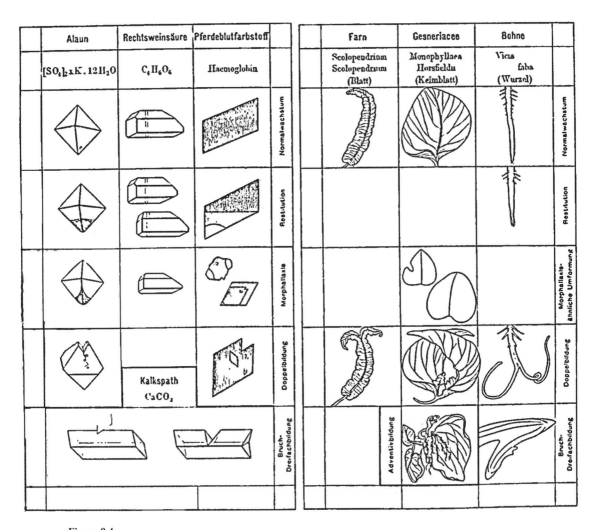

Figure 9.4
The tabulated arrangement of the sketches was to demonstrate the parallel patterns in crystals and plants. The terms on the right margin are (from top to bottom): normal growth, restitution, morphallaxis, doubling and, finally, crack or tripling. Reproduced from Przibram (1907), p. 619.

as phenomena due to vital forces.[59]) This explained why adult animals, which no longer grow, are not as capable of regeneration as young forms of the same species; and why plants regenerate only if the meristematic parts of their organs are involved, such as the tips of shoot, root, and leaves or the leaves' bases. Interestingly, almost no mention was made in this context of the renowned embryologist (later, geneticist) Thomas H. Morgan, who had published prominently on the issue. Przibram cited some of Morgan's factual findings—on the phenomenon of morphallaxis, for example—that is, the regeneration of a full organism from fragments. But Przibram failed to mention that Morgan had already conceptualized regeneration as belonging to "the general category of growth phenomena" and that Morgan had also rejected the idea that it could be explained "as the outcome of the selective agency of the environment."[60] The ultimate reason for regenerative processes, Przibram argued—and this is where he went far beyond Morgan—was that every natural form, whether mineral, plant, or animal, had a morphological equilibrium which it tried to regain after loss or lesion. The farther its form moved from this state (by losing an organ, for instance), the faster the organism would try to rebalance its shape.[61] In places where the original form had been injured, therefore, growth processes were accelerated to catch up with the rest of the organism and reestablish the equilibrium state. The process of regeneration, thus, was part of the individual's self-regulation, which Przibram considered as one of the fundamental properties of nature.

This example demonstrates how a very specific, plant biological study, such as the investigation of regenerative reactions in the cotyledons of tropical Gesneriaceae, became an integral, even indispensable part of the larger research program of the BVA, contributing to its claims on nature's fundamentals. It was Przibram who in his books, papers, talks, and institutional reports brought forward these bold, comparative syntheses. Figdor, in contrast, seems to have been little interested in generalization. However, the regeneration project reveals that his contributions were crucial. As mentioned earlier, the examples presented by Przibram to demonstrate that cleavage and doubled organs were general patterns of regeneration were taken almost exclusively from work that had been done in the BVA—down to the tabulated sketches. Further, there is no mention of "cleavage" as a pattern of regeneration in Przibram's papers before 1906, not even in the paper on injured crystals from which the figures for the 1906 tables were taken.[62] The phenomenon was first mentioned by Przibram, as far as I can see, in an extended article of 1906 dealing with the analogy between developmental mechanics in organisms and in crystals. In this paper, Przibram reported instances of "doubled formation" (*Doppelbildung*) in animals, as a reaction to certain kinds of injury, and compared them with similar processes in crystals.[63] On this issue he immediately cited Goebel's and Figdor's work on cleavage and regeneration in plants. It seems very plausible that Przibram, who became this idea's great proponent, had picked up the concept from plant biology—and then started to develop it further and use it as an interpretative approach for a number of phenomena in crystals and animals. The study of regeneration, thus, ought to remind us how closely the members of the BVA followed each other's work, even where this is not apparent on the surface. This is particularly true for Przibram, who depended on his colleagues for his cross-disciplinarian program.

Conclusion

The plant biological work undertaken in the BVA is, unfortunately, a sorely underexplored topic. One of the aims of this chapter, therefore, has been to demonstrate that some of the BVA's most original features cannot be adequately understood without taking into account the work on plants. However, the BVA's plant biological projects were designed, in the first place, to contribute to some of the highly contested research questions at the time in botany and plant physiology. It is important to keep in mind that the experimental investigation of growth and development as well as the interplay of internal and external factors in shaping the organism were central themes within plant biology in general, far beyond the BVA. Figdor did not choose to work on anisophylly and regeneration in order to serve the institution's superordinate goals; but, nevertheless, the goals were promoted. Despite the organizational and, perhaps, personal differences between the BVA's heads of departments, they were in close communication with one another, and they followed, enriched, and supported each other's work in ways that are not always obvious. Figdor's work, in particular, was crucial for Przibram's daring syntheses and theoretical approaches, for which the BVA became so famous; while at the same time, Figdor himself benefited from the unique interdisciplinary environment. After all, how else would Figdor have known about cleavages in a catfish's whiskers?

Notes

1. See, e.g., Hirschmüller (1991), Reiter (1999), Coen (2006), Gliboff (2006), Logan (2013). Even reports written at the time of the BVA mostly focused on the zoological work, such as Edwards (1911).

2. For references to the history of plant physiology, see, e.g., the respective chapters in Jahn (2004).

3. Pringsheim (1932), p. 15; German: "[Sachs] möge etwas anderes als Pflanzenphysiologie vortragen, weil er sonst in 2 bis 3 Stunden fertig sein würde."

4. The lengthy discussions preceding his employment are documented in Wiesner's personal files at the university. The favorite candidates were Anton de Bary and Julius Sachs. Both had declined the offer, while the third person on the list was rejected by the ministry; it was only at this point that Wiesner's name was brought up and agreed upon by the committee. See Archive of the University of Vienna, PH PA 3727 (Julius Wiesner).

5. For Wiesner's biography, see Wurzbach (1888), Molisch (1916), Winninger (1932).

6. The same was true for Richard von Wettstein, professor of systematic botany, and the professors of zoology at the University of Vienna. The report by von Wettstein (1902), p. 14, uses this example to emphasize the programmatically close relation of the BVA to the university.

7. Coen (2006), p. 494, suggested that the BVA was at "a sharp contrast to the tradition-bound university across town"; and Hofer (2002), p. 171, mentioned "animosities" between the university and the BVA. This may be true in view of regularities and institutionalized anti-Semitism at the Vienna university. It cannot be confirmed, however, in terms of biographical ties and research orientation.

8. On Figdor's biography, see Anonymous (1956).

9. See Molisch (1934) for his account of this journey.

10. See on Portheim, e.g., Metcalfe (1947).

11. The other two editors were the half-brothers Karl and Ludwig Linsbauer. For the Festschrift, see Linsbauer et al. (1903).

12. "Die ganze Lehre vom Bau, von der Entwicklung und vom Leben der Pflanze." Wiesner (1910), p. 106.

13. Wiesner (1884), p. 360.

14. Wiesner (1889). In the preface to this work (p. 1), Wiesner comments on the ambiguity of the term. The British used it, Wiesner reported, to refer to the study of living organisms in general; while in Germany, it was mostly used in a very restricted sense, referring only to the way of lives of plants and animals (similar to today's "ecology"). Alternatively, as Wiesner himself preferred, it implied something in between (in the sense outlined above).

15. See Przibram (1908), p. 235 German: "Während es dem Physiologen genügen mag, seine Versuchsobjekte so lange am Leben zu erhalten, als er eine bestimmte Funktion verfolgen will, und dann zu weiterer Beobachtung frische Exemplare zu nehmen, kommt es dem Biologen meist auf Durchverfolgung der Formänderungen während einer längeren Versuchszeit an."

16. It is striking that this original organization—a zoological, botanical, physiological and chemical department—was identical with the organization of the well-known Zoological Station in Naples, Italy, that also in many other respects served as a model for the BVA (e.g., in view of the table system). For the station in Naples, see, e.g., Partsch (1980), Simon (1980), Groeben and Müller (1975).

17. See ÖAW [Österreichische Akademie der Wissenschaften] Archiv, Biologische Versuchsanstalt (Vivarium). Promemoria an das Hohe Präsidium der Kaiserlichen Akademie der Wissenschaften in Wien, 1911; K1/M1.

18. See Reiter (1999), p. 592, and Hofer (2002), pp. 156–158. On the history of the families Portheim and Przibram in the nineteenth century, see also Niedhammer (2013).

19. ÖAW Archiv, Biologische Versuchsanstalt (Vivarium), K1/M4: Erster Bericht des Kuratoriums der biologischen Versuchsanstalt der Kais. Akademie der Wissenschaften, 25.3.1914. In contrast to the favorable reception of Figdor's request (as he was one of the institution's founders), the proposal to introduce a sixth department (for radiobiology, to be headed by the pharmacologist Walther Hausmann) was declined at the same meeting. See also, for the minutes of the debate: ÖAW Archiv, Biologische Versuchsanstalt (Vivarium), K2/M3: Protokoll der Sitzung des Kuratorium der Biologischen Versuchsanstalt. 5. März 1914, 4 Uhr nachmittags (ordentliche Jahressitzung).

20. This was emphasized again in 1914, when Figdor insisted that, in the new statute of the BVA, the outdoor areas for experimentation granted to the several departments ought to be clearly defined and demarcated: ÖAW Archiv, Biologische Versuchsanstalt (Vivarium), K2/M3: Protokoll betreff Kenntnisnahme der Leitungsordnung; anwesend: Prof. W. Figdor, Generalsekretär Becke, 8. Juni 1914.

21. See the BVA's report of December 4, 1903, to the Academic Board: Aktenvermerk 40926. Oesta. AVA. Phil: Biol. Versuchsstation, 1902–1919. Ktn. 128. Sign. 2A.

22. Notably absent here are questions of heredity and genetics in plants. Przibram (1908), p. 239, explained that these were already studied by Erich Tschermak-Seysenegg (1871–1962) at the Vienna Hochschule für Bodenkultur. As there was neither need nor desire for a duplication of efforts, Przibram maintained, the BVA would not pursue issues related to heredity.

23. Programm für die Entwicklung einer biologischen Forschungsanstalt in Wien; ÖAW, Biologische Versuchsanstalt K1./M.1. German: "[Ein Institut, das] durch Vereinigung beider Zweige [der experimentellen Biologie] in einem Raume die Wechselbeziehungen zwischen Tieren und Pflanzen weiterzuverfolgen gestatten würde."

24. One of the very few exceptions was a study of the immunological reaction of plant species to horse serum; see Kraus et al. (1907).

25. First contributions to the topic include Frank (1868) and Wiesner (1868).

26. Przibram (1908), p. 237.

27. See, e.g., Przibram (1904a). The term "experimental morphology" had been used earlier by, for example, the American zoologist Charles Davenport, albeit in a narrower sense, referring only to the investigation of the external causes of organic form, while excluding others. See Davenport (1897), p. viii. For an introduction to the "experimental morphology" of plants, in the line of Przibram, see Goebel (1908).

28. Przibram (1904a), p. 3: "Für ihn [den Neo-Darwinisten] ist die natürliche Zuchtwahl geradezu 'allein-seligmachend' und von wirklichen, realen Ursachen ist keine Rede mehr."

29. Figdor (1909).

30. Goebel (1880), pp. 817, 839. See, for the report, Figdor (1909), p. 109. Wiesner explicitly acknowledged his change of opinion in Wiesner (1892b), p. 552.

31. Figdor (1909), p. 110. German: "Ich werde mich nur darauf beschränken, ich sage dies hier ein für allemal, jene Tatsachen anzuführen, welche auf Grund von Experimenten bekannt geworden sind oder auch durch Beobachtung von in der freien Natur vorkommenden Erscheinungen, die, soviel wir heute wissen, überhaupt nur in einem Sinne gedeutet werden können."

32. Figdor (1909), pp. 111–112. German: "Bezüglich der Versuchsanordnung möchte ich nur noch kurz erwähnen, daß ein einwandfreies Resultat stets nur dann zu erwarten ist, wenn der Einfluß eines einzigen Faktors auf das betreffende Untersuchungsobjekt bei möglichster Konstanz aller übrigen äußeren Bedingungen zur Geltung kommt. [...] [E]igentlich ist all dies ganz klar, gerade so, wie daß bei derartigen Untersuchungen Alterszustände der Organe, Induktions- und Korrelationsverhältnisse u.a. mehr Berücksichtigung finden müssen. Ich halte es jedoch nicht für überflüssig auf diese Verhältnisse speziell hinzuweisen, weil gerade auf dem uns hier interessierenden Gebiete der Formgestaltung nach dieser Richtung hin viel gesündigt worden ist."

33. See, e.g., Levere (1996) and, more recently, Leber and Nickelsen (2016). See also Goebel (1898), Preface, in which morphology was framed as a field in transition, as slowly the shape of plants had been acknowledged to be part of the phenomena of life (*Lebenserscheinungen*), which, consequently, had to be studied in their interaction with the environment, and not a merely conceptual matter (*Begriffskonstruktion*).

34. The species were: *Selaginella spec.* (p. 117), *Abies nordmannia* (p. 120f), *Abies pectinata* (p. 122), *Acer platanoides* (p. 129), *Aesculus hippocastaneum* (p. 132), *Centradenia floribunda* (p. 141), *Anagallis grandiflora* (p. 143), *Streptocarpus wendlandii* (p. 144) and *Monophyllea horsfieldii* (p. 146)—and many more.

35. Figdor (1909), p. 164.

36. Figdor (1909), p. 163: "Die primäre Stellung der Blattanlagen am pflanzlichen Organismus."

37. He emphasized this point again in Figdor (1912), p. 136.

38. See, e.g., Wiesner (1892a).

39. Figdor sketched a procedure in consonance with John S. Mill's method of difference, see Mill (1973); the reception of Mill's methods of experimental research has been discussed in Scholl (2013). An analytical approach to experimentation as "difference tests," in a more refined meaning, was suggested in Graßhoff et al. (2000). For episodes that have been analyzed from this perspective, see Nickelsen and Graßhoff (2011), Nickelsen and Graßhoff (2008).

40. See, e.g., Coen (2006).

41. See, e.g., the contributions to Dinsmore (1991).

42. For publications on this theme, see, e.g., Figdor (1903, 1906, 1907, 1910, 1918, 1926); Portheim (1926); for further work by their associates, see, e.g., Stingl (1905, 1909).

43. See, e.g., Goebel (1902), pp. 503–505; a photograph of the frond is printed on p. 504. The example was also presented in Goebel (1908), p. 215.

44. Goebel (1905), pp. 224–228.

45. Figdor (1906).

46. Roth (1905). In Figdor's paper the title is cited as dealing with the "Tangerwels," while, in actual fact, it is a study of the "Panzerwels."

47. Figdor (1903).

48. Figdor (1907), p. 56. Both genera had been studied by Pischinger (1902), while Figdor criticized this study's methods and approach.

49. Stingl (1905); Figdor's inspiration is acknowledged on p. 221.

50. Figdor (1910).

51. Przibram (1907).

52. Weismann famously defended this position in Weismann (1892), book II, chapter 2. Regeneration, he claimed, was not a general property but a special adaptation produced by natural selection; see p. 140 and pp. 152–168. See also, e.g., Martins (2011), section 7.2.

53. A comprehensive critique of Weismann's position is given in, e.g., Przibram (1914).

54. German: "1. das allgemeine Vorkommen dieser Eigenschaft bei wachstumsfähigen Formen, 2. die Erklärung der scheinbaren oder wirklichen Ausnahmen und endlich 3. eine gemeinsame Ursache als Grund dieser Erscheinung in den drei Reichen." Przibram (1907), p. 619.

55. See Nemec (1905), for the latter's book on regeneration phenomena in leaves.

56. German: "Sie [die Spaltungen] gestatten, *an allen Organen der Pflanzen* Regenerationsfähigkeit nachzuweisen." Przibram (1907), p. 620; my emphasis.

57. In Przibram (1914), p. 349, he pointed out, for example, that olms were unable to regenerate the loss of organs in their natural, rather cool environment; while they were perfectly able to regenerate the same organs when brought into the warmth.

58. German: "Ich könnte darauf hinweisen, daß die höheren Tiere […] auf Kosten ihrer Regenerationsfähigkeit eine günstigere Ökonomie und die Möglichkeit eines bleibenden Gedächtnisses eingetauscht haben." Przibram (1907), p. 648.

59. Crystals became one of the favorite examples in the debate around 1900 between mechanists and vitalists; see, e.g., Brandstetter (2011), Allen (2005).

60. Morgan (1901), p. 292.

61. Przibram (1907), p. 646; the German term he uses is *Gleichgewichtszustand*. The importance of a morphological equilibrium for all organisms became a central assumption of Przibram's. It figures prominently, e.g., in Przibram (1922). In his summarizing chapter, Portheim picked up this notion without, however, developing it further or discussing its innovative potential in the plant sciences. See Portheim (1926), p. 1135.

62. See Przibram (1904b).

63. See Przibram (1906), p. 250.

References

Allen, Garland E. 2005. Mechanism, vitalism and organicism in late nineteenth and twentieth-century biology: The importance of historical context. *Studies in History and Philosophy of Biological and Biomedical Sciences* 36:261–283.

Anonymous. 1956. Wilhelm Figdor. In *Österreichisches Biographisches Lexikon 1815–1950*, vol. 1, 313. Vienna: Verlag der Österreichischen Akademie der Wissenschaften.

Brandstetter, Thomas. 2011. Lebhafte Kristalle. *Rheinsprung. Zeitschrift für Bildkritik* 11:128–137.

Coen, Deborah R. 2006. Living precisely in fin-de-siècle Vienna. *Journal of the History of Biology* 39:493–523.

Davenport, Charles. 1897. *Experimental Morphology*. London: Macmillan.

Dinsmore, Charles E, ed. 1991. *A History of Regeneration Research. Milestones in the Evolution of a Science*. Cambridge: Cambridge University Press.

Edwards, Charles L. 1911. The Vienna Institution for Experimental Biology. *Popular Science Monthly* 78:584–601.

Figdor, Wilhelm. 1903. Über Regeneration bei Monophyllaea Horsfieldii. *Österreichische Botanische Zeitschrift* 53:393–396.

Figdor, Wilhelm. 1906. Über Regeneration der Blattspreite bei Scolopendrium scolopendrium. *Berichte der Deutschen Botanischen Gesellschaft* 24:13–16.

Figdor, Wilhelm. 1907. Über Restitutionserscheinungen an Blättern von Gesneriaceen. *Jahrbücher für Wissenschaftliche Botanik* 44:41–56.

Figdor, Wilhelm. 1909. *Die Erscheinungen der Anisophyllie*. Leipzig, Vienna: Deuticke.

Figdor, Wilhelm. 1910. Über Restitutionserscheinungen bei Dasycladus clavaeformis. *Berichte der Deutschen Botanischen Gesellschaft* 28:224–227.

Figdor, Wilhelm. 1912. Zu den Untersuchungen über das Anisophyllie-Phaenomen. *Berichte der Deutschen Botanischen Gesellschaft* 30:134–139.

Figdor, Wilhelm. 1918. Zur Kenntnis des Regenerationsvermögens von Crassula multicava. *Berichte der Deutschen Botanischen Gesellschaft* 36:241–246.

Figdor, Wilhelm. 1926. Über das Restitutionsvermögen der Blätter von Bryophyllum calycinum Salisb. *Planta* 2:424–428.

Frank, Albert Bernhard. 1868. Über die Einwirkung der Gravitation auf das Wachsthum einiger Pflanzentheile. *Botanische Zeitung* 26: 873–882.

Gliboff, Sander. 2006. The case of Paul Kammerer: Evolution and experimentation in the early 20th century. *Journal of the History of Biology* 39:525–563.

Goebel, Karl. 1880. Beiträge zur Morphologie und Physiologie des Blattes. *Botanische Zeitung* 38:753–759; 769–778; 785–795; 800–826; 833–845.

Goebel, Karl. 1898. *Organographie der Pflanzen, insbesondere der Archegoniaten und der Samenpflanzen*. Jena: Fischer.

Goebel, Karl. 1902. Über Regeneration im Pflanzenreich. *Biologisches Centralblatt* 32:385–397; 417–438; 481–505.

Goebel, Karl. 1905. Allgemeine Regenerationsprobleme. *Flora* 95:223–241.

Goebel, Karl. 1908. *Einleitung in die Experimentelle Morphologie der Pflanzen*. Leipzig: Teubner.

Graßhoff, Gerd, Robert Casties, and Kärin Nickelsen. 2000. *Zur Theorie des Experiments. Untersuchungen am Beispiel der Entdeckung des Harnstoffzyklus*. Bern: Bern Studies for the History and Philosophy of Science.

Groeben, Christiane, and Irmgard Müller. 1975. *The Naples Zoological Station at the Time of Anton Dohrn*. Naples: Stazione Zoologica di Napoli.

Hirschmüller, Albrecht. 1991. Paul Kammerer und die Vererbung erworbener Eigenschaften. *Medizinhistorisches Journal* 26:26–77.

Hofer, Veronika. 2002. Rudolf Goldscheid, Paul Kammerer und die Biologen des Prater-Vivariums in der liberalen Volksbildung der Wiener Moderne. In *Wissenschaft, Politik und Öffentlichkeit. Von der Wiener Moderne bis zur Gegenwart, ed. Mitchell G. Ash and Christian H. Stifter, 149–184*. Vienna: WUV Universitätsverlag.

Jahn, Ilse, ed. 2004. *Geschichte der Biologie*. 3rd ed. Hamburg: Nikol.

Kraus, R., Leopold von Portheim, and T. Yamanouchi. 1907. Biologische Studien über Immunität bei Pflanzen. 1. Untersuchungen über die Aufnahme präcipitierbarer Substanzen durch höhere Tiere. *Berichte der Deutschen Botanischen Gesellschaft* 25:383–388.

Leber, Christoffer, and Kärin Nickelsen. 2016. Wissenschaft im Glaubenskampf: Geschichte als Argument in den akademischen Festreden Emil DuBois-Reymonds (1818–1896). *Berichte zur Wissenschaftsgeschichte* 39:143–164.

Levere, Trevor. 1996. Romanticism, natural philosophy, and the sciences: A review and bibliographical essay. *Perspectives on Science* 4:463–488.

Linsbauer, Karl, Ludwig Linsbauer, and Leopold von Portheim, eds. 1903. *Wiesner und seine Schule: Ein Beitrag zur Geschichte der Botanik. Festschrift anlässlich des dreissigjährigen Bestandes des Pflanzenphysiologischen Institutes der Wiener Universität*. Vienna: Hölder.

Logan, Cheryl A. 2013. *Hormones, Heredity, and Race: Spectacular Failure in Interwar Vienna*. New Brunswick, NJ: Rutgers University Press.

Martins, Lilian al Chueyr Pereira. 2011. Regeneration as a difficulty for the theory of natural selection: Morgan's changing attitudes, 1897–1932. In *Brazilian Studies in Philosophy and History of Science: An Account of Recent Works, ed. D. Krause and A. Videira, 119–129*. Dordrecht: Springer.

Metcalfe, C. R. 1947. Dr. Leopold von Portheim. *Nature* 159:835–835.

Mill, John Stuart. 1973. *A System of Logic, Ratiocinative and Inductive: Being a Connected View of the Principles of Evidence and the Methods of Scientific Investigation. Books I–III (Reprint)*. Toronto: University of Toronto Press.

Molisch, Hans. 1916. Julius Wiesner 1838–1916. *Berichte der Deutschen Botanischen Gesellschaft* 34:71–99.

Molisch, Hans. 1934. *Erinnerungen und Welteindrücke eines Naturforschers*. Wien: Haim.

Morgan, Thomas Hunt. 1901. *Regeneration*. New York: Macmillan.

Nemec, Bohumil. 1905. *Studien über die Regeneration*. Berlin: Borntraeger.

Nickelsen, Kärin, and Gerd Graßhoff. 2008. Concepts from the bench: Krebs and the urea cycle. In *Going Amiss in Experimental Research, ed. Giora Hon, Jutta Schickore, and Friedrich Steinle, 91–117*. Dordrecht: Springer.

Nickelsen, Kärin, and Gerd Graßhoff. 2011. In pursuit of formaldehyde: Causally explanatory models and falsification. *Studies in History and Philosophy of the Biological and Biomedical Sciences* 42:297–305.

Niedhammer, Martina. 2013. *Nur eine "Geld-Emancipation"? Loyalitäten und Lebenswelten des Prager jüdischen Großbürgertums 1800–1867*. Göttingen: Vandenhoeck & Ruprecht.

Partsch, Karl Josef. 1980. *Die Zoologische Station in Neapel. Modell internationaler Wissenschaftszusammenarbeit*. Göttingen: Vandenhoeck & Ruprecht.

Pischinger, Ferdinand. 1902. Über Bau und Regeneration des Assimilationsapparates von Streptocarpus und Monophyllaea. *Sitzungsberichte der Kaiserlichen Akademie der Wissenschaften* 111:278-302.

Portheim, Leopold. 1926. Regeneration bei Pflanzen. In *Handbuch der Normalen und Pathologischen Physiologie* 14(1): 1114–1140. Berlin: Springer.

Pringsheim, Ernst G. 1932. *Julius Sachs. Begründer der neueren Pflanzenphysiologie (1832–1897)*. Jena: Fischer.

Przibram, Hans L. 1904a. *Einleitung in die Experimentelle Morphologie der Tiere*. Leipzig: Deuticke.

Przibram, Hans L. 1904b. Formregulationen verletzter Krystalle. *Zeitschrift für Kristallographie* 39:576–582.

Przibram, Hans L. 1906. Kristall-Analogien zur Entwicklungsmechanik der Organismen. *Archiv für Entwicklungsmechanik der Organismen* 22:207–287.

Przibram, Hans L. 1907. Die Regeneration als allgemeine Erscheinung in den drei Reichen. *Naturwissenschaftliche Rundschau* 22:619–622; 633–634; 645–648.

Przibram, Hans L. 1908. Die Biologische Versuchsanstalt in Wien. Zweck, Einrichtung und Tätigkeit während der ersten fünf Jahre ihres Bestandes (1902-1907). Bericht der Zoologischen, Botanischen und Physikalisch-Chemischen Abteilung. *Zeitschrift für biologische Technik und Methodik* 1:234–264.

Przibram, Hans L. 1914. Regeneration und Transplantation im Tierreich. In *Allgemeine Biologie (= Die Kultur der Gegenwart Teil 3, Abt.4, Bd.1)*, ed. C. Chun and Wilhelm Johannsen, 343–377. Leipzig: Teubner.

Przibram, Hans L. 1922. *Form und Formel im Tierreich. Beiträge zu einer Quantitativen Biologie*. Leipzig, Wien: Deuticke.

Reiter, Wolfgang L. 1999. Zerstört und Vergessen: Die Biologische Versuchsanstalt und ihre Wissenschaftler/Innen. *Österreichische Zeitschrift für Geschichtswissenschaften* 10:585–614.

Roth, Wilhelm. 1905. Doppelte Regeneration eines Bartfadens bei einem Panzerwelse. *Blätter für Aquarien- und Terrarienkunde* 16:408–410.

Scholl, Raphael. 2013. Causal inference, mechanisms, and the Semmelweis case. *Studies in History and Philosophy of Science* 44:66–76.

Simon, Hans-Rainer, ed. 1980. *Anton Dohrn und die Zoologische Station Neapel*. Frankfurt Main: Erbrich.

Stingl, Georg. 1905. Untersuchungen "Über Doppelbildung und Regeneration bei Wurzeln." *Österreichische Botanische Zeitschrift* 68:219–225; 260–263.

Weismann, August. 1892. *Das Keimplasma*. Jena: Eine Theorie der Vererbung.

Wettstein, Richard von. 1902. Österreichische Biologische Stationen. *Neue Freie Presse* no. 13647 (August 21): 14–15.

Wiesner, Julius. 1868. Beobachtungen über den Einfluss der Erdschwere auf Grössen- und Formverhältnisse der Blätter. *Sitzungsberichte der Kaiserlichen Akademie der Wissenschaften* 58:369–389.

Wiesner, Julius. 1884. *Elemente der wissenschaftlichen Botanik II. Elemente der Organographie, Systematik und Biologie der Pflanzen. Mit einem Anhang: Die historische Entwicklung der Botanik*. Vienna: Hölder.

Wiesner, Julius. 1889. *Biologie der Pflanzen*. Vienna: Hölder.

Wiesner, Julius. 1891. *Organographie und Systematik der Pflanzen*. Vienna: Hölder.

Wiesner, Julius. 1892a. *Die Elementarstruktur und das Wachstum der lebenden Substanz.* Vienna: Hölder.

Wiesner, Julius. 1892b. Vorläufige Mitteilung über die Erscheinung der Exotrophie. *Berichte der Deutschen Botanischen Gesellschaft* 10:552–561.

Wiesner, Julius. 1910. Die Beziehungen der Pflanzenphysiologie zu den anderen Wissenschaften. In *Natur– Geist–Technik. Ausgewählte Reden, Vorträge, Essays*, ed. Julius Wiesner, 103–138. Leipzig: Engelmann.

Winninger, Salomon. 1932. Julius Wiesner. In *Große Jüdische National-Biographie* 6:282–283. Czernowitz: Orient.

Wurzbach, Constant von. 1888. Julius Wiesner. In *Biographisches Lexikon des Kaiserthums Österreich* 56. Bd.:88–93. Vienna: K.u.K. Hof- und Staatsdruckerei.

10 Wolfgang Pauli and Colloid Science at the Biologische Versuchsanstalt

Heiko Stoff

Colloid chemistry currently has the status of a normal science; it is globally institutional-ized and produces important knowledge for pure and applied chemistry. As Graeme K. Hunter outlines in a valid physicochemical definition, colloids are stable mixed phases in which one substance is nonhomogeneously dispersed within another (Hunter 2000, 156). But in the first three decades of the twentieth century colloid science was one of the most discussed new fields of chemical research, praised by its most outspoken protagonists as a new key to the secrets of life and the phenomena of vitality, by more moderate ones as a new way of doing electro- and physicochemical research. Skeptics denounced colloid science as mystical, speculative, and misleading a whole generation of chemists away from the path of classical and structural chemistry.

A colloid chemistry of organic and inorganic compounds was established in the mid-nineteenth century and joined in the 1890s with new electrochemical methods for research on protein solutions. Its renaissance around 1900 coincided with and rivaled the beginning of a highly successful era of structural chemistry and the isolation and standardization of biologically active substances such as small molecules. This is the reason we now have two histories of colloid science: one that highlights the continuity of a rather successful field of research, and another that focuses on the deficiencies of colloid science in com-parison to the triumphant history of molecular biology. Colloid science was not only in competition with biochemistry and natural products chemistry, but it also seemed to enable research on macromolecules like proteins, which until the 1920s was an insoluble task for classical chemistry. Whereas colloidal theory treated proteins as colloidal aggregates of a changing composition, as the historian of science Ute Deichmann points out, classical chemistry conceptualized large molecules with physical and chemical individualities. According to Deichmann, the controversy between the molecular and the colloidal concep-tion of the microstructure of cell components had the status of a "pre-history of what was later called molecular biology" (Deichmann 2007a, 105). But if the success story of molecular biology depended on the concept of structural chemistry, colloid chemistry, in the words of John W. Servos, would belong to a "history of failures" (Servos 1985, 139). According to this opinion, indeed rejected by Servos himself, colloid chemistry wasn't able to grasp the structural identity of proteins as macromolecules, and it was open to speculations about protoplasmic substance that were intermingled with and rather disguised by complex data. As Deichmann comments, colloidal chemistry seemed to neglect questions of structure and function for experimental research and theories (Deichmann 2007a, 105).

A persuasive argument for the importance of colloid science is that its main protagonists between 1890 and 1930, people like Wilder Bancroft, Eli Burton, Heinrich Bechhold, Herbert Freundlich, William B. Hardy, Franz Hofmeister, Jacques Loeb, Theodor Svedberg, and even Otto Warburg, were all highly respected scientists. Therefore Marcel Florkin's witticism of a "dark age of biocolloidology" has been rejected by most historians of science, because colloid chemistry indeed produced valuable knowledge in the field of physical chemistry (Florkin 1972; Bechtel 2006, 94). Any hasty condemnation of colloid chemistry in favor of classical chemistry would, as Joseph S. Fruton states, "dismiss the importance of the physical-chemical approach ... at a time when the chemical nature of the catalytic agents in enzymatic reactions was in doubt" (Fruton 1999, 158). Or to use William Bechtel's metaphor: "One avenue toward complexity was offered by colloid chemistry" (Bechtel 2006, 94). To denounce colloid chemistry as a rather ideological if not reactionary neo-vitalist pseudoscience misses out on the opportunity for historical analysis of the complex junction of technology, science, and approaches to life in the early twentieth century.

One of the most devoted and interesting colloid chemists was Wolfgang Pauli (figure 10.1), who between 1907 and 1914—a crucial time for colloid science—resided as head of the Physicochemical Department of the Biologische Versuchsanstalt (BVA, or Institute for Experimental Biology). During this period Pauli collaborated with Karl Landsteiner, who gained fame for distinguishing among the main blood groups, and distanced himself from classical chemistry, i.e., the school of Paul Ehrlich, a general concept of specific molecular entities, and the receptor theory. It is my aim to show not only how the concept of colloid chemistry developed and how colloid chemistry, as a new experimental branch of chemistry with the goal of developing novel explanations for vital life processes, fit into the BVA, but also how it merged speculations, data, and concepts with new techniques and experimental systems. Pauli combined the missionary stance of a prophet of a new way of chemical thinking with highly accurate experimental practices, based on the innovative method of electrophoresis, for studying the electrochemistry of proteins.

Pauli and Colloid Chemistry at the BVA

Wolfgang Pauli is nowadays better known for being the father of the physicist Wolfgang Ernst Pauli (or Wolfgang Pauli Jr.), the 1945 Nobel laureate for the exclusion principle and the spin theory. But "Wolfgang Pauli" wasn't even the original name of the elder

Figure 10.1
Wolfgang Joseph Pauli, head of the Physicochemical Department of the BVA from 1907 to 1914 and father of the Nobel Prize recipient in physics in 1945, Wolfgang Ernst Pauli. Courtesy of the Leopoldina Archiv, M1 3177.

scientist, who came from a Jewish family of publishers and booksellers and was born on September 11, 1869, in Prague as Wolfgang Josef (Wolf) Pascheles. Probably to foster his academic career in the anti-Semitic climate of Austria but also to enable a marriage with a Catholic woman, he changed his name to Wolfgang Josef Pauli in 1898 and converted to Catholicism. Pauli studied medicine at the Karl-Ferdinand University in Prague under the physicist and philosopher of science Ernst Mach and the physiologist Ewald Hering. He received his MD in April 1893. About this time he was invited to work at the Rudolf-spital in Vienna, where he stayed as a departmental assistant at the hospital until 1898. In 1899—shortly after he changed his name—Pauli earned his habilitation with research on swelling processes, gained the position of an assistant at the university clinic of Vienna, and finally married the intellectual Bertha Camilla Schütz, who wrote articles on different cultural, social, historical, and political—particularly feminist and pacifist—topics. In 1911 both converted to Protestantism. The children of this ingenious couple—the already mentioned physicist Wolfgang Ernst (1900–1958), but also the writer Hertha Pauli

(1906–1973)—proved to be highly creative themselves. Wolfgang Pauli Jr. got his second name from his godfather, Ernst Mach, the famous philosopher of science and a family friend of the senior Wolfgang Pauli. Bertha Pauli died in 1927, and Pauli later married the young artist Maria Rottler. In 1922 he was appointed as professor for biochemistry at the Institute for Medical Colloid Chemistry at the University of Vienna. When in March 1938 political power in Austria was handed over to the Nazis, Pauli left for Zurich as a guest at Paul Karrer's chemical institute of the University of Zurich. Pauli stayed in Zurich until his death, on November 4, 1955 (Enz 2010, 1–24; Jacobi 2000; Smutný 1990).

Under the influence of the Prague pharmacologist and protein expert Franz Hofmeister, Pauli in the mid-1890s gave up his career as a physician and switched to physical and colloid chemistry. He published his first research findings in the field of colloid chemistry, emphasizing that physical chemistry was in the service of medicine. In the first years of the new century he began collaborating not only with Hans Handovsky but also with Karl Landsteiner, who at the time famously worked on the identification of blood groups. Pauli focused his research on the parallelism of the colloid state in vitro and in the living organism, and the transitions between the living and the nonliving (Mazumdar 1995, 218). Both Landsteiner's and Pauli's research programs stated, in a rather revolutionary tone, that from now on physical chemistry had to deal with both the organic and inorganic. Life processes had to be examined with the methods of colloid chemistry. Living substances, Landsteiner explained, are mostly made up of colloids. The protoplasm itself is a variable system of colloids, always in a state of a versatile equilibrium and disturbed by intoxication (Dold 1922, 290; Mazumdar 1995, 218–220).[1]

In 1861 the British chemist Thomas Graham—Pauli called him the "the father of colloid chemistry"—had undertaken a fundamental distinction between colloids and crystalloids (Pauli 1922, 2). Graham identified substances marked by the absence of the power to crystallize, slow in the extreme, of low diffusibility and distinguished by the gelatinous character of their hydrates, as colloids. He defined the peculiar form of aggregation "as the colloidal condition of matter" and delineated a distinction "of intimate molecular constitution": colloidal condition as opposed to crystalline condition; the colloidal is a dynamical state of matter, while the crystalloidal is the statical condition (Graham 1861, 183–184). In the words of the biologist and skeptical biocolloid chemist Jacques Loeb, Graham characterized crystalloids by a tendency to form crystals when separating from a watery solution and colloids by a tendency to separate out in the form of "gelatinous" masses. Crystalloids and colloids differ in their diffusive mobility and in a peculiar physical aggregation: the crystalloids diffuse readily through different kinds of membranes through which colloids can diffuse not at all or only very slowly. Colloids form aggregates when in solution; this property is lacking or less pronounced in crystalloids (Loeb 1922, 1; Tanford and Reynolds 2001, 42–60).

Pauli, who distanced himself subtly from Graham's fundamental assertions, described Graham's differentiation of colloids and crystalloids as "a distinction as wide as that between an organized substance and a lifeless mineral" (Pauli 1922, 2). All three major colloidal chemists, Pauli, Bechhold, and Loeb, who otherwise often disagreed, qualified Graham's definition as an exaggeration. Crystalloids and colloids do not represent "two distinct worlds of matter," as Graham had proclaimed. In contrast there exist all sorts of transition stages, as Bechhold concluded; whereas Pauli stated that "all substances show a greater or less difference in the rate or ease with which they pass from the solid or liquid state into the state of vapour" (Bechhold 1919, 2; Pauli 1922, 2).

Colloidal Controversies from around 1900 to the 1920s

The major focus in the colloid science of the first decades of the twentieth century was the significance of conditions and states, of "variations in stability": "There is no fixed nomenclature for these changes, which lie on the border line of the physical and the chemical, and they are termed in general the alteration of state of colloids" (Pauli 1922, 16). Colloids were simply a state and not a type of matter (Hunter 2000, 155). While Graham was sure that the difference between colloids and crystalloids was based on their chemical nature, Pauli defined the colloid or crystalloid state rather as a "physical manifestation of a dependent condition of matter." Therefore the same chemical substance can be obtained, according to circumstances, with colloid or crystalloid characteristics (Pauli 1922, 3, 5). The main point of this new colloid chemistry was that depending on the solution, the same substance may behave either as a crystalloid or as a colloid. Loeb summarized with reference to the American chemist Eli Burton that the distinction between colloidal and crystalloidal substances was dropped in favor of a distinction between the colloidal and the crystalloidal state of matter (Loeb 1922, 275). Still, for colloid chemists there existed an important difference between filterable crystalloids and the "glue-like" colloids, which were held back by parchment membranes (Bechhold 1919, 3).

In 1902 Richard Zsigmondy and Henry Siedentopf, by using their highly innovative ultramicroscope, showed that there is indeed a difference between colloidal solutions, fine dispersions that pass easily through a filter, and true molecular solutions (Siedentopf, Zsigmondy 1902; Bechhold 1919, 6; Pauli 1922, 4). Pauli again defined colloid solutions as built or structured by the smallest, definite, and consistent particles in a solution whose size can be clearly measured. The size of the particles of a colloid solution has an upper boundary to preserve a consistent dispersion. Therefore colloid solutions must be differentiated from suspensions and emulsions. But, according to the laws of surface tension, there is also a boundary for the smallness of the colloid particles. This means that a colloid solution can be distinguished from an actually homogenous solution (Pauli 1908, 2; 1922, 5). According to Pauli there were two general methods of preparing a colloidal system: dispersion methods of breaking up large particles or aggregates (like electrical dispersion), and condensation methods (Pauli 1922, 4–5). Pauli differentiated between hydrated colloids and colloids with nonsolvated particles, as well as between colloids with neutral particles that are unstable in solution (e.g., suspensoids) and those with stable neutral particles (hydrated or solvated colloids) (Pauli 1922, 11, 13). In 1924, Pauli summarized that colloids could be measured with the ultracentrifuge as having a size between 1 and 100 micromicron They are bigger than crystalloids but smaller than raw dissipations. Colloids are either aggregates of molecules, and therefore products of condensation, or they are an atomization of raw mass, the product of dispersion, meaning they represent a process coming to a halt. But there also exist high-molecular compounds that from the outset are in a colloidal condition. These are proteins, some carbohydrates, starch, and glycogen (Pauli 1924a, 1).

Bechhold explained that many colloids "form with liquids, especially with water, a more or less fluid solution," called *sol* (a neologism which originates from "solution"). This sol could by various means be transformed into a more waterless form called *gel* (from "gelatin"). According to Bechhold, sols are substances that possess very large molecules, which are unable to pass through the pores of an animal skin or a parchment membrane. But this was not true for every substance. Some substances such as albumin, starch, and glycogen seemed to occur exclusively in the colloidal condition. In those substances, every further subdivision of the colloidally dissolved particles would have to be associated with a splitting of the molecule (Bechhold 1919, 2–3). For Pauli, therefore, a colloid system was "dispersion within certain limits of subdivision" (Pauli 1922, 10). Between colloids like albumin and crystalloids like amino acids there seemed to exist all kinds of transition forms, which pass through the same membranes more or less rapidly. As Bechhold, trying to provide the final word on the discussion, proclaimed: "There is, indeed, no sharp line of demarcation between colloids and crystalloids" (Bechhold 1919, 8).

Fundamental basic research in the field of colloid science took place in the first decade of the twentieth century and culminated around 1905 in a fierce debate on the chemistry of immunity reactions. Karl Landsteiner's "electrochemical theory of immune affinity" suggested that electrical charges constituted the affinity between antigen and antibody, and not Paul Ehrlich's receptors. Landsteiner, who was far from being a colloidal extremist but, after all, was a disciple of Max von Gruber, who was an opponent of Ehrlich's side-chain theory and the idea of chemical bonding (Rees 2014, 36–38; Silverstein 2009, 107–108; Deichmann 2007a, 108). Pauli admired Landsteiner's colloid-chemical interpretation of the antigen-antibody reaction and his immunological research on nonspecific component reactions and variations in charge. The two cooperated to disprove Ehrlich's concept of immunological specificity. Pauli and Landsteiner clearly felt that they stood at the beginning of a new era of chemistry, one that defined the problem of immunity in strict colloidal terms and heralded the close of structural or classical chemistry (Mazumdar 1995, 221–222). Bechhold, himself a central figure in colloid chemistry but also a disciple of Ehrlich, was not pleased with this attack on his teacher and defended organic chemistry against physical chemistry with Emil Fischer's unassailable lock-and-key argument. Pauli combatively remarked that antibodies are colloids and that their reactions show the same general laws as colloidal substances, especially the fact that precipitations need the presence of neutral salts (Pauli 1905b). Even if the colloidal standpoint in the case of immunology would be dropped in the 1920s, the collaboration of Pauli and Landsteiner left its mark on modern chemistry because it resulted in the further improvement of an electrophoretic apparatus that was of great importance for biochemical research in the twentieth century and enabled Landsteiner to separate protein from blood sera (Pauli 1922, 19–36; Fruton 1999, 228; Mazumdar 1995, 233–235, 236n70). Colloidal science was highly technical anyway; one might even say the success of colloidal chemistry in the early twentieth century was based on technical inventions, experimental practices, technological visualizations, and a whole new set of data, meticulously registered by Pauli.

Besides Pauli's and Landsteiner's improvement of electrophoresis, also notable was Zsigmondy and Bechhold's linking of the ultramicroscope to colloidal research and Wolfgang Ostwald's viscosimeter, which served as a means to measure the viscosity of suspensions and solutions. Finally, the data Theodor Svedberg gained by the use of the ultracentrifuge put an end to a colloidal theory of proteins. Colloid chemistry was a technical problem; technical inventions like filtration methods, the ultramicroscope, electrodialysis, and electrophoresis were deeply intertwined with colloidal research (Deichmann, 2001, 260). In his chapter on methods of colloid research, Bechhold included, among others, dialysis, ultrafiltration, gauging of ultrafilters, adsorption by filters, diffusion in aqueous solution, osmotic compensation methods, interferometry, ultramicroscopy for colloidal solution, and ultramicroscopy for organized material (Bechhold 1919, 89–126).

Pauline M. H. Mazumdar has outlined in great detail the scientific battle that took place in 1905 between Bechhold and Pauli and that led to a sharp distinction between structural and colloidal chemistry. Mazumdar concludes that for Pauli everything in nature could be explained by electrochemistry, for Landsteiner electrochemical forces were responsible for specific affinity, and for the Ehrlich group only precipitation could be explained by colloid chemistry. Pauli and Landsteiner neglected the role of chemical structure, while Bechhold upheld it as a central concept (Mazumdar 1995, 228–230). As Deichmann summarizes, "Pauli considered Ehrlich's structural chemistry to be outdated and claimed that there was a colloid-chemical explanation for every single phenomenon in biology and medicine, including immunology" (Deichmann 2007a, 108). But it shouldn't be neglected that Bechhold himself qualified Ehrlich's receptor theory in proclaiming that the substances involved in immunity reactions are all dissolved or suspended colloids. This was easy to say, because it had so far been impossible to produce immune bodies by means of a crystalloid. In 1919 Bechhold closed his argument with the authoritative words that "the proof of the colloid character of antigens and immune bodies has been demonstrated in numerous cases," and that according to Pauli and Landsteiner "it is well to regard antigens and antibodies as amphoteric electrolytes" (Bechhold 1919, 196, 205; Fruton 1972, 140–144).

Until the early 1920s not many scientists really doubted the "Landsteiner-Pauli model." An important exception was Leonor Michaelis, who clearly objected to the idea that specific affinity is chemical affinity independent of electrical affinity. But in 1922 the hygienist Hermann Dold could still speak of a colloidal school of immunity that included among others Landsteiner, Pauli, and Friedmann, as well as Michaelis and Bechhold (Dold 1922, 291; Rees 2014, 38). While Pauli dropped the subject of immunology as a colloidal effect in the 1910s, Landsteiner, now influenced by the experiments of Ernst Peter Pick and Friedrich Obermayer in producing chemically modified immunizing serum, in the 1920s generally gave up the colloidal standpoint and elaborated a concept of the "chemical specificity of the antigen-antibody reaction" based on a "a combination of various weak bonds." This in some ways slowly rehabilitated Ehrlich's theory of chemical bonds between antigen and antibody and strengthened ideas of molecules and specificity rather than colloids and nonspecificity, even if Pick himself had insisted on a colloidal explanation of his experiments (Deichmann 2007a, 108, Rees 2014, 38–40).

Pauli's Department at the BVA

Exactly at the climax of this controversy, in March 1907, the physiologist Sigmund Exner made an application for the approval of a physicochemical department at the BVA under the direction of Pauli. There was no word of colloid science, but Exner highlighted the value of physical chemistry for pharmacology and physiology.[2] The approval was granted, and the 38-year-old Pauli was without further ado appointed director of the Physicochemical Department, which was opened the next month and soon housed around 20 scientists and 10 students. The new department was supposed to collaborate with the Zoology Department and the Department of Botany and Plant Physiology. Indeed, Pauli discussed with Leonore Brecher from the Zoology Department parts of her important work on the coloring of pupae (Brecher 1922, 258; see also chapter 8 in this volume). The department's main fields of work were physical chemistry, the chemistry of biocolloids, and the application of physical and chemical methods for biological problems. Once it had been established on April 1, 1907, students of Vienna University were able to do a practical course at the Physicochemical Department, and a work space could be offered each year to a student from the Faculty of Medicine.

Hans Handovsky, Johann Matula, and the remarkable Max Samec, who besides being an expert in the colloid chemistry of starch was also an aviation pioneer (Ullmann 1961), are among those who made their academic careers in the field. The case of Handovsky—who was one of Pauli's earliest assistants, held the position of a professor for pharmacology at the University of Göttingen from 1926 to 1933, and was the author of an important textbook on colloid chemistry in 1923—was rather tragic (Handovsky 1923). Handovsky had to leave Germany with the rise to power of National Socialism and worked under very difficult circumstances at the University of Gent in Belgium, where he was again dismissed for anti-Semitic reasons (Louis and Verschooris 2012, 128; Szabó 2000, 398). He returned to Göttingen in the 1950s but never regained his former status (Szabó 2000, 399–403, 443–444, 507; Schäfer-Richter and Klein 1992, 90–91).

The main purpose of the department was to combine colloid science with physical chemistry and to apply physical and chemical methods for biological problems, especially in regard to the research undertaken at the departments of zoology and botany. The department itself had five working rooms, and Pauli himself took care of the arrangement.[3] Pauli's years at the BVA were highly productive, even if it seems that the directorship for him was just a stopover to a university position. With his assistants—besides Handovsky, Matula, and Samec there were also Oskar Falek, Leo Flecker, Erwin Strauss, and Regina Meller—he experimentally provoked fundamental chemical changes with proteins (albumin) through acid, base, and salt and published a huge amount of articles in professional journals, thereby sketching a colloid theory of proteins. Together with Handovsky he elaborated in numerous experimental investigations a new electrochemical colloid science of proteins. Both, during their time at the BVA, especially focused on the influence of alkali and acid on proteins (Pauli 1910; Fruton 1999, 158).

Pauli's textbook on the colloid chemistry of proteins, published in German in 1921 and in English in 1922, was actually based on his lectures of the years 1912 and 1913 (Pauli 1921, Pauli 1922). His highly accurate working style was praised by such admirers as the Swiss chemist and Nobel laureate Paul Karrer, but dismissed as pedantic and hiding a lack of method by skeptics like Michaelis (Deichmann 2007b, 120; von Meyenn 2008, 398). It seems that by around 1913 Pauli had already made plans to leave the BVA and start a university career. He had been appointed as honorary professor in 1907, which gave him one foot in the door of Vienna University. In June 1914 he wrote to Friedrich Becke, then general secretary of the BVA, that due to the worsened material situation of his department, the loss of time because of the long distance from his apartment to the institute, and finally his increased duties as a lecturer for physicochemical biology, he had to give up the position of director at the BVA. These reasons were rather trivial, and Pauli anticipated criticism by noting that only *after* his resignation did he get the chance to take on the directorship at the laboratory for physicochemical biology at the University of Vienna.[4]

The Austrian Academy of Science indeed had been tough on the directors at the BVA. In 1910 Pauli obtained 1,000 Kronen from the Academy to purchase apparatuses for research on the physical change of states of biocolloids. Three years later he ordered a water interferometer for the amount of 900 Kronen. And that was it. The financing of apparatuses was crucial for his experiments on colloid chemistry, and it is doubtful that the BVA really could have satisfied his needs (Anzeiger der kaiserlichen Akademie der

Wissenschaften in Wien 1910, 24).[5] Also, departments heads like Steinach and Pauli weren't allowed to do independent research, and their departments weren't as well equipped as those of zoology and botany. While Steinach received an additional 1,000 Kronen and therefore stayed at the BVA, Pauli left in 1914 (Walch 2016, 59–61).

The BVA was the place to be for new scientific and experimental approaches, and for theories of life associated with new ways of forming life. No wonder that colloid chemistry, which promised a "new chemistry," was part of it. In particular the application of the developmental-mechanical concept of the plasticity of the human material corresponded to Graham's assumption that the peculiar physical aggregation of colloids is essential to substances that can intervene in the organic processes of life and form "the plastic elements of the animal body" (Graham 1861, 183; Mocek 1998, 360). In his well-received 1905 festival address, "Wandlungen in der Pathologie durch die Fortschritte der allgemeinen Chemie," Pauli installed colloid chemistry as a new chemical master science (Pauli 1905a; Mazumdar 1995, 227–228). According to the colloid chemical doctrine, colloids are of much wider distribution and fundamental importance for life itself than crystalloids (Bechhold 1919, 3). Therefore, the much-debated question, whether proteins are crystalloids or colloids, was of central importance for the significance of colloid chemistry as a science.

Proteins, Biocolloids, and Micelles

In 1908 Pauli stated simply that the fact that proteins are of colloidal nature had been proved by their difficult permeation through dialysis membranes, their capacity for diffusion, and their actions observed in the ultramicroscope (Pauli 1908, 2). According to Pauli, who referred to Svedberg's kinetic studies, there is no sharp demarcation between the general chemistry of substances of high molecular weight like proteins and their colloid chemistry (Pauli 1922, 8). In the early twentieth century it was accepted knowledge that proteins held the central position in the organization of living matter. Pauli emphasized that they occur in nature in close connection with vital processes. They are completely irreplaceable in the living cell, and they alone display the specific properties of living matter. But while proteins have high molecular weight, physical changes are brought about by minute quantities of substances of low molecular weight, notably electrolytes (Pauli 1922, 1). According to colloid theory, the addition of bases, acids, and salts influences the electric charge of proteins, which can be made visible by their motion in an electric field (i.e., via electrophoresis) (Pauli 1906, 104; 1908, 3; 1922, 12, 14–15, 19; Deichmann 2001, 260). Proteins seemed to be colloids that exist as charged particles in solutions. While this field of research was highly interesting for mainstream chemists, it also led, as Pauli admitted, "into a field of various conflicting theoretical speculations." The "purely ionic conception" was soon to be in conflict with structural conceptions of proteins (Pauli 1922, 14).

Pauli interpreted his experimental findings and gathered data to the effect that some proteins are rather unstable hydrophobes, while others have hydrated particles and are stable in solution even when the particles are not electrically charged. But electric charge again is of the highest importance for the stability of the lyophile or hydrated proteins. The alteration of the state of colloids corresponds to variations in stability, that is, to reversible or irreversible coagulations (Pauli 1922, 16–17). With the electrophoresis apparatus Pauli and Landsteiner showed that proteins regularly indicate an electro-negative charge when in the natural position. Michaelis's electrophoresis experiments exposed that a certain acid concentration positively charged the negatively charged albumin particles. Therefore at a certain acid concentration "an indifferent point in the sign of the charge will occur" and albumin shows no electrophoresis in either direction. William B. Hardy had named this moment when there is no difference between the particles and the medium an "isoelectric point" (Pauli 1922, 22–23; Morgan 1990, 148; Fruton 1999, 204). To understand the alterations in state of the proteins, Pauli concluded, it is of great importance to have an accurate knowledge of their dissociation when in the isoelectric state. The alterations in properties which natural proteins display in the isoelectric region therefore can be explained by the relation between the alterations in natural proteins and their electric charge (Pauli 1922, 29, 37).

While it seems that between 1900 and 1920 Pauli worked exclusively on a colloid theory of proteins, unifying colloid chemistry and the physical chemistry of proteins, he also emphasized "every tendency to a relation between their colloid chemistry and their structure" (Pauli 1922, 9; 1920, 912). Most colloid scientists were much more scrupulous and interested in questions raised by electrochemical research than their critics apprehended.

Wolfgang Ostwald, son of the physical chemist and Nobel laureate Wilhelm Ostwald, was the most outspoken representative of colloid chemistry in Germany. During a lecture tour in North America in the winter of 1913/14, he spread the news of a revolutionary new way of chemical thinking, later published by Martin Fischer as *An Introduction to Theoretical and Applied Colloid Chemistry: The World of Neglected Dimensions* (Ostwald 1917; Servos 1985, 139; Servos 1990, 53–70). This monograph, in which Ostwald tried to explain in an accessible manner the rather complicated way of thinking that characterized colloid science, represented the state of the art in the science of biocolloids in the 1910s (Deichmann 2007a, 107). The phrase "world of neglected dimensions" thereby referred to the colloidal microsphere, which supposedly was the venue of all life processes (Olby 1985; Hunter 2000, 156). Ostwald was a mere propagandist of colloid chemistry who, like so many other scientists in the early twentieth century, merged scientific innovations with a new approach to life itself. But he also left his mark on colloid science in coining the terms "dispersion medium" and "disperse phase" for the component parts of a colloidal system. He devised a classification based on the state of aggregation of the disperse phase and differentiated between suspensoids (a solid disperse phase) and emulsoids (a disperse phase with liquid particles) (Pauli 1922, 10). Colloidal chemistry was not only a rather specialized, if not esoteric, new field of chemical research but also a new universal science of life, and Ostwald was its main herald. According to the historian of medicine Heiner Fangerau, colloid chemistry combined vitalistic and holistic with mechanist and reductionist concepts, modern chemistry with romantic ideas of the living (Fangerau 2010, 38–39). Pauli himself can be called a "mechanistic vitalist" (Mazumdar 1995, 218). The vague concept of colloids itself was open enough to make for a popular science of colloid chemistry (Hunter 2000, 156–157).

It was Carl Nägeli's mid-nineteenth century theory of micelles, the micella-hypothesis (*Mizellarhypothese*) that stood at the center of colloidal speculation. James William McBain, from the University of Bristol, transferred the concept into modern colloid science in 1913 (Vincent 2014). Loeb introduced his 1922 monograph *Proteins and the Theory of Colloidal Behaviour* by stating that colloid chemistry had been developed on the assumption that the ultimate unit in colloidal solutions is not the isolated molecule or ion but an aggregate of molecules or ions, the so-called micelle. Thereafter, in a colloid system micelles are "surrounded by hydration shells" and play a fundamental role in the protoplasm (Loeb 1922, v, 2). If protoplasm as a subcellular substance is indeed built of colloids, colloid chemistry must also give answers to the fundamental questions of life. Most colloid chemists combined their generally accurate findings with ambitious statements about life itself.

When Wolfgang Ostwald concluded that living matter seemed to be a meeting ground for adsorption effects and colloid catalysis, he even joined two of the most important concepts of the new chemical approach to life: colloids and catalysis (Ostwald 1917, 162). According to Ostwald's syllogism, catalysis is a characteristic of colloids, and therefore enzymes must be colloids (Hunter 2000, 156). For Bechhold, carbohydrates, lipoids, proteins, food, and condiments, as well as enzymes, were all colloids. The organism as a whole appeared as a "colloidal system" (Deichmann 2007a, 107). Pauli described the living substance as a complex of different dissolved or swelled matters. Every change in the functional actions of the living substance corresponds to changes in the physical-chemical characteristics, which are just manifestations of biocolloidal changes of state (Pauli 1906, 102; 1910, 483). He also recapitulated that the variations of proteins reflect the distinctions between different kinds of organisms and between individuals of the same kind, and stated that all cellular processes were in close relation to colloidal changes of state: the living substance is a complex of different dissolved or swelled and interacting materials—of lecithin, cholesterol, proteins, enzymes, and certain salts—in a specific relation. Besides these intracellular processes, extracellular intermediate substances (*Zwischensubstanzen*) play an eminent role in vital processes for higher organisms. Therefore, Pauli proclaimed, colloid science is an important aspect of physiology (Pauli 1906, 101).

One of Pauli's main arguments was that the colloid reaction of proteins (the changes of states of these substances) accompanies all manifestations of life. And not without a tinge of pathos, he pondered that through the colloid chemistry of proteins a straight path leads to the dark territory, which the researcher touches only with a silent longing like a land of promise: the physical chemistry of living substance (Pauli 1908, 13; 1922, 1). The most important life processes seemed to take place in a hitherto neglected and rather invisible microstructural sphere. Colloid chemists in this regard had a privileged access to the secrets of life. According to Pauline M. H. Mazumdar, Pauli "presented colloid chemistry as the new explanation of almost everything in medicine and biology" (Pauli 1922, 126; Mazumdar 1995, 227).

Colloid Science and Vital Processes

Pauli was well aware that—as with the new physiology or the new dietetics—it was the right time for scientific revolutions in the field of biomedicine, for the replacement of old knowledge with new facts. Therefore he constituted colloid chemistry as a change of paradigm, as an independent branch of science based on a number of observations that stood "in strong contrast to previous records" (Pauli 1922, 2). In the same tone Ostwald tried to establish "the right of modern colloid chemistry to existence as a separate and independent science." He proudly announced that colloid chemistry had already become so vast "that no one can master it in its entirety" (Ostwald 1917, xi, 3; Ede 2007, 21). Colloid science was inaugurated as a new master science, a promising approach to the secrets of life, and a new discipline of chemistry. In the first two decades of the twentieth century, at least in Germany, colloid science outdid biochemistry in institutionalization (professional journals, professorships) and academic discussion. This is what Florkin, with a tendency to scapegoat colloid chemistry, called the "dark age of biocolloidology" (Florkin 1972).

Around 1900 colloid science was no exception if it aroused miraculous expectations. Indeed, the BVA was the worldwide epicenter for the promise of a cure for all: for rejuvenation and improved performance, transplantations, Lamarckism, eugenics, and more. And maybe it should be recalled that even in the 1950s the research of Feodor Lynen, later a Nobel laureate, on coenzyme A was embedded in the rather unscientific hopes of finding a magic bullet (Stoff 2004, 392–404; 2012, 165–170). In 1921, an authority like Svedberg—five years before he switched sides after gaining data through the use of the ultracentrifuge, which demonstrated that proteins were very large macromolecules of high molecular weight—still would start an article on colloids with this remarkable praise: "All living beings are built up of colloids; almost all our food, our articles of clothing, our building materials, are colloids. Or, to mention some special systems, protoplasm, proteins, glue, starch, all kinds of fibres, wood, brick, mortar, cement, certain kinds of glass, rubber, celluloid, etc. The importance of colloid science for many industrial questions is, therefore, beyond all doubt" (Svedberg 1921, 2).

Colloid science was another promising junction for science and industry; the latter around 1900 was already well attuned to organic and biochemistry, but especially to natural products chemistry. So it is no wonder—or, rather, normal, as Ute Deichmann highlights with critical undertones—that in Germany the institutionalization of colloid chemistry at universities was in part funded by industry (Deichmann 2007a, 107). But even if most of the colloidal chemists referred to far-reaching speculations about living substance, in general they stuck to highly specific problems. Pauli especially reached deep into physicochemical questions: the emancipation of proteins from electrolytes, the relation between proteins and salts, electrical conductivity, and the precipitation of proteins (Pauli 1906, 102). While during the course of the 1910s colloid chemistry became more and more institutionalized, it was also increasingly criticized. In particular, Søren Sørensen, Leonor Michaelis, and Jacques Loeb rejected Ostwald's and Pauli's speculations as vitalistic and metaphysical explanations (Deichmann 2007a, 107, 109–110).

In his obituary for Loeb in 1924, the colloid chemist Herbert Freundlich underlined that Loeb had sensed in colloid science a return to something mystical. Loeb disliked that, instead of the clear concepts of chemical affinity or osmotic pressure, researchers like Bechhold, Ostwald, and Pauli used terms like *Oberflächenkräfte* or *Quellungsdruck* (Freundlich 1924, 602). Loeb didn't doubt the existence of colloids themselves; in general he appreciated colloid science, but neglected the extrapolation of a specific science of colloids, any priority of colloids over crystalloids, the colloid chemistry of proteins, and the idea of colloidal association: "As long as chemists continue to believe in the existence of a special colloid chemistry differing from the chemistry of crystalloids, it will remain impossible to explain the physical behavior of colloids in general and of proteins in particular" (Loeb 1922, v). Neil Morgan states, in his balanced article on the relation between biochemistry and colloid chemistry, that "Loeb placed the electrical properties of proteins on a classical, that is to say stoichiometric, footing" (Morgan 1990, 147). Loeb opposed the idea that colloids form aggregates as the fundamental property. For him, colloids were clearly defined molecules that could be described with the proper methods of organic or physical chemistry. Colloids therefore were protein solutions, whose chemical actions followed stoichiometric laws (Loeb 1922, 2; Freundlich 1924, 603). Loeb denounced one of the central dogmas of colloid science—Freundlich's adsorption formula, which stated that electrolytes were adsorbed on the surface of colloidal particles according to a purely empirical formula—as a methodical error. He was convinced that proteins combine with acids, alkalis, or salts according to the stoichiometric laws of classical chemistry and that the chemistry of proteins does not differ from the chemistry of crystalloids (Loeb 1922, v).

On the basis of his own research Loeb came to the conclusion that "the parallelism between the concentration of protein ions and the physical properties of proteins demanded by Pauli's theory could not be demonstrated." Throughout his monograph Loeb repeatedly criticized Pauli's hydration theory, but he never attacked him as unscientific (Loeb 1922, 19). In concrete terms he doubted Pauli's assertion that the viscosity of protein solution depends primarily upon the protein ion. For Pauli each protein ion is hydrated, and each individual protein ion is surrounded by a considerable shell of water. Loeb strikingly disproved Pauli's assumption that the effect of alterations to pH value on the variable swelling and viscosity of a protein solution had something to do with the ionization and hydration of proteins (Loeb 1922, 18; Fangerau 2010, 38–40, 113, 147). Even if Loeb never rejected Pauli, whom he accepted personally and scientifically (much as he did his former pupil Ostwald, whom he despised as a nationalist and reactionary fraud) Loeb nevertheless questioned the validity of fundamental assumptions underlying Pauli's colloidal concept (Hunter 2000, 158; Deichmann 2007a, 110; Fangerau 2010, 40, 113).

When Svedberg praised the importance of colloid science in 1921, he also pointed out that colloid chemists were genuinely interested in research on the problem of the microstructure of matter and the problem of structure in its full extent: "In the great science of the structure of matter, the science of colloids forms the domain that lies above molecular dimensions and beneath macroscopic dimensions. In this domain we have a great number of those systems which are the basis of our material culture and the basis of life as a whole" (Svedberg 1921, 2). Colloid science raised the right questions but tended to insist on wrong answers in the controversies over antibodies and proteins. In the early 1920s the problem of the status of enzymes and proteins was at the top of the scientific agenda. In Germany, enzymes, together with hormones and vitamins, were defined as biologically active substances, even if their chemical structure was enigmatic (Stoff 2012, 152–170). In contrast to hormones and vitamins, which were defined in terms of their biophysiologic effects, enzymes were chemical agents and discussed in the context of the debate on the chemistry of life (Fruton 1999; Kornberg 1989; Kohler 1973, 1982).

While the history of enzymes is written in terms of debates about vitalism and mechanism, the interplay of regulation, specificity, and catalysis is crucial to the concept of biologically active substances. Emil Fischer's famous and momentous analogy, which constituted the relationship between enzyme and substrate in terms of a lock-and-key model, established the principal of complementarity as a highly productive hypothesis for enzymatic reactions. The axiom of catalytic specificity linked structure to efficiency. According to Alwin Mittasch, the main theoretician of the concept of catalysis in Germany in the 1920s and 1930s, specificity and selectivity meant the reduction of a certain substance to a certain performance (Mittasch 1936). When in 1926 James B. Sumner with a rather simple procedure isolated urease as a protein, he was met with disbelief by the German experts who were partial to a colloidal theory of enzymes and followed Richard Willstätter's axiom of a catalysis free of proteins. Michaelis's doubts about the effect of an ionic environment and his research concerning the relation between enzyme and substrate; the successful isolation of enzymes by Sumners and John Howard Northrop; and last but not least Loeb's monograph *Proteins and the Theory of Colloidal Behaviour* all put colloid chemistry in perspective in the early 1920s. Svedberg's research on the macromolecular nature of proteins and Hermann Staudinger's proof, in the second half of the 1920s, that macromolecules were not aggregates but bondings of molecules finally discredited the role of colloidal chemistry for the structure of proteins, which finally were shown to be well-defined physical entities and not mixtures of particles (Fruton 1972, 138; Deichmann 2001, 243–255, 259–265; 2007a, 110–111).

Loeb's proof of the stoichiometric character of the reactions of proteins and his mathematical and quantitative theory of colloidal behavior, on the basis of Frederick George Donnan's theory of membrane equilibria, were thus a rejection of colloidal theory with regard to proteins (Loeb 1922, vi). All the characteristics of colloidal behavior, according to Loeb, could be explained mathematically from the difference in diffusibility between colloids and crystalloids (Loeb 1922, 2). New methods of measuring hydrogen ion concentration proved that, concerning proteins, the chemistry of colloids did not differ from the chemistry of crystalloids. According to Loeb, who offered a purely chemical and mathematical view of the significance of the isoelectric point and of the cause of the influence of acids and alkalis on the direction of the migration of colloidal particles, the development of an exact theory of colloidal behavior was prevented by the methodical error of failing to measure the hydrogen ion concentration of colloidal solutions and of gels (Loeb 1922, 5, 7). Even if Loeb introduced another paper in 1923 with the simple statement that living substance is of colloidal nature, colloid chemistry was by then old–fashioned, having been ousted by a biochemistry based on molecules (Loeb 1923, 21). While in the 1940s biochemistry, molecular biology, and polymer chemistry took their places as new master sciences, colloid chemistry—as economically and industrially important as it was—regressed into a subdiscipline of chemistry.

Conclusion

Pauli, who had focused his whole academic career on the new science of colloid chemistry, tried to adapt his colloid concept to his colleagues' objections (Pauli 1929, 679). In the late 1920s he started to classify colloid chemistry in a history of new concepts and ideas in general: As a rather new science, colloid science grew fast in a typical way—the terms fermented, and the adoption of thoughts as facts was premature. First, differences were overstated, then they were bypassed, and finally reconciled with a new consensus (Pauli 1924b, 421). Colloid chemistry was special, he insisted, but it had to be affiliated with chemistry, complex chemistry, and crystal physics (Pauli 1924c, 557). From the late 1920s, Pauli would not hold colloid science in an untenable position, so he adjusted his concept to accord with the now-strong position of structural chemistry. He therefore reverted to the main questions of experimental research. The historian of science Andrew Ede, author of a book on the history of colloid science in the United States, elaborated that while the scope claimed for colloid science covered almost every aspect of the physical world, in practice research was carried out only in a few areas: adsorption, electrical behavior, cell function, and the structure of proteins (Ede 2007, 177). While any colloid science concerning the structure of protein and antibodies was discredited in the 1920s, colloid chemistry in general remained as an important but not a spectacular chemical discipline. And in this rather comfortable and productive niche, colloid chemistry has survived until today.

Pauli's time as head of the Physicochemical Department at the BVA was short—he stayed for just seven years—but crucial for the history of colloid research. The important dispute about Ehrlich's receptor theory as well as the collaboration with Landsteiner took place during his stay at the BVA. But the department wasn't called the "Institute for Colloidal Science"; Pauli was hired as an expert in physicochemistry, and it was expected that he would collaborate with the Zoology Department and the Department of Botany and Plant Physiology. Exner had commended Pauli's research as having value for pharmacology and physiology, but it seems that in this regard, he didn't come up to expectations. On the other hand, the department's rather poor equipment must have been an obstacle for Pauli's own experimental studies. Colloid research was based on high technology, and Pauli needed expensive apparatuses for his electrochemical experiments. While it is hard to find evidence that colloid science played a crucial role at the BVA, it fitted well into the institute's field of research.

As a new chemical master science (as Pauli called it two years before he acceded to the Physicochemical Department) it seemed to give answers to the vital questions of life, especially to the main topic of developmental mechanics: the plasticity of the elements of the animal body. Pauli's chief activity during his time at the department, therefore, was to find evidence and data for the colloidal worldview. To achieve this goal he indeed established a well-trained working group, consisting of excellent researchers like Handovsky and Samec. Like other research topics at the BVA that were surrounded by mystery or even accusations of fraud, including Paul Kammerer's case of the midwife toad and Walter Finkler's head transplants with insects, colloid science had to be cleaned of speculations to retain the status of a respected but far from new science. But history has shown that Pauli was unable to achieve a compromise of his chemical research with the expectations of a new concept of life.

Notes

1. A thesis which, since the 1930s, was in accordance with other developing theories, like Hans Selye's stress model and a virulent concept of poisoning noxa (Stoff 2015).

2. Akten18, Chapter 2. BVA Chronologie, 1903{1948 Vermerk 10325. Oesta. AVA. Phil: Biol. Versuchsstation, 1902{1919. Ktn. 128. Sign. 2A.

3. Aktenvermerk 2877, 11802, 11803. Oesta. AVA. Phil: Biol. Versuchsstation, 1902{1919. Ktn. 128. Sign. 2A.

4. 06.06.1914 letter from Pauli to Friedrich Becke, Secretary General of the Austrian Academy of Sciences, Wien ÖAW Archiv. "Biologische Versuchsanstalt (Vivarium)." G 1914. K3./M1.

5. See also the archival material on "Subventionen" at the Österreichische Akademie der Wissenschaften (http://www.oeaw.ac.at/basis/).

References

Anzeiger der kaiserlichen Akademie der Wissenschaften, Mathematisch-naturwissenschaftliche Classe. 1910.

Bechhold, Heinrich. 1919. *Colloids in Biology and Medicine*. New York: Van Nostrand.

Bechtel, William. 2006. *Discovering Cell Mechanisms: The Creation of Modern Cell Biology*. Cambridge: Cambridge University Press.

Brecher, Leonore. 1922. Die Puppenfärbungen der Vanessiden (*Vanessa io, V. urticae, Pyrameis cardui, P. atalanta*). *Archiv für Entwicklungsmechanik* 50 (1): 209–308.

Deichmann, Ute. 2001. *Flüchten, Mitmachen, Vergessen: Chemiker und Biochemiker in der NS-Zeit*. Weinheim: Wiley/VCH.

Deichmann, Ute. 2007a. "Molecular" versus "colloidal": Controversies in biology and biochemistry, 1900–1940. *Bulletin for the History of Chemistry* 32 (2): 105–118.

Deichmann, Ute. 2007b. "I detest his way of working": Leonor Michaelis (1875–1949), Emil Abderhalden (1877–1950), and Jewish and non-Jewish biochemists in Germany. In *Jews and Sciences in German Contexts: Case Studies from the 19th and 20th Centuries*, ed. Ulrich Charpa and Ute Deichmann, 101–126. Tübingen: Mohr-Siebeck.

Dold, Hermann. 1922. Kolloidchemie und Immunitätsforschung. *Kolloid-Zeitschrift* 31 (5): 290–292.

Ede, Andrew. 2007. *The Rise and Decline of Colloid Science in North America, 1900–1935: The Neglected Dimension*. Farnham: Ashgate.

Enz, Charles P. 2010. *No Time to Be Brief: A Scientific Biography of Wolfgang Pauli*. Oxford: Oxford University Press.

Fangerau, Heiner. 2010. *Spinning the Scientific Web: Jacques Loeb (1859–1924) und sein Programm einer internationalen biomedizinischen Grundlagenforschung*. Berlin: Akademie.

Florkin, Marcel. 1972. The dark age of biocolloidology. In *Comprehensive Biochemistry*, vol. 30, ed. Marcel Florkin and Elmer H. Stoltz, 279–283. New York, Amsterdam: Elsevier.

Freundlich, Herbert. 1924. Jacques Loeb und die Kolloidchemie. *Naturwissenschaften* 12 (30): 602–603.

Fruton, Joseph S. 1972. *Molecules and Life. Historical Essays on the Interplay of Chemistry and Biology*. New York, London: Wiley.

Fruton, Joseph S. 1999. *Proteins, Enzymes, Genes. The Interplay of Chemistry and Biology*. New Haven, London: Yale University Press.

Graham, Thomas. 1861. Liquid diffusion applied to analysis. *Philosophical Transactions of the Royal Society of London* 151:183–224.

Handovsky, Hans. 1923. *Grundbegriffe der Kolloidchemie und ihrer Anwendung in Biologie und Medizin*. Berlin: Springer.

Hunter, Graeme K. 2000. *Vital Forces: The Discovery of the Molecular Basis of Life*. London, San Diego: Academic Press.

Jacobi, Manfred. 2000. "Von antimetaphysischer Herkunft": Zum 100. Geburtstag von Wolfgang Ernst Pauli. *Physikalische Blätter* 56 (4): 57–60.

Kohler, Robert E. 1973. The enzyme theory and the origin of biochemistry. *Isis* 64:181–196.

Kohler, Robert E. 1982. *From Medical Chemistry to Biochemistry*. Cambridge: Cambridge University Press.

Kornberg, Arthur. 1989. *For the Love of Enzymes: The Odyssey of a Biochemist*. Cambridge: Cambridge University Press.

Loeb, Jacques. 1922. *Proteins and the Theory of Colloidal Behaviour*. New York: McGraw-Hill.

Loeb, Jacques. 1923. Die Erklärung für das kolloidale Verhalten der Eiweißkörper. *Naturwissenschaften* 11 (12): 213–221.

Louis, Yves, and Marc Verschooris. 2012. Leonardo Conti et ses rapports avec les médecins belges pendant la Seconde Guerre mondiale. *Témoigner– Entre Histoire et Mémoire* 112:124–136.

Mazumdar, Pauline M. H. 1995. *Species and Specificity: An Interpretation of the History of Immunology*. Cambridge: Cambridge University Press.

Mittasch, Alwin. 1936. Über Katalyse und Katalysatoren in Chemie und Biologie. *Naturwissenschaften* 24 (49): 770–777, 785–790.

Mocek, Reinhard. 1998. *Die werdende Form. Eine Geschichte der kausalen Morphologie*. Marburg: Basilisken-Presse.

Morgan, Neil. 1990. The strategy of biological research programmes: Reassessing the "dark age" of biochemistry, 1910–1930. *Annals of Science* 47 (2): 139–150.

Olby, Robert C. 1985. Structural and dynamical explanations in the world of neglected dimensions. In *A History of Embryology*, ed. T. J. Horder, J. A. Witkowski, and C. C. Wylie, 275–308. Cambridge: Cambridge University Press.

Ostwald, Wolfgang. 1917. *An Introduction to Theoretical and Applied Colloid Chemistry: The World of Neglected Dimensions*. New York: Wiley.

Pauli, Wolfgang. 1905a. *Wandlungen in der Pathologie durch die Fortschritte der allgemeinen Chemie. Festvortrag, gehalten in der Jahressitzung der k.k. Gesellschaft der Ärzte in Wien am 24. März 1905*. Wien: Perles.

Pauli, Wolfgang. 1905b. Ueber den Anteil der Kolloidchemie an der Immunitätsforschung. *Wiener Klinische Wochenschrift* 18:665–666.

Pauli, Wolfgang. 1906. Beziehungen der Kolloidchemie zur Physiologie. *Zeitschrift für Chemie und Industrie der Kolloide* 1 (4): 101–107.

Pauli, Wolfgang. 1908. Kolloidchemische Studien am Eiweiss. *Zeitschrift für Chemie und Industrie der Kolloide* 3 (1): 2–13.

Pauli, Wolfgang. 1910. Die kolloiden Zustandsänderungen von Eiweiss und ihre physiologische Bedeutung. *Pflügers Archiv für die Gesamte Physiologie des Menschen und der Tiere* 136 (1): 483–501.

Pauli, Wolfgang. 1920. Fortschritte der allgemeinen Eiweißchemie. *Naturwissenschaften* 8 (47): 911–917.

Pauli, Wolfgang. 1921. *Kolloidchemie der Eiweißkörper*. Dresden, Leipzig: Steinkopff.

Pauli, Wolfgang. 1922. *Colloid Chemistry of the Proteins*. Philadelphia: Blakiston's.

Pauli, Wolfgang. 1924a. Die Neuere Entwicklung der Kolloidchemie und die Medizin. *Klinische Wochenschrift* 3 (1): 1–5.

Pauli, Wolfgang. 1924b. Neuere Untersuchungen über den Aufbau der Kolloide). I. *Naturwissenschaften* 12 (22): 421–429.

Pauli, Wolfgang. 1924c. Neuere Untersuchungen über den Aufbau der Kolloide. II. *Naturwissenschaften* 12 (27): 548–557.

Pauli, Wolfgang. 1929. Einige Fortschritte in der Kolloidchemie der Eiweisskörper und Ihre Biologische Bedeutung. *Klinische Wochenschrift* 8 (15): 673–679.

Rees, Anthony R. 2014. *The Antibody Molecule: From Antitoxins to Therapeutic Antibodies.* Oxford: Oxford University Press.

Schäfer-Richter, Uta, and Jörg Klein. 1992. *Die jüdischen Bürger im Kreis Göttingen 1933–1945. Ein Gedenkbuch.* Göttingen: Wallstein.

Servos, John W. 1985. History of chemistry. *Osiris* 1: 132–146.

Servos, John W. 1990. *Physical Chemistry from Ostwald to Pauling: The Making of a Science in America.* Princeton: Princeton University Press.

Siedentopf, Henry, and Richard Zsigmondy. 1902. Über Sichtbarmachung und Größenbestimmung ultramikroskopischer Teilchen, mit besonderer Anwendung auf Goldrubingläser. *Annalen der Physik* 315 (1): 1–39.

Silverstein, Arthur M. 2009. *A History of Immunology.* London, San Diego: Academic Press.

Smutný, František. 1990. Ernst Mach and Wolfgang Pauli's ancestors in Prague. *European Journal of Physics* 11:257–261.

Stoff, Heiko. 2004. *Ewige Jugend. Konzepte der Verjüngung vom späten 19. Jahrhundert bis ins Dritte Reich.* Köln, Weimar: Böhlau.

Stoff, Heiko. 2012. *Wirkstoffe. Eine Wissenschaftsgeschichte der Hormone, Vitamine und Enzyme, 1920–1970.* Stuttgart: Steiner.

Stoff, Heiko. 2015. *Gift in der Nahrung. Eine Genealogie der Verbraucherpolitik in Deutschland Mitte des 20. Jahrhunderts.* Stuttgart: Steiner.

Svedberg, The. 1921. A short survey of the physics and chemistry of colloids. *Transactions of the Faraday Society* 16:A002–A013.

Szabó, Anikó. 2000. *Vertreibung, Rückkehr, Wiedergutmachung: Göttinger Hochschullehrer im Schatten des Nationalsozialismus.* Göttingen: Wallstein.

Tanford, Charles, and Jacqueline Reynolds. 2001. *Nature's Robots: A History of Proteins.* Oxford: Oxford University Press.

Ullmann, M. 1961. Max Samec zum achtzigsten Geburtstag. *Stärke* 13:195–196.

Vincent, Brian. 2014. McBain and the centenary of the micelle. *Advances in Colloid and Interface Science* 203:51–54.

Von Meyenn, Karl, ed. 2008. *Teil III: 1955–1956.* Volume 4, Wolfgang Pauli. Wissenschaftlicher Briefwechsel mit Bohr, Einstein, Heisenberg u.a. Berlin: Springer.

Walch, Sonja. 2016. Triebe, Reize und Signale. Eugen Steinachs Physiologie der Sexualhormone. Vom biologischen Konzept zum Pharmapräparat, 1894–1938. Wien: Böhlau.

11 The Physiology of Erotization: Comparative Neuroendocrinology in Eugen Steinach's Physiology Department

Cheryl A. Logan

Eugen Steinach was a pioneer in reproductive endocrinology. He devised gonadal transplants that vividly demonstrated reversals in sexual anatomy and behavior in mammals that had been castrated in infancy and given gonads from the opposite sex. Infant males given ovaries became feminized and could, as adults, even suckle young. In male guinea pigs with ovaries, feminized attributes and behavior could persist for up to three and a half years. Yet this was Steinach's second career. His first specialty was neurophysiology. He had spent many years exploring the nervous system at the German University in Prague, initially as an assistant to the renowned physiologist Ewald Hering. But as he assumed the Directorship of the BVA's Physiology Department, the newer field of endocrinology dominated his research. His department was perhaps best known for its demonstrations of the secretory effects of gonadal interstitial cells and the impact of their secretions on the development of mammalian reproductive behavior. But Steinach's concept of "erotization"—framed as early as 1910—shows that he was also a neuroendocrinologist. And the laboratory's scientists recognized early the importance of the anterior pituitary gland and its relationship with the brain. In 1929, Steinach (figure 11.1) even proposed the existence of brain hormones.

When Steinach's transplant findings were first announced in 1912, just before he arrived at the Physiology Department, they raised eyebrows across Europe. When his subsequent demonstrations (Steinach 1916) also showed that individual infants given both male and female gonads developed as functional and anatomical hermaphrodites, the shock was even greater. By 1920, Steinach and his students had consistently shown that in rodents masculine and feminine characteristics develop flexibly in *either* sex under the influence of secretions acting independently of neural controls. Such studies and their clinical application in Steinach's concept of "rejuvenation"—his surgical technique designed to reactivate glands in the elderly (Hirshbein 2000; Stoff 2004)—gave Steinach a sensational reputation across Europe and in the United States. That notoriety, however, sometimes clouds the recognition that the laboratory also pioneered an understanding of the cellular foundation of the secretions that shape sexual development in vertebrates.

Figure 11.1
Eugen Steinach (1861–1944) at about age 70, reproduced with the permission of the Austrian National Library (ÖNB) Vienna, #222543B.

Since the 1880s, it had been known that there are three cell types in the mammalian testes. One is the interstitial cells, large cells containing fat globules; the second is the Sertoli cells, which reside inside the tubes in which sperm grow; the third is the generative cells: the sperm and their forebears. Until 1900, the interstitial cells were considered innocuous connective tissue that perhaps functioned to bring the nourishment needed for spermatogenesis, while the gonad's secretions (first termed "hormones" in 1905) were considered the product of sperm. That view began to change in 1903, when the French anatomists Paul Ancel and Pol Bouin proposed that the gonad was actually a double gland, in which the interstitial cells functioned as a separate endocrine structure within the testis. Possibly influenced by the Viennese anatomist Julius Tandler, Steinach took up the interstitial cell question. And between 1910 and 1930, Steinach and his collaborators repeatedly demonstrated that in male and female rodents the interstitial cells indeed produced the chemicals promoting the development of sexual anatomy and shaping sexual behavior. Transformations and reversals of both kinds of sexual attributes were almost always accompanied by the proliferation of interstitial cells. "Not simply the maintenance, but the entire development of masculinity and the growth of the secondary sexual characteristics is dominated by the activity of the inner secretory [interstitial] tissue" (Steinach 1910, 564).

The idea of a double gland was not new. But applied to the gonads, its implications were huge. If the hypothesis was accurate, it would mean that the development of sexual characteristics and the sex drive were physiologically separate from the processes of reproduction. Sexuality and reproduction could be understood as independent functions maintained by distinct cellular processes. This meant that medically and biologically, one function could change without altering the other. And, because the research showed that the development of sexual behavior was flexible, Steinach's findings violated prevailing social norms: Giving female glands to a male could produce milk-secreting mammary glands that transformed that male into an individual with "more than the ability, also the *inclination* to suckle" and instilled "nurturing in the manner of a *true* mother" (Steinach and Holzknecht 1916, 498, my stress; figure 11.2). Such provocative results sometimes brought strong criticism (e.g., Stieve 1921). But these and other discoveries, including the demonstration that estrogen is active in male bodies, which can convert "female" hormones into "male" hormones, garnered Steinach seven Nobel Prize nominations and two Lieben Prizes (Logan 2010; Södersten 2012; Södersten et al. 2014).

But neither historians nor endocrinologists usually think of Steinach as a neuroendocrinologist. I argue that he was: His background in neurophysiology, his concepts of "erotization" and "new erotization (*Neuerotisierung*)," his work on the anterior pituitary gland, and the pituitary research continued by his students and colleagues all indicate that the anatomy and histology of sexuality were just part of the Physiology Department's focus. Scientists in Steinach's laboratory regularly measured sexual behavior in an effort to link endocrine processes to neural processes. Further, they did so using a comparative and evolutionary frame that eventually contrasted with the more reductionistic approach that began to dominate hormone research around 1930. I present the comparative framework that guided their approach to the brain, stressing their use of behavior and their research on the pituitary. I also suggest that Steinach's career-long attention to the brain, understood as neuropituitary-gonadal interactions altering behavior, contributed to the unusual resistance to his research in America. Exploring the "feedback" hypothesis of pituitary control—developed in the United States by Carl Moore and Dorothy Price and in Europe by Steinach, Heinrich Kun, and Walter Hohlweg—I argue that, though they converged on similar solutions, the two teams were guided by distinct cultures of experimentation and, by the 1930s, these differences helped produce American endocrinology's negative reaction to the pioneering research done in Steinach's department.

Brain, Behavior, and Context

Steinach had worked at the BVA occasionally from 1903, when he helped fabricate apparatuses for physiological research. In 1907, while still employed at the German University in Prague, he was assigned a research station at the BVA (Brauckmann 2013). In 1913, he finally left Prague to become director of the BVA's newly formed Animal Physiology

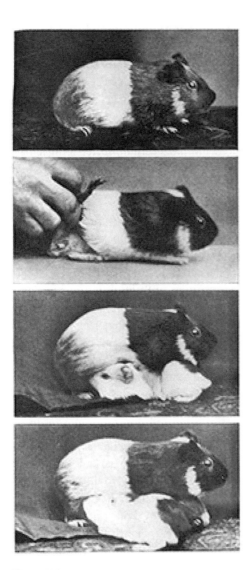

Figure 11.2
A male guinea pig feminized by ovarian implants that produced an underdeveloped penis (second panel) and milk-secreting mammary glands as well as the inclination to suckle one (third panel) and even two (fourth panel) pups. The inclination to suckle returned periodically and could be sustained for life (Steinach and Holzknecht, 1916, reproduced with the permission of Springer Scientific).

Figure 11.3
An undated photo of the Animal Physiology Department's laboratory staff pictured in about 1924. From E. Steinach, *Sex and Life* (1940). Steinach is third in the second row.

Department (figure 11.3). Though much effort was devoted to exploring sexual anatomy and gonad histology, the work consistently involved behavioral measures that could reveal the evolutionary functions of hormones. The department's scientists viewed these measures as reliable indices of brain action.

In 1910, Steinach published the first of many studies on the effects of successful gonadal transplants in castrated infant mammals, mostly rats and guinea pigs. The publication was his first on the topic of hormones in sixteen years, and it marked the beginning of his career as an endocrinologist. The paper includes data gathered during Steinach's research stays at the BVA; and though the Physiology Department had not yet been founded, the paper's key concepts defined the foundation of much of its research. In addition, the work reveals important subtleties of hormone action viewed comparatively and in relation to the brain. Steinach (1910) challenged those who still held that the mate-clasping reflex in male frogs is controlled exclusively by nerve impulses coming from the gonads (or seminal vesicles), arguing instead for the independent effect of hormones produced in the testes. But the paper begins oddly, with Steinach noting that the clasping reflex can also occur in castrated frogs; in some invertebrates, he stressed, castration produces no disruption at all in the expression of sexual reflexes. To Steinach these comparisons implied that, as neural control waned, the dependence of sexual characteristics on gonadal hormones had emerged gradually during the course of evolution, becoming more pronounced in mammals.

Noting that research had already established the effect of hormones, he stressed a functional integration that explored "the reproductively necessary increase and fulfillment of the functional, that is the psychic, breeding attributes" that occur *before* breeding begins. Of mammals he asked, how is "the central organ [brain] tuned so that the sexual senses serving it call forth on the one hand, the drive-like longing for the opposite sex, and, on the other, produce that special disposition of the reflex apparatus which makes possible the goal-directed activation of specific sexual functions ... ?" (Steinach 1910, 553). Explaining how such neural tuning coordinates sensory and motor processes with a functioning sex drive was the foundation of his concept of "erotization (*Erotisierung*)," a term that Steinach introduced in this paper.

In Prague, Steinach had already located nerve centers that inhibit the clasping reflex in male frogs. These new experiments confirmed that within 24 to 48 hours, injected extracts of testes restored the reflex, which had been eliminated in castrated and decapitated males. The studies also traced the waxing and waning of the potential to release the reflex with changing seasons. Steinach tested animals when that tendency was lowest, but took tissues from animals producing the secretions when it was highest; he then pretested experimental and control animals to confirm the absence of the reaction. He explored both the specificity and the selective nature of the restored response to the injected tissue—injected gonadal substances did not alter other reflexes. Finally, to confirm that the secretory tissue had reached the central nervous system, Steinach also showed that the reflex could be restored by injected extracts of brain and spinal cord taken from breeding males, but not by brain extracts from castrated males or females. The work was a model of methodological precision in early behavioral neuroendocrinology.

Steinach had no neural data on mammals (this was true throughout his career). However, in the second part of the paper he inferred that the development of normal mating behavior in mammals, which he produced by implanting gonads into infant castrates, also requires the action of hormones on the central nervous system. When the implanted testes healed, male rats showed full masculine somatic development and, as adults, behaved like intact breeding males. As hormones act on the brain, "ganglia," Steinach wrote, are "set in working order" so that sensory impressions emanating from the female are, in the male, "re-evaluated into feelings of lust" (1910, 565). Even at this early point in his endocrinological career, therefore, Steinach was integrating brain mechanisms with hormone action. Internal secretions were the department's focus; but its scientists consistently stressed how the effect of secretions on the brain produced "the erotization of the central nervous system" (566).

Early on, Steinach worried that his demonstrations of flexible sexual development shaped by interstitial cell secretions might lead others to oversimplify the action of hormones. Their power was great; but that did not mean that the nervous system played no role. In part to explore the effect of neural control on secretions, Steinach (1936) later conducted isolation/stimulation experiments (under way since 1924) to show the power of the central nervous system. Beginning with four-week-old male rats, Steinach housed males in sexual isolation lasting up to six months. During the isolation period, repeated short tests with receptive females showed that the intensity of males' sexual responses became weaker and weaker. By the end of six months, the males were "like eunuchs": They had lost their "libidos" and their glands had atrophied, a finding confirmed by examination of their sexual anatomy and by the absence in the testes of spermatogenesis.

Steinach then added small compartments to the males' cages to give them olfactory exposure to a series of receptive females, but no direct contact with them. He even blinded some males to rule out visual influence. After two or three weeks, the barriers between compartments were raised. Extended exposure had reversed the males' apathy; they actively pursued the females, the females responded indicating their receptivity, and some pairs even mated. Histology showed that both live sperm and interstitial cells full of secretory granules once again filled the males' testes. These once apathetic "eunuchs" had undergone a "new erotization" (*Neuerotisierung*) presumably triggered by the odors of receptive females. Steinach concluded: "The nervous, that is to say psychic, processes exercise a powerful controlling influence on the inner-secretory activity of the gonads through which bodily and psychological maturity can be automatically protected from regression or possibly from persistent depression" (Steinach 1936, 165). He speculated that the mechanism involved increased blood flow as the olfactory cortex altered the nervous system's effect on the gonads. Erotization and new erotization were hypothetical neurohumoral processes underlying the complex sensory and behavioral manifestations of masculinity and femininity. As a sensitive and technically feasible measure of brain action, the animals' behavior confirmed this.

Comparative Neuroendocrinology and the Pituitary

Steinach's early comparative and evolutionary approach is not surprising, given the comparative perspective that guided research at the BVA. Later research on frogs confirms that this comparative frame continued to shape the laboratory's work, and it kept the brain front and center. In 1929 and 1930, Steinach published two papers presenting work done with his student Heinrich Kun. Again using the frog's clasping reflex, the authors explored the impact of two kinds of brain tissue injections (a pulp and a more purified extract) on triggering the reflex. In the 1929 paper, instead of triggering an all-or-none reflex (as in 1910), Steinach employed graded quantitative measures of responses that were produced using chemical stimulation. Chemical stimuli permitted the determination of how threshold concentrations of an effective stimulus could be altered by different kinds of injections.

The work involved about 50 experimental series, using more than 800 *Rana esculenta*. For each, Steinach calculated the average threshold concentration needed to evoke a reaction. He stressed three findings. Compared to control injections, both brain tissue injections produced reactions at much lower concentrations. The intensity of the reflexes produced at the lowest thresholds was stronger and the latency to produce them shorter with the brain injections. Steinach estimated that brain extracts had increased reflex activity by 400 to 600 percent over control substances. Finally, the effects were specific to brain tissue: They could not be produced by extracts from liver or muscle, or by other hormones, including the sex hormones. The central point of the second paper was to show the impact of temperature on the threshold concentrations at which brain extracts produced the reflex in frogs that naturally lay eggs in winter. At high temperatures (mimicking summer), injections had no effect. But brain extracts injected at lower temperatures were effective at relatively low concentrations (Steinach and Kun 1930).

These demonstrations did not involve typical glandular action; rather, Steinach asserted, he had introduced the problem of the "incretory function of the central organ"—the brain. There was a "central stimulating substance" (*centrale Reizstoff*)—a CNS hormone—at work. And it had wide phylogenetic distribution: Extracts from the brains of frogs, rats, dogs, cattle, and even a human were effective. Steinach stressed that he had devised new methods that were necessary to explore "the experimental foundation for [brain] hormone research" (Steinach 1929, 1275). To illustrate the power of the brain's secretions, Steinach and Kun (1930) added a paragraph describing work on tree frogs. Tree frogs feed on flies; and to test them the authors used a fly-catching response that permitted them to leave the animals' brains intact. In this way they could test the extracts' effects on higher neural processes, such as those involved in attention. Indeed, frogs that received brain injections caught many more flies in the same period than those given control injections. The authors added incidentally that the department had been doing such studies on tree frogs during the years 1918 to 1925; the new experiments just replicated older findings using more purified extracts.

There are no frog publications that I know of from the earlier period they mention; but the series shows that Steinach had been taking a comparative and evolutionary approach all along. The work revealed that in different seasonal contexts, using extracts from several animal sources, and with different responses triggered by distinct sensory modalities in different species, the brain was not just storing or reacting to hormones—it produced them. Steinach called the substance they had demonstrated "*Centronervin*."[1] He wrote that without a doubt there existed "relationships or interactions with other incretions" (1929, 1276), and he acknowledged that centronervin might be one (or several) of many CNS hormones. He indicated that they were isolating it chemically—work that was entrusted to the Berlin pharmaceutical corporation Schering-Kahlbaum, AG.[2]

Given his emphasis on brain interactions, it is not surprising that early in his endocrine career Steinach also explored the significance of the pituitary gland. As his colleague Paul Kammerer wrote, Steinach assumed that "the central nervous system so sensitized ('erotized') by the secretions of the breeding period would therefore be a regulator that intervenes between the producing glands (gonads) and the receiving organs (physiological and morphological sexual characteristics); their release and development occur through innersecretory not neural influence; [the latter] however plays the role of an organizing mediator" (Kammerer 1919, 305). Lying just below the base of the brain and hanging off of it on a stalk, the pituitary gland had long been thought to be involved in neural control. The Viennese gynecologist Bernhard Aschner suggested as early as 1909 that the gland's influence over the ovaries depended on sexual centers in the brain, probably in the hypothalamus (see Hohlweg 1974).

Steinach's research on the pituitary is oddly distributed over his career. A short piece by his student Joseph Schleidt (who was killed in World War I) appeared in 1914. But it was followed only in 1928 and thereafter by several papers that placed Steinach and his students Kun and Hohlweg in the thick of the battle for the holy grail of endocrinology: the relationship between the brain, the endocrine glands, and what for a time was called the master gland—the anterior pituitary. Clinical work had long shown the importance of the pituitary. But it was only in the late 1920s and early 1930s that the complexities linking regulation and control to anterior pituitary secretions began to be clear. The 1928 study by Steinach and Kun was groundbreaking in part because they used injections rather than the less precise implants to assess the role of the anterior pituitary. But beyond their methods, the question they posed was why, given signs of sexual development early in infancy, do gonadal secretions exert their effects on male sexual maturity so late: Why is there a pause, an inactive period that defines "a biological law of puberty" (Steinach and Kun 1928, 529)?

To explain this late "awakening" of the testes, Steinach and Kun turned to the anterior pituitary. They had, they wrote, begun their study in 1926; but several important papers appeared in 1927, just before theirs. That year Philip Smith successfully completed the first pituitary implants at the University of California, and Bernhard Zondek and Selmar Aschheim at Berlin's Charité Hospital showed the impact of pituitary implants on activating the ovaries of immature female mice. Citing both, Steinach and Kun stressed the advantages of their injection method: With it they could precisely dose and vary the amount of putative hormone involved. The paper's greater significance, however, is the power their interpretation gave the gonads *over* the pituitary.

Steinach and Kun injected infant male rats daily or every other day for up to 18 days with 0.2–0.3 cm^3 extracts of cattle pituitary and compared them to littermates matched for size. After 12 days of injections, the development of sexual anatomy (penis, prostate, and seminal vesicles) was complete in the injected animals. In 4 more days, the animals showed full "psychosexual" maturity (at ages 38 to 45 days). This was twice as rapidly as their control brothers reached puberty. The findings were confirmed by microscopic examination of the testes, which, in injected animals, were full of granulated interstitial cells—a sign of intense secretory activity. This suggested, they wrote, "that the closest relationship exists between this awakening of the interstitial cells and the increased testicular secretions activated by the anterior pituitary hormone." The endocrine function of the testes depended on secretions from the anterior pituitary, as did, they argued (using data from Steinach's student Berthold Wiesner) the endocrine function of the ovary. And because pituitary extracts from either male or female cattle were effective, the pituitary hormone was probably "not sexually specific." Contrary to what Aschheim and Zondek concluded, therefore, pituitary secretions were probably not sex hormones; they were, rather, "activators of gonadal incretions" (Steinach and Kun 1928, 526).

Steinach and Kun then reported the effects of similar injections on adult males that naturally failed to mature sexually and on senile males that no longer showed sexual interest in females. Anatomically and "psychically," pituitary injections had the same effect; in both groups, injections triggered the growth of the interstitial cells. And pituitary reactivation of gonadal secretions then reinstilled masculine sexual behavior. Unlike with infants, however, in the "eunuchs," after the injections ended, sexual attributes again regressed. This difference led Steinach and Kun to conclude that in addition to the effect of the pituitary at sexual maturity, anterior pituitary hormones also regulated gonadal secretions in adulthood to protect masculinity and femininity. This idea then led them to reinterpret Steinach's controversial demonstrations of rejuvenation in humans: Rejuvenation was probably produced by the effect of surgically restored gonadal hormones acting on a "senile hypophysis [pituitary]." They wrote: "One can more easily consider [rejuvenation] as a retroaction (*Rückwirkung*) of the gonads on the hypophysis" (1928, 529). Collectively the data suggested that the actions of the two glands were reciprocal: "It appears after all to justify the assumption that under normal conditions the sexual hormone balance (*Gleichgewicht*) in individual life is assured through the reciprocal interaction of hypophysis and gonads" (529).

Both Steinach's pituitary research and his suggestion of secretions produced in the brain itself had far-reaching effects. As Sonja Walch (2010) has shown, although he was trained as a biochemist, Walter Hohlweg was deeply influenced by the years he spent in Steinach's department.[3] This is nowhere clearer than in his approach to the anterior pituitary, which included pathbreaking work done in the 1930s. In a paper with Max Dohrn, head of Schering's physiology laboratory, the authors explored the "castration cells" that had been described in rat pituitaries by Schleidt in 1914. Hohlweg and Dohrn (1931) explored the timing of the castration cells' development, showing that in both adults and infants, castration produced the cellular changes in pituitary in 18 to 24 days. The authors then showed that normal pituitary histology could be restored when castrated rats were given daily injections of Progynon (Schering's placental extract).[4] The minimally effective dose required for degeneration of the castration cells in infant females, however, was much lower than that needed in either infant males or adult females. And that same low dose induced premature ovarian development in infant females. In all animals, to be effective the doses had to be administered over a period of about three weeks.

The effectiveness of the Progynon injections meant that during development pituitary activation was being inhibited by the retroactive effect of small amounts of "innersecretory activity" from infant females' gonads. To confirm this, the authors first castrated infants and gave some injections of Progynon; then after three weeks—enough time for castration cells to develop—they implanted both control and Progynon pituitaries into other infant females. The ovaries of infants given "castration pituitaries" from untreated infants showed clear signs of early sexual maturity—large ovarian follicles and corpora lutea; secretions of the overactive anterior pituitary had acted on the immature gonad. But females given pituitaries from Progynon-treated castrates remained infantile, confirming a "functional dependence of the anterior pituitary" on gonadal secretions. The ovary clearly depended on the anterior pituitary; but the reverse was also the case. It was "a beautiful example of the hormonal balance of the two glands. The supraordinate gland A produces hormone a, which in the subordinate gland B affects the production of hormone b. Hormone b directly or indirectly reciprocally influences the supraordinate gland A to such an extent that there a specific amount of secretion of hormone a cannot be exceeded" (Hohlweg and Dohrn 1931, 346). The authors dedicated their paper to Eugen Steinach on the occasion of his 70th birthday.

Hohlweg and Dohrn (1931, 1932) also discovered an important difference between the sexually mature versus the immature females. In infants, the implanted pituitaries produced only one ovarian cycle; the ovaries then returned to their immature state. Only once the infants reached the age of puberty did animals in both implant groups show reproductive cycles: Only then did the "anterior pituitary-gonadal balance shift ... [in ways] that caused, in the interplay between the two glands, the rhythmic course of the female sexual cycle" (Hohlweg and Dohrn 1932, 234). Hohlweg reasoned that the mature cyclical interdependence of the glands required involvement by the brain, which somehow mediated gonadal influence on the anterior pituitary. To show this, Hohlweg and Junkmann (1932) implanted excised anterior pituitaries into the kidneys of infants (males and females) and then castrated the animals. Four weeks later—again time for castration cells to appear—the authors compared the histology of implanted versus in situ (native) pituitaries. Only the native pituitaries showed the typical cellular changes. Implantation had cut all neural connections to the repositioned glands, preventing the castration-induced changes in cell structure. Because the native glands retained their neural connections, the difference suggested that the development of oversecreting pituitaries required neural input.

To confirm neural control, the authors implanted already developed "castration pituitaries" into the kidneys of both normal and castrated males and females. The histology showed that after three weeks the anterior pituitaries of both groups lacked castration cells: In castrates, despite the lack of gonadal hormones, no castration cells developed in the absence of neural connections. Whereas anterior pituitary control of the gonads was purely chemical, gonadal control of the anterior pituitary involved hormonal and neural input. Hohlweg and Junkmann proposed a model in which gonadal secretions that were first triggered by the anterior pituitary affected a "neural sexual center" (1932, 322), which in turn limited further pituitary output.

Herbert Evans and Miriam Simpson at the University of California had already shown that secretions from the "castration anterior pituitary" were more hormonally effective than those of the normal pituitary, an effect that they too linked to the castration cells. Evans, however, argued that this anterior pituitary overeffectiveness was due to the storage of excess pituitary secretion. All of Hohlweg's evidence, however, suggested a process of regulation in which pituitary production was limited or went unchecked by the presence or absence of gonadal estrogens. In all three of the reports cited above, Hohlweg explicitly contrasted the two competing interpretations, finally concluding that the changes in castration cells produced by gonadal hormones overwhelmingly favored not storage, but inhibition by gonadal hormones on the production of further anterior pituitary secretions. Hohlweg had discovered negative feedback as an alternative to the less dynamic process proposed by Herbert Evans.

Less direct but perhaps no less significant is Steinach's impact on the field of neurosecretion. Among the pioneers of that field were the German-American scientific couple Ernst and Berta Scharrer. While working as students of Karl von Frisch in Munich, the Scharrers developed the first successful concept of neurosecretion. Using the same animals that von Frisch studied in his doctoral research at the BVA (see chapter 12), Ernst Scharrer (1928) discovered gland-like nerve cells in fish hypothalamus. Ernst later directed Frankfurt's Edinger Neurological Institute (Kleft 1997), and in 1937, integrating work on vertebrates and invertebrates, the Scharrers published the first review of the neurosecretion concept (Scharrer and Scharrer 1937). Their evidence was based in cell histology. But they acknowledged the importance of exploring the physiological significance of the proposed neurosecretions, referencing (among others) Steinach's centronervin paper. In the same year, moved by moral indignation at the Nazis' treatment of Jews, the Scharrers emigrated to the United States, where they were among the leaders in establishing the field of neuroendocrinology (Kreier and Swaab 2010).

Von Frisch's dissertation explored the neural processes by which light influenced color change in minnows, and he speculated that the mechanisms could involve light receptors in the pineal gland or brain ventricles. Von Frisch had left the BVA before Steinach became director of the Physiology Department. But while at Prague, Steinach—an award-winning neurophysiologist—had himself explored light and color change in minnows. And by 1909—the year von Frisch began his doctoral research—Steinach's endocrine transplants were already under way at the BVA. Von Frisch's 1911 report of his dissertation is regarded

as a pioneering advance in neuroendocrinology because it established the power of extero-
ceptive sensory processes to act on neuroglandular activity (Watts 2011). This concept was
central to Steinach's notion of erotization, and the two men may have discussed the sensory
engagement of neuroglandular activation during the short period in which they overlapped
in the BVA's Zoology Department. Von Frisch even cited Steinach's two papers on color
change in fish in the 1911 report of his research. The men shared common interests in
color change, light sensitivity, the brain, and the stimulation of brain glands; and von Frisch
may well have engaged the senior visiting scientist in ways that indirectly influenced the
Scharrers' later work on neurosecretion.

The American Rejection of Steinach

Despite the wide-ranging scope of his and his students' contributions, Steinach's work was
widely criticized in America. Some of the most dismissive criticism came from the pituitary endocrinologist Herbert Evans. In a 1933 review of progress in pituitary research,
Evans wrote:

Endocrinology, which suffered obstetric deformation in its very birth by the extravagant claim
of the septuagenarian Brown-Sequard that he had magically restored his youth with testicular
substance, has continued to suffer the same sort of obloquy through similar claims of the modern
Steinach school, whereas to be an endocrinologist among the practicing profession today means too
often to be primarily [sic] concerned with making fat ladies thin. In research matters are hardly better.
(1933, 425)

Evans was referring to the rejuvenation controversy—an honest scientific debate. And
Steinach's claims there were extreme. But why is Evans so dismissive of basic research?
Why "in research" are matters "hardly better"?

Other points in the review suggest that Evans was distancing himself from several
conclusions of the "Steinach school." In discussing pituitary control of the gonads, Evans
referred to a "true sex hormone" secreted by the anterior pituitary (1933, 428), a view that
Steinach and Hohlweg rejected in favor of pituitary regulators. Evans even distanced
himself from Steinach's conclusion about the centrality of the interstitial cells. Referring
to the origin of the hormones shaping sexual characteristics, Evans added: "—from whatever gonadal tissue they spring" (1933, 427). The phrase seems designed to evade endorse-
ment of Steinach's account. And despite his vast knowledge of the pituitary control of
metabolism, Evans's review omitted any reference to the groundbreaking work on pituitary
regulation of the thyroid gland completed by the BVA-trained scientist Eduard Uhlenhuth,
then at the University of Maryland. Finally, Evans did not reference either Kun's or Hohlweg's anterior pituitary work. Hohlweg's account of gonadal control of the pituitary both
challenged Evans's interpretation and endorsed what Evans saw as Steinach's misguided
treatment of senility. Though his enormously successful career in endocrinology gave him

ample opportunity to answer Steinach empirically, Evans never did.[5] Instead, he used rejuvenation as an excuse to ignore 20 years of groundbreaking research from the BVA's Physiology Department. Rejuvenation gave him an easy out, which he used to avoid the more difficult task.

Evans was one of several elites of American endocrinology and medicine who strongly criticized Steinach. The most principled rejoinder, however, came from Evans's younger colleague, the University of Chicago zoologist Carl Moore. Moore confronted Steinach with scientific argument and empirical evidence, which led Moore to endorse different interpretations of hormone action. Because his challenge was scientifically based, it is more revealing of the underlying differences in approaches that shaped the conflict. I use Moore's most effective challenge—his rejection of Steinach's concept of hormone "antagonism"—to explore influences beyond the rejuvenation controversy that shaped the reaction of American scientists to the work of Steinach's group. Steinach and Kun's 1928 reinterpretation indeed linked rejuvenation to the anterior pituitary, and that would have been heresy to Evans and to Moore. But beginning in the 1930s, just as important was the contrast between Steinach's comparative evolutionary physiology and the growing value of the mechanistic precision and efficiency of chemistry, a difference that was aggravated by the intense international competition to identify the structure and function of those anterior pituitary hormones.

America and the Pituitary Wars

Steinach was of an older generation, trained before newer disciplinary boundaries and narrower research perspectives had foreclosed the broad integrations across animal groups, and between physiology and evolution, that he preferred. As a comparative physiologist, he had adjusted his approach to the chemical revolution, incorporating chemical methods into his evolutionary framework through his work with the Schering Corporation and his affiliation with Hohlweg (a biochemist). Despite this, younger men like Evans and Moore may have seen him as an "old codger," whose time, like (they hoped) the hegemony of German-speaking science, was past, but who was trying too late to "get in" the pituitary game. In fact, he had been in it all along, exploring neuroendocrine processes framed by a comparative perspective that stressed evolution, function, and developmental flexibility.

Carl Moore and his student Dorothy Price famously published their interpretation of the relationship between gonadal hormones and the anterior pituitary in 1932. Many endocrinologists credit Moore and Price with discovering that gonadal secretions feed back onto the anterior pituitary to decrease pituitary secretions and therefore reduce the amount of gonadal hormone acting on sexual attributes. Less well known, however, is that within days of one another in July 1930, at the Second International Congress for Sexual Research, Hohlweg, collaborating with Dohrn, and Moore both presented papers proposing the feedback hypothesis, and each published an account of it in the conference proceedings. Both then published more detailed accounts in 1931 and 1932 (see Simmer and Suß 1993). Years later, in 1975, presenting their reminiscences of the period in the same volume, Dorothy Price and Walter Hohlweg each claimed priority as the first to propose the concept. Simmer and Suß (1993) convincingly show that neither Hohlweg nor Price was original; several authors had already suggested gonadal influence over the pituitary. Instead, the two groups had converged on an exciting idea that each enriched with important new and clarifying evidence. In addition, however, their efforts reflect different cultures of science: One was linked to the integration of adaptation, brain, and behavior; while the other, narrower and more mechanistic, relied primarily on the swift power of chemistry.

Moore's research had challenged Steinach's ideas since 1921. In 1924, he successfully transplanted opposite-sex gonads into male and female rats without first castrating the animals. For this reason, the finding contradicted Steinach's reports of glandular "antagonism." Moore (1924) also challenged Steinach's rejuvenation surgery and his conclusion that the postsurgical degeneration of sperm in the testes was due to increases in the interstitial cells and their hormones. Moore suggested instead that degeneration was caused by the excessive heat to which the testes were exposed when transplant or surgery removed them from the thermally insulating properties of the scrotum. The 1932 paper with Price was the culmination of Moore's challenge to Steinach. It incorporated a great deal of information derived from data gathered with new methods—again rejecting Steinach's concept of antagonism. But this time Moore and Price replaced antagonism with a "new conception of hormone interactions."

That conception was based on evidence that the gonadal secretions of either sex "have a depressing effect on the hypophysis [pituitary]" (Moore and Price 1932, 19, 20), effectively decreasing circulating levels of gonadal hormones. The authors' Chicago colleagues in F. C. Koch's Department of Physiological Chemistry and Pharmacology had just successfully purified "the male hormone" from extracts of bull testis, and Moore operationalized a dosing system adapted from bird research for use in mammals. This permitted quantitative variation in injections of the "testis hormone," which could then be compared to or mixed with varying quantities of "oestrin" to directly assess the occurrence of antagonism. Moore and Price stressed the chemicals' effects on "accessory structures" and on the gonads in males and females. Among others, they reported three key findings that challenged antagonism. First, neither the testis hormone nor oestrin (nor their mixture) damaged the accessory structures of the other sex: Hormones "stimulate homologous accessories, but have no effect on heterologous characters" (22)—no direct antagonism. Second, although they replicated Steinach, showing that injections of oestrin could badly damage the testes, high doses of the purified testis hormone *also* damaged the testes. This surely could not be antagonism, and the finding led them to explore the pituitary. Finally, they showed that either pituitary implants or injections of "hebin"—taken from the urine of pregnant females and containing high levels of pituitary hormones—eliminated the damage produced by injections of either oestrin or the testis hormone. Steinach's "antagonism" was not a direct effect of sex hormone interaction. Rather, it resulted from the absence of a quantity of homologous gonadal secretion that was governed by that hormone's "depression" of anterior pituitary secretions.

Moore and Price did not clearly define Steinach's idea of antagonism; they implied both a struggle between gonads and a "chemical" antagonism before rejecting what they termed "a 'sex antagonism' phenomenon" (1932, 39). But Steinach's concept (1916) dealt with two different problems. The "Kampf," which he described in 1916, he interpreted as an artifact of the transplant procedures in which one gland "cannot establish roots" (310) after the other begins to function. In referring to the natural process, he implied chemical antagonism between male and female hormones, which he too described as "inhibition" (*Hemmung, Wachstumshemmung*). His usages were sometimes unclear, and the Moore-Price data were convincing. But, why would a former neurophysiologist, steeped in the concept of inhibition, introduce the term "antagonism" at all?

The idea was becoming prominent in neurophysiology just as Steinach completed his transition to endocrinology. John Langley, Michael Foster's successor in physiology at Cambridge University, explored the mutual chemical antagonism between atropine and pilocarpine in salivary secretions and in adrenal extracts that stimulated sympathetic nerves in the autonomic nervous system. In 1903, he proposed that mutually antagonistic chemicals acted neither on muscles nor on nerves directly, but on special "receptive substances," the deep forerunners of modern receptors (Maehle 2004). Langley proposed that similar processes could occur with other internally secreted chemicals, such as "thyroidin" and the "chemical bodies formed by the generative organs" (quoted in Maehle 2004, 166). In 1906, he spoke on the topic in Vienna, and his presentation was published in the *Zentralblatt für Physiologie*. While in Prague, Steinach had done research on smooth muscles, on inhibitory effects in frog stomachs, and on contractility in capillaries—all phenomena involving the autonomic nervous system. Steinach surely knew of Langley's work, and he may well have employed Langley's concept of antagonism—usually measured by varying the doses of opposing chemicals—in his analysis of the interactions among chemicals being secreted by the gonads.

Laboratory Cultures in Endocrinology

By 1940, the Stanford University psychologist Calvin Stone had completed much research on reproductive behavior and the sex drive in male rodents. Stone's papers consistently reference the results of Steinach's transplants and his acceptance of the interstitial cells as the source of the secretions. Stone even identified Steinach's "erotization of the nervous system" as the kind of idea needed to understand the neurohumoral mechanisms underlying sexual behavior (Stone 1939). But the fascination with identifying the chemicals themselves was great, and the promised precision of chemistry began to dominate endocrinology in the late 1920s, especially with the discovery of large amounts of putative hormones in mammalian urine (Clarke 1998; Oudshoorn 1994). One result, as the American behavioral endocrinologist Frank Beach noted, was the growing distrust accorded behavioral studies in Europe and America: Unlike "pure-science techniques" (Beach 1981, 332, citing Turner and Bagnara 1976), behavioral studies took too much time and were too variable. "Pure-science" methods were preferred, especially to standardize the quality and clinical value of the proposed substances.

It was known that Carl Moore rejected behavioral measures (Beach 1981; Crews 2014). In 1924, he expressed concern that the "only morphological basis on which Steinach's regeneration rests" (Moore 1924, 503) was the proliferation of interstitial cells, whose secretory function Moore doubted. His phrase implies that Steinach's other measures—many of them behavioral—were unreliable. Moore and Price made this quite explicit. They stressed new measurements, reiterating Moore's dissatisfaction with other indices of sex difference, including "disposition," body length, and coat—each a measure that Steinach had employed—among the indices that could not be used "with any certainty of gaining an idea of the relative sex-hormone state of the animal within a reasonable experimental period" (Moore and Price 1932, 20). Moore regarded as unworkable long-term chronic experiments, like Steinach's, in which behavior revealed an adaptive developmental integration of hormones and brain. In the 1930s, Moore even warned his student William Young that behavior was "unordered by hormonal events, and unrelated to variables of significance to reproductive biology" (Goy 1967, 5).

Moore's rejection of antagonism was empirically based. But his refusal to consider Steinach's measures, preferring those that could be used within "a reasonable experimental period," shows that by the 1930s, endocrinology placed a premium on the speed of scientific production, the precision of chemistry, and circumscribed changes occurring in cellular anatomy. Relying on the collaboration with Koch's group (which Frank Lillie had arranged), Moore adopted what he saw as a more objective biochemical approach, devaluing a broader biological integration of adaptation, evolution, and the experiential triggering of neuroendocrine processes. Chemical methods were by that time clearly producing huge advances in endocrinology; but why must their success preclude biological integration involving brain and behavior as part of "reproductive biology"?

At the time, "feedback" had rather different meanings for the Steinach-Kun-Hohlweg team versus the Moore-Price team.[6] Steinach's group viewed it as part of a physiological balance occurring within a neuroendocrine system that coordinated adaptation. It helped establish a hormonal equilibrium that assured chemical balance, but as part of a system that also included the reshaping of the brain as well as female olfactory cues that could reinvigorate males with regressed testes. The short-term biochemical control of Moore's "feedback," by contrast, referred to an internal process that simply replaced antagonism with a mechanism of reciprocal inhibitory control. It did not reach to the neural regulation of adaptations in "masculine" or "feminine" behavior or to the sensory effects of behavior on the other sex. For this reason, Hohlweg and Junkmann could readily accommodate Moore's finding, which allegedly challenged antagonism. Citing Moore's paper from the International Congress, they wrote: "The antagonistic effect, as Moore has already explained, probably occurs through an inhibition of the anterior lobe" (Hohlweg and Junkmann 1932, 321). Because Steinach had already proposed retroactive gonadal control of the anterior pituitary, the new data changed little, and Hohlweg could without pause explore whether the effect involved neural processes. Ironically, Moore was using "feedback" to disprove Steinach's antagonism, while Hohlweg was using it to confirm the idea. Kun even noted that Hohlweg's proposed brain sexual center could explain the "inhibitory mechanism" (*Hemmungsmechanismus*) (Kun 1934, 321) underlying Moore's result. Steinach's own summary (1936) of his career's work cites several studies by the Chicago biochemists on the biochemistry of androgens—he praised them as important advances. But it does not refer to the Moore and Price paper. For Steinach, Hohlweg, and Kun, "feedback" integrated gonadal control into an adaptive system of sensory, seasonal, secretory, behavioral, and neuroglandular processes—including experience as well as chemicals—in which balanced neuroendocrine cycles could even be altered by social action.

Moore and Price's paper was admittedly reactive. But with the narrower precision favored as biochemically oriented physiologists and biochemists began to dominate research on endocrine processes, American reproductive endocrinology moved away from evolutionary physiology and behavior, especially when it was cast—as Steinach often did—in mental terms. These changes were fed by the potentially huge corporate profits that would result from the availability of chemically purified and standardized compounds (Oudshoorn 1994; Stoff 2010, 2013). Adele Clarke describes the "reproductive sciences" as a distinct scholarly specialty in the United States that diverged from physiology after 1910 to rapidly become its own discipline. Using "disciplining" to refer both to the development of distinct boundaries and the "policing and enforcing" (Clarke 1998, 7) of reproduction, she describes the loss of a prewar integration of evolutionary, developmental, and genetic problems, especially in the period between 1925 and 1940. She presents several dimensions of US reproductive science—beyond its separation from evolutionary and physiological concerns—that created lucrative new markets, including a bias among basic researchers against clinical work and the pressures for "Fordist mass production" (10). These too, likely affected the American attitude toward Steinach. In the United States, money was flowing, competition was intense, and results had to come fast.

Even as Steinach (1936) applauded the inclusion of chemistry, he too voiced objection to the estrangement among physiological, evolutionary, and chemical approaches to hormone action, a view probably fed by great disappointment that his pioneering efforts were being misrepresented. Harry Benjamin, a pioneer in the humane treatment of inter-sexuality in the United States and Steinach's American defender, inadvertently confirmed the growing emphasis on chemistry. Alone and exiled at almost 80 years old, Steinach finally sought to come to America. When Benjamin contacted some American institutions to explore the prospects, several objections were raised. Benjamin (1945, 442) anonymously quoted one respondent: "If Steinach would be a biochemist, we would go to any length to bring him here." But by the 1940s, his integrative and comparative approach was irrelevant to the rush to purify marketable chemicals that dominated American reproductive science.

Steinach's connections with Swiss scientists may have been what enabled him to cross the border into Switzerland on the day the Nazis entered Vienna in 1938. Witnessing from afar the fate of his BVA colleagues, he never returned. He died there in 1944, an unhappy man. Although he could be arrogant and very difficult, Steinach also had an unusual scientific and life spirit (Benjamin 1945). Nearing the age of 70, he had adjusted his approach and persevered; but his conceptual and historical preference and the failing Austrian economy impeded any greater alignment with chemistry—a move that for him could never have been complete.

Acknowledgments

I am very grateful for informative discussions with and insightful comments provided by Sabine Brauckmann, David Crews, George Michel, Per Södersten, and Rudolf Soukup.

Notes

1. *Time* magazine highlighted the discovery: "Brain Juice," *Time*, vol. 14, no. 14 (September 30, 1929).

2. Steinach also pioneered hormone replacement therapy. In 1923, he collaborated with Schering-Kahlbaum to develop the first commercially available sex hormone preparation, Progynon (Soukup 2010).

3. Hohlweg left Steinach's laboratory in 1928 to work at the Schering Corporation in Berlin. He remained there until 1945, when he joined the Institute for Experimental Endocrinology at Berlin Charité (Walch 2010). Other Steinach students who left Austria and became notable endocrinologists include Berthold Wiesner and Oskar Pezcenik, both of whom emigrated to the United Kingdom.

4. Hohlweg sent a crude oil of the extract to Adolf Butenandt, who in 1939 won the Nobel Prize in chemistry for purifying it as estrone; see Soukup 2010.

5. Evans chaired the Department of Anatomy at the University of California, Berkeley, where he directed the largest endocrinology lab in the United States before World War II (Raacke 1976). While in graduate school, he spent summers doing research in Germany, and the 1933 paper contains many citations in German.

6. In the 1930s, neither group used the word *feedback* or its German equivalent, *Rückkopplung*.

References

Beach, F. 1981. Historical origins of modern research on hormones and behavior. *Hormones and Behavior* 15: 325–376.

Benjamin, H. 1945. Eugen Steinach, 1861–1944: A life of research. *Scientific Monthly* 61 (6): 427–442.

Brauckmann, S. 2013. BVA Chronologie. Unpublished manuscript.

Clarke, A. 1998. *Disciplining Reproduction: Modernity, American Life Sciences, and the Problem of Sex.* Berkeley: University of California Press.

Crews, D. 2014. Personal communication, June 5.

Evans, H. 1933. Present position of our knowledge of anterior pituitary function. *Journal of the American Medical Association* 101: 425–432.

Goy, R. 1967. William Caldwell Young—a biographical sketch. *Anatomical Record* 157: 3–12.

Hirshbein, L. 2000. The glandular solution: Sex, masculinity, and aging in the 1920s. *Journal of the History of Sexuality* 9 (3): 277–304.

Hohlweg, W. 1974. Rückblick auf 60 Jahre Forschung zur neuroendocrinen Regulation im Hypothalamus-Hypophysenvorderlappen-Keimdrüsen-System. In *Endocrinology of Sex*, ed. G. Dörner, 159–165. Leipzig: Barth.

Hohlweg, W., and M. Dohrn. 1931. Beziehungen zwischen Hypophysenvorderlappen und Keimdrüsen. *Wiener Archiv für innere Medizin* 21: 337–350.

Hohlweg, W., and M. Dohrn. 1932. Über die Beziehungen zwischen Hypophysenvorderlappen und Keimdrüsen. *Klinische Wochenschrift* 11: 233–235.

Hohlweg, W., and K. Junkmann. 1932. Die Hormonal-Nervöse Regulierung der Funktion des Hypophysen-vorderlappens. *Klinische Wochenschrift* 11: 321–323.

Kammerer, P. 1919. Steinachs Forschungen über Entwicklung, Beherrschung und Wandlung der Pubertät. *Ergebnisse der inneren Medizin und Heilkunde* 17: 295–398.

Kleft, G. 1997. The work of Ludwig Edinger and his neurology institute. In *Neuroendocrinology: Retrospect and Prospectives*, ed. H.-W. Korf and K.-H. Usadel, 407–423. Berlin: Springer.

Kreier, F., and D. Swaab. 2010. History of neuroendocrinology: "The spring of primitive existence." In *Handbook of Clinical Neurology*, vol. 95, *History of Neurology*, ed. S. Finger, F. Boller, and K. Tyler, 335–360. Amsterdam: Elsevier.

Kun, H. 1934. Psychische Feminierung und Hermaphrodisierung von Männchen durch weibliches Sexualhormon. *Endokrinologie* 13: 311–320.

Logan, C. 2010. Reproduction versus endocrine bisexuality: Eugen Steinach and the mammalian puberty gland. In *Pioniere der Sexualhormonforschung*, ed. R. W. Soukup and C. Noe, 35–54. Vienna: Book of Abstracts.

Maehle, A.-H. 2004. "Receptive substances": John Newport Langley (1852–1925) and his path to a receptor theory of drug action. *Medical History* 48: 153–174.

Moore, C. 1924. The behavior of the testis in transplantation, experimental cryptorchidism, vasectomy, scrotal insulation and heat application. *Endocrinology* 8: 493–508.

Moore, C., and D. Price. 1932. Gonad hormone functions, and the reciprocal influence between gonads and hypophysis with its bearing on the problem of sex hormone antagonism. *American Journal of Anatomy* 50: 13–67.

Oudshoorn, N. 1994. *Beyond the Natural Body: An Archeology of Sex Hormones*. New York: Routledge.

Raacke, I. 1976. "The die is cast"—"I am going home": The appointment of Herbert McLean Evans as head of anatomy at Berkeley. *Journal of the History of Biology* 9: 301–322.

Scharrer, E. 1928. Die Lichtempfindlichkeit blinder Elritzen (Untersuchungen über das Zwischenhirn der Fische. I). *Zeitschrift für Vergleichende Physiologie* 7: 1–38.

Scharrer, E., and B. Scharrer. 1937. Über Drüsen-Nervenzellen und neuro-sekretorische Organe bei Wirbellosen und Wirbeltieren. *Biological Reviews of the Cambridge Philosophical Society* 12: 185–216.

Simmer, H., and J. Suß. 1993. Zur Frühgeschichte des negativen Feedbacks der Östrogene auf die Gonadotropine des Hypophysenvorderlappens. Der Prioritätsstreit zwischen Dorothy Price und Walter Hohlweg. *Geburtshilfe und Frauenheilkunde. Ergebnisse der Forschung für die Praxis* 53: 425–432.

Södersten, P. 2012. A historical and personal perspective on the aromatization revolution: Steinach confirmed. In *Brain Aromatase, Estrogens, and Behavior*, ed. J. Balthazart and G. Ball, 281–314. Oxford: Oxford University Press.

Södersten, P., D. Crews, C. Logan, and R. W Soukup. 2014. Eugen Steinach: The first neuroendocrinologist. *Endocrinology* 155: 688–702.

Soukup, R. W. 2010. Eugen Steinach: Spiritus rector des ersten zyklusregulierenden Hormonpräparates. In *Pioniere der Sexualhormonforschung*, ed. R. W. Soukup and C. Noe, 16–34. Vienna: Book of Abstracts.

Steinach, E. 1910. Geschlechtstrieb und echt sekundäre Geschlechtsmerkmale als Folge der innersekretorischen Funktion der Keimdrüsen. *Zentralblatt für Physiologie* 24: 551–566.

Steinach, E. 1916. Pubertätsdrüsen und Zwitterbildung. *Archiv für Entwicklungsmechanik* 42: 307–332.

Steinach, E. 1929. Ein Reizstoff des Centralorgans und die centrale Funktion. *Medizinische Klinik* 25 (33): 1273–1276.

Steinach, E. 1936. Zur Geschichte des männlichen Sexualhormons und seiner Wirkungen am Säugetiere und beim Menschen. Im Anschluß an neue eigene Forschung. *Wiener klinische Wochenschrift* 49: 161–172; Schluß: 196–205.

Steinach, E., and G. Holzknecht. 1916. Erhöhte Wirkungen der inneren Sekretion bei Hypertrophie der Pubertätsdrüsen. *Archiv für Entwicklungsmechanik* 42: 490–507.

Steinach, E., and H. Kun. 1928. Die entwicklungsmechanische Bedeutung der Hypophysis als Aktivator der Keimdrüseninkretion. Versuche an infantilen, eunuchoiden und senilen Männchen. *Medizinische Klinik* 24 (14): 524–529.

Steinach, E., and H. Kun. 1930. Notiz zur biologische Prüfung eines Hirnreizstoffes. *Medizinische Klinik* 26: 119–121.

Steinach, E. 1940. *Sex and Life. Forty Years of Biological and Medical Experiments*. New York: The Viking Press.

Stieve, H. 1921. Entwicklung, Bau und Bedeutung der Keimdrüsenzwischenzellen. Eine Kritik der Steinachschen "Pubertätsdrüsenlehre." *Ergebnisse der Anatomie und Entwicklungsgeschichte* 23: 1–249.

Stoff, H. 2004. *Ewige Jugend. Konzept der Verjüngung vom spaten 19. Jahrhundert bis ins Dritten Reich*. Köln: Böhlau.

Stoff, H. 2010. Wirkstoffe als Regulatoren des Leitsungsgetriebes, 1889-1950. In *Pillen und Pipetten—Wie Chemie und Pharmazie unser Leben bestimmen*, ed. V. Koesling and F. Schülke, 118–139. Leipzig: Koehler & Amelang.

Stoff, H. 2013. Vital regulators of efficiency: The German concept of *Wirkstoffe*, 1900–1950. In *Biologics: A History of Agents Made from Living Organisms in the Twentieth Century*, ed. A. Von Schwerin, H. Stoff, and B. Wahrig, 89–104. London: Pickering and Chatto.

Stone, C. 1939. Sex drive. In *Sex and Internal Secretions*, 2nd ed., ed. E. Allen, 1213–1262. Baltimore: Williams and Wilkens.

Turner, C. and J. Bagnara. 1976. *General Endocrinology*, 6th ed. Philadelphia: Saunders.

Walch, S. 2010. Walter Hohlwegs Laborexperimente zwischen Chemie und Physiologie—Sexualhormone in pharmazeutischen Tierversuch. In *Pioniere der Sexualhormonforschung*, ed. R. W. Soukup and C. Noe, 102–111. Vienna: Book of Abstracts.

Watts, A. G. 2011. Structure and function in the conceptual development of mammalian neuroendocrinology. *Brain Research. Brain Research Reviews* 66: 174–204.

V RAMIFICATIONS

12 Fish in the Prater: Karl von Frisch's Early Work at the Biologische Versuchsanstalt, 1909–1910

Tania Munz

In his memoirs, Karl von Frisch remembered the Biologische Versuchsanstalt (BVA) fondly: "There it did not reek of clove oil and denatured alcohol, there the living animal reigned—and that's where I felt myself drawn" (von Frisch 1962, 32). Karl von Frisch (figure 12.1) is remembered today as one of the winners of the 1973 Nobel Prize in Physiology or Medicine, together with his fellow Viennese, Konrad Lorenz, and Niko Tinbergen of the Netherlands. The three men were recognized for their pioneering work in animal behavior studies, and von Frisch was awarded a share of the prize for his discovery of the honeybee dance language. He determined that bees communicate precise information about the distance and direction of food sources to their hive mates after they return from foraging flights by means of the "dances" they run in the hive (Munz 2016; von Frisch 1962; Kreutzer 2010). Less well known today is his work on fish. But throughout most of his career, von Frisch performed sensory physiological work on fish in the winter months, when it was too cold for the bees to leave the hive. Indeed, von Frisch performed some of his earliest work on fish at the BVA. During the years 1909–1910, he conducted the research that would culminate in his doctoral dissertation on the nervous control of pigmentation changes in minnows and trout under the supervision of Hans Przibram as well as the unofficial guidance of his uncle, Sigmund Exner.

In this paper, I focus on von Frisch's earliest work with animals, including his work on fish at the BVA, where "the living animal reigned." I show that despite this assertion, in his early experiments, animals reigned in a highly constrained manner, if the verb even applies. We see a budding young scientist who consumed animals freely for his work: from cutting, to lobotomizing, to killing them by the dozen, von Frisch evidenced an aggressive and short-term approach to his experimental animals. It is only in his later work that we see an approach that relied on keeping animals longer-term and that emphasized their proper care. I show this transition in his 1923 work "A Catfish That Comes When Summoned by Whistling," in which he trained a catfish to respond to sound. In a final section, I examine the construction of his new Zoological Institute in Munich, completed in the early 1930s, and argue that the building marks the culmination of this shift—careful consideration was given, in its planning and construction, not just to the needs of the

Figure 12.1
Karl von Frisch in 1932. (Nachlaß Karl von Frisch, Bayerische Staatsbibliothek, Munich, ANA 540)

scientists but also to its many and varied animal inhabitants. In taking seriously von Frisch's characterization of the BVA as a place that focused on the living animal, we gain valuable insight into the development of this approach. The sensibility of *artgerechte Haltung*—the keeping of an animal under circumstances that are appropriate and specific to its kind—as it was practiced at the BVA, would become more important over the course of von Frisch's career and would have a lasting practical and conceptual effect for animal behavior studies more generally.

Early Science

Karl von Frisch was born in Vienna in 1886 as the fourth son of Marie and Anton Ritter von Frisch. Young Karl's academic credentials were auspicious, as both sides of the family boasted impressive medical and academic pedigrees. His paternal grandfather, Anton, had earned himself a knighthood through his reforms of the Austrian Imperial Army medical Corps (Burkhardt 1990; von Frisch 1962). Including Karl's father, the family had already produced three generations of physicians. Karl's mother, Marie, came from no less impressive stock, hailing from the multitalented and influential

Exner family, where all three of her brothers would become accomplished academics (Coen 2007, 2004, 2006).

Though the von Frisches stressed the importance of a well-rounded, humanist education, Karl struggled with the formal aspects of his schooling. He had little patience or talent for the more theoretical subjects, especially mathematics and foreign languages, that kept him indoors and took him far from the living animals he came to love (von Frisch 1962, 19–20). In contrast, his home menagerie flourished—by the time he entered secondary school, he had accumulated an astonishing 123 different species of animals (von Frisch 1962, 16). He also became an avid aquarist and collector of the local fauna and flora that surrounded his family's summer home in the hamlet of Brunnwinkl in the Salzkammergut of Austria (figure 12.2). Over the years, he gathered and preserved approximately 5,000 specimens for his local natural history museum. He was captivated by the abundance of natural life and recalled having "wanted to collect everything, not only butterflies or another specific group, as is customary" (von Frisch 1962, 20).

Figure 12.2
Karl von Frisch with his uncle, the experimental physiologist Sigmund Exner, in the private museum that was located in the attic of the von Frisches' summer residence in Brunnwinkl. Von Frisch started to collect the animals that were native to the region surrounding the mill when he was a young boy. Parts of the collection are still on display today in the nearby city museum of St. Gilgen. (Nachlaß Karl von Frisch, Bayerische Staatsbibliothek, Munich, ANA 540)

In 1905, von Frisch enrolled as a medical student at the University of Vienna. He embarked on the path with reservations, since he continued to be more interested in zoology. But his physician father had convinced him that medicine could provide a more certain future. Despite his misgivings, von Frisch didn't squander his time at the University of Vienna. He acquired a firm grounding in anatomy, zoology, and histology. And most important, he worked with his uncle, the renowned experimental physiologist Sigmund Exner. Many years later, he would recall his uncle's course with enthusiasm. Exner had "conveyed the function of human organs with exemplary clarity and free of unnecessary baggage." And he praised his uncle for his ability to "compellingly support his claims with well-thought-out demonstrations" (von Frisch 1962, 26).

In addition to his coursework, von Frisch also took up his first independent research project under Exner's guidance during this time. In these experiments, he expanded on his uncle's own investigations of compound eyes to study the pigments in the eyes of moths, lobsters, and shrimp (von Frisch 1908).[1]

In vertebrate eyes, the iris changes shape to regulate the amount of light that passes through the pupil to the retina. Invertebrate eyes lack this structure and instead depend on moving pigments to regulate light influx. Von Frisch set out to uncover the nature and cause of these pigment changes. To this end, he moved invertebrate animals from dark to light and vice versa and examined their eyes. The two different pigment configurations that resulted from these differing light conditions, he termed "light eyes" and "dark eyes." He performed these examinations both on living animals, in various stages of light and darkness, and by killing the animals at regular time intervals in the light-dark transition. In a subsequent set of experiments, he exposed living animals to lights of different wave-lengths and determined that the pigments were most sensitive to blue-violet and violet light (von Frisch 1908).

After studying the speed of transition of the pigments, von Frisch went on to examine whether light itself or nervous stimulation triggered the pigment changes. He tested all the usual suspects for nervous stimulation—electrical stimuli, chemical stimuli in the form of acids and bases, different temperatures, and even radium and x-rays—but found that none of them caused a change in the animals' eye pigments.

The young scientist was learning to perform delicate surgeries, and the resulting publi-cation gives us a vivid sense of what it was like to work with these animals (von Frisch 1908, 662). In a group of shrimp, for example, he severed the hair-like fibers of the animals' optic nerves. The text does not always make clear whether the animals survived these procedures, or indeed, whether that was desired for the experiment. But we can glean that, at least in some cases, he performed experiments with postoperative animals that had survived the earlier procedures.

Even when von Frisch was not wielding a scalpel, manipulations of the animals required considerable skill and patience. To test the effects of partial illumination, he smeared a tacky alcohol-soot mixture on the animals' eyes. He described the procedure in some detail and deemed it "not entirely easy. Their shape, smooth nature, and great sensitivity to even the smallest amounts of light are challenging circumstances." The effort was made even more difficult by the animals' active resistance to the treatment. "Because the animals know to deftly remove the annoying cap, one also has to include the eyestalk." But "after several failed experiments," he settled on a combination of soot, celluloid, and ethanol that made a "quick-drying paste and with this the eyes are pasted over" (von Frisch 1908, 701).

Although von Frisch's scientific publication maintained a degree of detachment befitting a budding young scientist, his private reflections on these experiments give us a sense of how he felt about the work at the time. He remembered how, although he "went to work with enthusiasm," he "very quickly faced a conflict." He continued, "I had to stimulate the living crustaceans' eyes with electrical currents, which clearly was uncomfortable for them" (von Frisch 1962, 26). His aversion was so strong, he remembered later that "every experiment required that I overcome my reluctance." Although in the end, his scientific drive was stronger than his compassion, even in later years he conceded that he most likely would not have been able to perform such experiments "had they involved birds or mammals with their more highly developed and certainly more sensitive nervous systems" (26). Von Frisch's sympathies were visceral but also strongly mediated by his scientific understanding of kinship and an evolutionary hierarchy that placed humans and mammals at the top of the pain-perceiving scale. Although the study was ultimately inconclusive, he later recognized that the work awoke in him an abiding interest in experiment and sensory physiology (27).

After passing his medical exams with distinction in all subjects, von Frisch decided that he no longer wanted to endure the medical curriculum, which had become increasingly clinical. In 1908, he left medicine and the University of Vienna for Munich to pursue a doctorate in zoology at Richard Hertwig's Zoological Institute. But in 1909, after only two semesters in Munich, von Frisch returned to Vienna to be closer to his family while researching and writing his dissertation.[2]

Von Frisch's Research at the BVA

Back in Vienna, Karl von Frisch approached Hans Przibram for a spot at the BVA to conduct research for his doctoral thesis. Przibram led the biological research facility near the Prater amusement park on the outskirts of Vienna (see chapter 8). While the original structure had been built in ornate style in 1873 for the World Expo, the research institute was founded in 1903, only six years prior to von Frisch's arrival. Turn-of-the-century Vienna would come to be remembered as the home of Freud and Klimt, Mahler, and members of the Secession. The natural sciences at the University of Vienna also enjoyed a high point around this time (Coen 2007). And yet some disciplines fared less well in this setting. Biology, in particular, was underfunded and clung to a traditional approach. Matters were not helped by the fact that it was forbidden to keep live animals in the new science building of the university. These factors made it virtually impossible to pursue the new experimental biology with which Przibram had become enamored.

Much of the nineteenth century had been dominated by studies of morphology of dead and often extinct animals, but the second half of the nineteenth century saw the ascendance of a new kind of life science, one that turned its attentions to living processes such as inheritance, regeneration, and reproduction. Its practitioners increasingly looked to the experimental and quantitative methods of the physical sciences and physiology as models. For the success of these pursuits, it became critically important that the animals and plants under investigation be able to endure scientists' probings long enough to reveal their behaviors and structures over time and across generations.

Flush with familial wealth, highly capable and trained, Prizbram and two colleagues—Wilhelm Figdor and Leopold von Portheim—dedicated their energies toward creating an institute to house this kind of experimental and interdisciplinary life science (see chapter 3). Four departments at the new BVA—zoology, botany, physical chemistry, and physiology—were dedicated to investigating the problems and processes of life. The work at the BVA, according to Przibram, "should not limit itself to only certain problems but should rather pull into its domain all large questions of biology." He went on to write expansively about the organisms that would be called upon to serve these ends: "Animals, plants, inhabitants of fresh water, the seas as well as terrestrial parts are all equally welcome" (Przibram 1908, 234).

While the building's exterior exuded the stately grandeur of its past (see chapters 3 and 4), its interior told a different story. Hans Przibram and his colleagues had completely gutted the place and installed state-of-the-art equipment (chapter 7, this volume; Przibram 1908, 243–264). Salt- and freshwater tanks with elaborate aeration and circulation systems housed aquatic organisms, and the former were supplied with seawater that was brought in from Trieste by railcar (Przibram 1908, 246). The facility also boasted outdoor ponds, terraria, cages, and gardens to grow experimental plants and animal feed. Cages and pens held larger animals, while high-moisture rooms mimicked tropical conditions. A cave built five meters below ground ensured absolute darkness and a steady 12 degrees Celsius (258). Special contraptions served to alter the conditions of gravity and atmospheric pressure, while light and dark rooms accommodated experimental protocols that depended on special light conditions (261).

The very structures of the institute all converged on a single purpose: the maintenance and study of living organisms. Where a previous generation of natural scientists had pored over the remnants of death—be they in the form of fossils, skeletons, or taxidermy—the new biology focused on the processes of life. In Pzibram's institute, immense resources were poured into the breeding and keeping of experimental organisms. And indeed, Przibram considered the ability to keep animals alive—not just for the duration of an experiment, but long enough to study their development, reproduction, and life histories over generations—one of the key features of experimental biology: "While it may be sufficient for the physiologist to keep his experimental objects alive [only] for as long as he wants to pursue a particular function, and then for further observation to use fresh specimens, the biologist usually cares about the continuous observation of form changes over a longer experimental period" (Przibram 1908, 253).

In keeping with the larger goals of the institute, Przibram assigned to von Frisch a study of the developmental history of the praying mantis, an organism on which Przibram himself had been working. But von Frisch was soon bored with the topic and happened to observe a fellow student working on pigmentation changes in minnows (see also chapter 11). The animals underwent rapid color changes when swimming in differently colored environments. Von Frisch was hooked (von Frisch 1962, 32).

He felt the work was a natural outgrowth of the earlier study he had performed under his uncle on pigment changes in the compound eyes of invertebrates. Przibram gave his blessing to his student's change of heart, and von Frisch promptly began to study how the color changes were regulated in fish. Von Frisch wrote of how he had once again "fallen for a comparative physiological project," and noted that, when he needed advice, he would once again visit his uncle Sigmund Exner at the nearby physiological institute (von Frisch 1962, 33).

Von Frisch took an 1876 publication by Georges Pouchet (son of the famous Félix) on the nervous control of pigment cells as a touchstone for these investigations (Pouchet 1876). He began by cutting the sympathetic trunk in the fish's body just below its dorsal fin.[3] Almost immediately the fish's tail turned dark, starting from the point of the incision and proceeding down its body. When von Frisch stimulated the fish's sympathetic trunk with electricity, its skin once again turned light, suggesting that contraction caused the cells to appear lighter, while the relaxation caused by the cutting of the nerves prompted the pigments to move to the skin's surface, thereby giving the darker coloration. When he gradually moved his point of incision in different fish to successively higher positions along the animals' bodies, more and more of the animals changed color posterior to the cut. That is, until he reached a point just above the dorsal fin: then suddenly the animal's head turned black, while the posterior part below the incision remained unchanged (figure 12.3). Cutting part of the animals' optic nerve (*Ramus opthalmicus veini trigemini*) at the source behind the roof of the eye socket also caused the head to turn dark from the eye to the snout (von Frisch 1910, 18–20).

In a set of observations on dead fish, von Frisch noticed that the spinal cord (part of the central nervous system) also seemed to play a role. In a freshly dead fish, the animal's tail underwent a lightening, much as it had when he stimulated the sympathetic trunk electrically. But this lightening could be manipulated by cutting the spinal cord, rather than the sympathetic trunk, suggesting to von Frisch that the central nervous system was involved (von Frisch 1910, 20). This meant that coloration was not just controlled by the

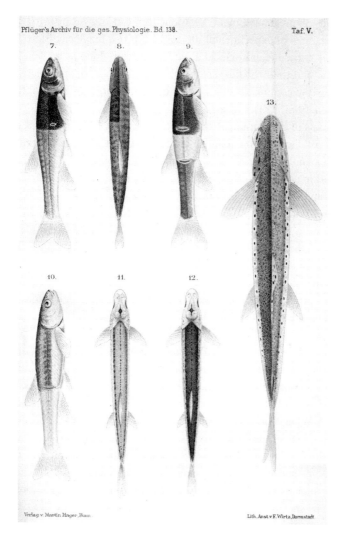

Figure 12.3
An illustration of the colorations fish displayed after von Frisch severed different parts of their nervous systems. He performed these experiments to study the causes of pigment changes in the skin of minnows. Karl von Frisch, "Beiträge zur Physiologie der Pigmentzellen in der Fischhaut," *Pflügers Archiv für Physiologie* 138 (1911), Plate V.

sympathetic—that is, involuntary—part of the nervous system, but that there was also a voluntary component.

By changing the position of incisions in live and dead fish, he was able to locate an area just above the dorsal fin where the nerve fibers that innervate the pigment cells branch from the spinal cord into the sympathetic trunk. He found the critical juncture to lie somewhere in the region of the fifteenth vertebra in minnows (von Frisch 1911, 331). He also determined that the light entering through the animal's eyes influenced its coloration, as could direct pressure to its skin (von Frisch 1911, 339–346, 348–365). In addition, some fish (including minnows) had the capacity to sense light that passed through an area of skin and skull on the fish's "forehead" area. While he was unable to localize the exact region of the brain responsible for this light perception, the finding was nonetheless of great interest (von Frisch 1911, 375–380).

But how exactly did von Frisch arrive at these results? In other words, what were his methods? To pinpoint the spinal region in question, he had taken what he somewhat vaguely called "a large number of minnows," some live and some dead, and destroyed their spinal cords at various locations (von Frisch 1911, 324). He then immediately plunged the animals into boiling water so that their flesh could be removed and the skeletal remains examined for the exact location of the surgical intervention. In other trials, he punctured the animal's skull and brain with a sharp platinum electrode (333–334). A second pole was grounded on the wet surface on which the fish lay to complete the circuit with its body. To keep the fish alive during this procedure, he attached a rubber hose to the animal's mouth, through which he forced aerated water (332).

In a subsequent trial aimed at determining the role of the forebrain, he removed a portion of the fish's brain and then replaced the skull flap with soft paraffin wax. The flap "stuck well," he wrote, "when the skin was scraped from the bone and the bone dried" (von Frisch 1911, 334). He further reported that the fish "tolerated the intervention very well," although when he so much as approached their tank, "they displayed a noticeable startle reaction and flitted about like crazy" (von Frisch 1910, 25). In another experiment, he scooped out the animals' eyes, so that he could stimulate the optic nerve that runs behind the animal's eye sockets (von Frisch 1911, 335). Clearly these experiments were not for the faint of heart.

Von Frisch used dozens of fish in these trials, and did with them as he saw fit—from boiling them alive or mutilating them by partial and full lobotomies to watching them die as their tails became necrotic and eventually fell off. In addition to minnows, he also experimented on trout at various stages of development, crucian carp, char, bass, eels, and fire salamander larvae. The animals were kept in sparse conditions—a glass aquarium "without plants or ground cover"—and were used as the work demanded (von Frisch 1911, 350). Thus, the work was of an experimental-physiological bent; it was unnecessary to keep the animals alive past a brief period of investigation.

Xaverl the Catfish

Let us shift now to a later work by von Frisch that was published in 1923 under the title "A Catfish that Comes When Summoned by Whistling" (Ein Zwergwels der kommt wenn man ihm pfeift). The paper was based on experiments von Frisch performed in the early 1920s, after he had taken his first professorship at the University of Rostock on the Baltic Sea. In the piece, von Frisch described experiments in which he tested whether fish could hear sounds. By the time he weighed in on the question, the dispute was already in full swing with the catfish swimming center stage. Although the fishes lack a cochlea—the snail-like structure that propagates sound in the mammalian ear—some observers reported that they react to sound. At least six scientists in the 1910s claimed that catfish reacted to whistling. But just as many found the catfish impervious to their "whistling and screaming … singing and clapping." One researcher even found that "the scales and trills of a celebrated singer" failed to elicit a response in his placid little fish (von Frisch 1923, 442).

Von Frisch argued that previous researchers had not managed to settle the question, because they had failed to expose the laboratory fish to the types of sounds that were likely to have played a role in the animals' evolutionary history. In his autobiography, he remembered thinking: "If I were a catfish, I would be interested in earth worms and other such tasty bites, but hardly in the trills of a celebrated singer. One cannot expect a fish to react to sounds that hold no meaning in its life" (von Frisch 1962, 71–72).

For his own experiment with the catfish he named Xaverl, von Frisch surgically removed its eyes so that it could not be cued visually. A few days later he began to train the animal; each day he whistled and then proffered a morsel of meat on the end of a stick. On the sixth day of the experiment, the animal reacted. When von Frisch whistled, it jerked from its shelter and swam about, apparently in search of food (figure 12.4).

The papers discussed here and in the previous section are separated by fifteen years and a war. These are major chronological and methodological breaks, and the latter piece much more closely resembles the project von Frisch would pursue in the decades to come—to show that an animal possessed greater sensory capacity than was previously believed by demonstrating its ability to perceive stimuli through conditioning. These experiments relied on the animals' implied ability to either react or not react. Thus, they depended on a kind of (albeit constrained) animal agency unlike the earlier dissertation work.

But what of Xaverl's missing eyes? In the otherwise playful piece, a single line lets on that von Frisch's investigation of the fish's one sense was preceded by his willful extinction of another: "To remove any objection from the outset that optical stimuli could have played a part, I had extirpated both of the animal's eyes a few days earlier" (von Frisch 1923, 443). The event is notable for being treated as utterly unremarkable.

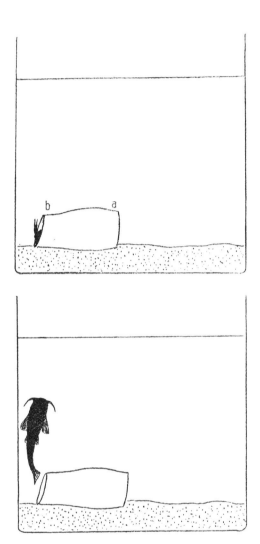

Figure 12.4
These images show Xaverl, the blind catfish von Frisch trained with food to respond to the sound of whistling. In the top image, Xaverl is shown resting in a tube von Frisch made for him by breaking off the ends of a candlestick. The second image shows the fish swimming from his shelter, presumably after von Frisch whistled. Karl von Frisch, "Ein Zwergwels der kommt, wenn man ihm pfeift," *Biologisches Zentralblatt* 43 (1923), p. 443, figures 1 and 2.

We get a bit more elaboration on the experiment from von Frisch's autobiography, in which he again talked about the removal of the fish's eyes. For his lay readership, he now minimized the importance of the organs to the animal's well-being: "A catfish has minute eyes, which mean very little to him. They can easily be removed without affecting his normal way of life and this was done." By simply declaring, "this was done," von Frisch stripped the deed of its agent. And as if to make amends, he hastened to assure readers that the fish was in good hands. He explained, "to make his glass container more comfortable, I took a hollow earthenware candlestick … smashed off its base so that its was open at both ends, and put it into the basin." He remembered that the little fish "eagerly availed himself" of its new home (von Frisch 1962, 72).

While the trade of two eyes for a broken candlestick presented at best a bad deal for Xaverl, the idea that an animal not only needed to be kept in conditions conducive to its longer-term survival—Xaverl lived long enough to be named—but also had to be kept in appropriate, species-specific conditions would stay with von Frisch over the course of his career. And, the tube that was to make Xaverl's tank more hospitable to the fish reflected a sensibility that would be evident in many of his later investigations. For his work with bees, for example, he would hire beekeepers to help ensure that the animals thrived and made it through the winter months.

A New Laboratory in Munich

In the summer of 1926, an official from the Rockefeller Foundation, Augustus Trowbridge, visited the old Zoological Institute in Munich, where von Frisch had studied under Hertwig. Von Frisch was now its director, having taken over in 1925 after Hertwig's retirement. Trowbridge's visit initiated lengthy negotiations that would eventually result in the Foundation donating 993,000 Reichsmark to the construction and operation of the new research facility (Rockefeller Foundation 1949). Von Frisch was on a lecture tour in the United States when he received the good news that the Rockefeller Foundation had finalized the funding. On that trip he visited more than 20 colleges, universities, and research organizations. The purpose of the trip had been twofold: first, to speak to his American colleagues about his research and findings on the sensory physiology of bees and fish. The second, and "main purpose of [his] visit," according to von Frisch, was "to inspect the American institutes for inspiration for the design of our new Zoological Institute." He was happy with the visit and felt its purpose had been "satisfied to the fullest degree" (von Frisch 1930). Later, he would credit this visit to the United States with having guided him in the planning of the new institute.

When it was completed in the fall of 1932, the new Zoological Institute in Munich was by all accounts spectacular. Set on three levels, every aspect had been carefully planned: from its laboratory rooms, to its general layout, down to its light fixtures, each and every feature was designed to meet the needs of its animal and human inhabitants (von Frisch and Kollmann 1935). When von Frisch first approached the architects, he stressed the importance of light, both for the scientists' working conditions and for the animals. In particular, he requested that the gardens and outdoor animal facilities be positioned to receive the most sunlight. Consequently, they were placed on the southern side of the Institute grounds. The outdoor spaces, moreover, were terraced to provide extended experimental access; large windows could be completely lowered so that indoor and outdoor areas merged into one continuous experimental space.

Although von Frisch's institute was dedicated to zoology, it was designed to accommodate different disciplinary areas of emphasis, as had been the case at the BVA. Some areas were meant for chemical work, while others were dedicated to physiology, bacteriology, or behavioral experiments. Here too the idea of emphasis was not to preclude cross-disciplinary exchange but rather to foster such interaction in the service of zoology. Von Frisch explained that while "morphology, physiology and physiological chemistry find certain rooms and building parts especially equipped for their needs, […] one cannot and should not say: here one specialty begins and the other ends" (von Frisch and Kollmann 1935, 7).

Separate rooms were dedicated to microscopy, where visual instruments were stored below large windows so they could easily be retrieved and used under optimal light conditions. Desk lamps were convertible to microscope lamps, and flat drawers accommodated slide collections. Culture rooms were installed next to where bacterial media were to be prepared, with special exhaust systems, centrifuges, and agitators also servicing those areas. A secure chemical room abutted a fireproof chamber in which flammable substances could be stored. A weighing room accommodated finely calibrated scales, while temperature-sensitive instruments were stored in yet another dedicated room. A "large and bright" photo and drawing studio received steady light throughout the day from its northern exposure, while a windowless cistern set five meters below ground offered a steady temperature of 13–14 degrees Celsius throughout the year. In addition to the varied offices and work stations, the institute also offered its inhabitants sleeping accommodations: a small apartment was to provide visiting scholars with "a friendly guestroom with adjacent bath," and another offered cots to weary scientists in case their experiments kept them at the institute overnight (von Frisch and Kollmann 1935, 13).

Throughout von Frisch's description of the facilities, he emphasized the building's sophisticated infrastructure: From electrically operated window shades to climate controls, to elevators, the facility was impressively modern. Power flowed from outlets at 110 or 220 volts as well as by direct and alternating current. The outdoor ponds, greenhouse, and cistern were supplied with groundwater but could be switched to the municipal system if the need arose. State-of-the-art telephone equipment offered two outside lines as well as an intercom system that could be used for Morse code communication throughout the building (von Frisch and Kollmann 1935, 23).

But perhaps even more impressive than the functional accommodations for humans was the extent to which the architecture was to meet the varied needs of its animal inhabitants. Separate facilities were dedicated to the breeding and care of small mammals that were to be used in experiments. Others offered fresh and saltwater aquariums as well as an elaborate system of outdoor ponds to house the institute's aquatic organisms. A large greenhouse facility was subdivided into three parts—an aquarium facility, a "warm house," and an insectarium that allowed for indoor bee work. A variety of plants could be grown in the adjacent garden plots: some for animal feed, others because they were of "particular biological significance," and still others for their propensity to attract bees (von Frisch and Kollmann 1935, 21).

In one of the outdoor ponds four islands were home to ant colonies. When experiments demanded that the insects intermingle, scientists could lay bridges to connect them across the water barrier (von Frisch and Kollmann 1935, 19). The ponds were constructed with different ground surfaces to provide a variety of aquatic environments. Indoors, a few of the saltwater tanks were to display seawater creatures of biological interest, so that "those students who aren't afforded a stay at the sea, won't be deprived of observing a selection of live, especially biologically interesting forms" (14). In addition, each floor had its own water tank facility, so animals could be kept near the lab spaces when they were needed in experiments.

Alongside the impressive practicality of the building, a notable decorative touch had also been created: beautiful masonry work by the artist E. A. Rauch framed the new institute's entrance and depicted animals that had played a significant role in the science practiced at the former institute: radiolaria, medusa, and a frog reminded of Hertwig's work, while a bee, minnow, and catfish represented von Frisch's favored animals. On purely aesthetic grounds the artist had also added a stag beetle to the menagerie. And finally, a phoenix graced the apex of the doorway, its wings outstretched above all who entered and left the building. Von Frisch explained that the mythical creature had been chosen as "a sign of hope." He continued that the depiction embodied the wish that "something of the work that happens here will endure and not be as fleeting as those figures that come and go through the door" (von Frisch and Kollmann 1935, 7).

The deliberate artistic choices framing the doorway told of the important role animals had played, and would continue to play, in the work conducted at the Zoological Institute in Munich. The living animal had come to "reign" in von Frisch's own work as well, and the new building was a testament to the importance he gave to the keeping and maintenance of animals under suitable conditions. The choice of the phoenix would also prove prescient in ways von Frisch could not yet know, for soon the beautiful new laboratory in Munich, just like the BVA in Vienna, would fall to the smoldering destruction of the Second World War.[4]

Conclusion

At first glance, von Frisch's early work seems puzzling and at odds with the ethos of the BVA. For his dissertation research, he used and maimed dozens of animals with little regard for their well-being or long-term survival. When we look at this work, the influence of experimental physiology as it was practiced by his uncle and mentor, Sigmund Exner, was very much in evidence. Przibram, in contrast, though certainly interested in physiology, looked at phenomena that spanned the course of an animal's development and across generations. This emphasis on long-term survival was inscribed into every aspect of the building on the Prater.

While von Frisch's work did not focus on the topics that most interested Przibram, his later experiments increasingly focused on the behavior and requirements of animals that were to survive at least from the training to the experimental phases (and in the case of bees, over the course of the winter and into the new experimental season). This later emphasis on the animal and its surroundings fit well with his theoretical commitments. Von Frisch studied the behaviors of animals with a close eye to their evolutionary history. The catfish did not respond to an opera singer, because the latter had been irrelevant to the animal's phylogenetic development. While this may beg the question of the importance of whistling to prehistoric fish, what matters is that von Frisch was keenly aware of the evolutionary history of the animals he studied. As his work matured, emphasis on long-term survival and behavior in relation to an animal's sensory physiological capacities would take on increasing significance. By the early 1930s, when he planned and constructed his Zoological Institute in Munich, form and function would be united to center on the care and research of living animals, much like the BVA had done three decades earlier. While in his postwar recollections von Frisch credited his trip to the United States with having served as inspiration for the new institute, there can be little doubt that the BVA also served as an early example of a research institute dedicated to the proper care and keeping of experimental animals. Well before von Frisch began the planning and construction of the new institute in Munich, he had cut his scientific teeth in an environment that sought to provide animals with precisely what they needed—light, temperature, moisture, food, and other aspects of proper keeping. And in this, the BVA stood at the very forefront of the direction animal studies would take so successfully in the decades to come.

Acknowledgments

With thanks to the Österreichische Akademie der Wissenschaften, Konrad Lorenz Institute for Evolution and Cognition Research, Gerd Müller, and especially Sabine Brauckmann for bringing us together. Also: Klaus Taschwer, Christian Reiß for useful help with sources.

Portions of this essay were previously published (Munz 2016) and are reproduced here with permission from the University of Chicago Press.

Notes

1. Von Frisch deemed Exner's work on the topic of such enduring significance that he initiated a republication and translation of the book in 1989, for which he wrote the introduction (Exner 1989).

2. For the purposes of this paper, I focus on his time in Vienna. For more on his time in Munich, see Munz 2016, chapter 1.

3. This discussion is based on two publications by von Frisch: a 1910 short preview of his work for a Festschrift for Hertwig and his more complete 1911 dissertation publication (von Frisch 1910, 1911).

4. On the tragic takeover of the BVA by the Nazis and its eventual destruction in the war, see Reiter 1999.

References

Burkhardt, R. W. 1990. Karl von Frisch. *Dictionary of Scientific Biography* 17 (Supplement II): 313–320.

Coen, D. 2004. Scientific dynasty: Probability, liberalism, and the Exner Family in Imperial Austria. PhD dissertation. Cambridge: Harvard University.

Coen, D. 2006. A lens of many facets: Science through a family's eyes. *Isis* 97 (3): 395–419.

Coen, D. 2007. *Vienna in the Age of Uncertainty: Science, Liberalism, and Private Life*. Chicago: University of Chicago Press.

Exner, S. 1989. *The Physiology of the Compound Eyes of Insects and Crustaceans*. Trans. R. C. Hardie. Berlin: Springer.

Kreutzer, U. 2010. *Karl von Frisch, Eine Biographie*. Munich: August Dreesbach Verlag.

Munz, T. 2016. *The Dancing Bees: Karl von Frisch and the Discovery of the Honeybee Language*. Chicago: University of Chicago Press.

Pouchet, Georges. 1876. *Des changements de coloration sous l'influence des nerfs*. Paris: Librairie Germer Baillière.

Przibram, H. 1908. Die Biologische Versuchsanstalt in Wien. Zweck, Einrichtung und Tätigkeit während der ersten fünf Jahre ihres Bestandes (1902–1907). Bericht der zoologischen, botanischen und physikalisch-chemischen Abteilung. (4 Parts). *Zeitschrift für biologische Technik und Methodik* 1 i: 234–264; ii: 329–362; iii: 409–433; Supplement. 1909. 1–34.

Reiter, W. L. 1999. Zerstört und Vergessen: Die Biologische Versuchsanstalt und ihre Wissenschaftler/innen. *Österreichische Zeitschrift für Geschichtswissenschaften* 4:585–614.

Rockefeller Foundation. 1949. Meeting Minutes, June 17. Rockefeller Foundation Archive. Sleepy Hollow, NY. RG 1.2, Series 705D, Box 6, Folder 56.

Von Frisch, K. 1908. Studien über Pigmentverschiebung im Facettenauge. *Biologisches Zentralblatt* 28:662–671, 698–704.

Von Frisch, K. 1910. *Über die Beziehung der Pigmentzellen in der Fischhaut zum sympathischen Nervensystem: Vorläufige Mitteilung. Festschrift zum sechzigsten Geburtstag Richard Hertwigs*, 17–26. Jena: Fischer.

Von Frisch, K. 1911. Beiträge zur Physiologie der Pigmentzellen in der Fischhaut. *Pflügers Archiv für Physiologie* 138:319–387.

Von Frisch, K. 1923. Ein Zwergwels der kommt, wenn man ihm pfeift. *Biologisches Zentralblatt* 43:439–446.

Von Frisch, K. 1930. Letter to the Foreign Office (Auswärtiges Amt) in Berlin, via the Bavarian State Ministry for Education and Cultural Affairs, May 23. Bayerisches Hauptstaatsarchiv, Munich, MK 54482. Personalakt Karl von Frisch.

Von Frisch, K., and Theodor Kollmann. 1935. *Der Neubau des zoologischen Instituts der Universität München.* Munich: Druck der graphischen Kunstanstalt A. Huber.

Von Frisch, K. 1962. *Erinnerungen eines Biologen.* Berlin: Springer.

13 The Biologische Versuchsanstalt in Transnational Perspective: The Cold Spring Harbor Connection

Kate E. Sohasky

In 1907, the Seventh International Congress of Zoology met in Boston. The congress attracted scientists from countries as distant as South Africa, Argentina, and Japan. Among the attendees was Hans Przibram, director of the Biologische Versuchsanstalt (BVA) of Vienna. Following the congress, Przibram traveled to New York's Long Island, where at Cold Spring Harbor he met Charles Davenport, director of the Carnegie Institution of Washington's Station for Experimental Evolution (SEE).[1] After his visit, Przibram expressed to Davenport his great interest in arranging for the exchange of experimental specimens between institutions, writing: "I am curious to see what one can do in sending living material & if you want to procure something from our country I would always be glad to endeavor to get the things forwarded. I would like to thank you for the great kindness shown to me during my visit and travel & trust there will soon be an opportunity of meeting again, perhaps for the next Congress."[2] This initial meeting marked the beginning of an intellectual exchange that lasted into the interwar years.

Przibram and Davenport's relationship challenges assumptions about the history of the acceptance of theories of heredity in the United States, and calls attention to the role of transnational networks in producing scientific knowledge in national contexts. Historians of science, including Alexandra M. Stern and Nathaniel Comfort, have broadened our view of the histories of eugenics and genetics in the United States by acknowledging a diversity of belief in the mechanisms of inheritance in scientists such as Irving Fisher and Harvey Kellogg in the early twentieth century (Stern 2005; Comfort 2012). Peter J. Bowler, in his discussions of neo-Lamarckism, orthogenesis, and contemporary theories in the first decades of the twentieth century, has similarly demonstrated the proliferation of a diversity of theories of inheritance, which repudiates the notion of a general consensus of belief in a hereditarian synthesis of the theories of Weismann and Mendel after 1900 (Bowler 1983). George M. Cook has suggested that Davenport harbored interest in the possibility of the inheritance of acquired traits around the turn of the century and that this interest subsequently shaped the models and form of early research at the SEE, arguing that the research on the inheritance of acquired traits by scientists including Francis B. Sumner and Herman C. Bumpus, conducted at Cold Spring in its early years, in fact led to the discrediting of that belief even while it contributed to the station's experimental methodology (Cook 1999). The research for this chapter builds on these contributions to paint a picture of the complexity of hereditarian thought in the United States among even those geneticists and eugenicists of the first three decades of the twentieth century who have been designated "mainline" by Daniel J. Kevles (Kevles 1985). An examination of the relationship between the SEE and the BVA indicates that Davenport maintained an interest in the inheritance of acquired traits well into the 1920s. That interest is marked in his attention to and encouragement of the institute's experimental forays into the influence of environment on heredity, as expressed in his letters to Przibram and other scientists in the United States.

Historians of biology recognize the experimental station as partly a response to internal criticism of what some early twentieth century scientists perceived to be a myopic laboratory approach to the study of evolutionary biology. These scientists sought to integrate field and laboratory research, applying modern laboratory technology to the study of live specimens in controlled laboratory conditions that sought to imitate the specimens' natural environments (Kohler 2002, 23–59; Maienschein 1991, 3). Biological research entered a new phase of experimentalism that transitioned away from a developmental approach to an increasingly hereditarian one, which focused on the "parts" of the organism as opposed to the whole. A hereditarian approach narrowed scientific practice by its focus on discriminate components of the organism over a view of the organism in its entirety. This transition fits within a broader historical trend away from "holism" in mainstream biological research over the course of the twentieth century (Maienschein 1991, 4–5; Lawrence and Weisz 1998, 1–24).

Research at international experimental stations was both deeply influenced by national context and internationally directed, at a time when many scientists perceived nationalism as a threatening and corrupting influence on the production and application of scientific knowledge. Though both the SEE and the BVA were officially established as independent of the state, the state was ever present in the daily operations of each institution: research tables were rented by foreign states at the BVA, and the SEE's researchers served as consultants to the United States legislature on major policy issues during the first quarter of the twentieth century, most notably including legislation concerning immigration. Political concerns likewise animated the scientific lives of the scientists of the BVA, which was established by Przibram in 1903, himself Jewish, as a research institution welcoming to Jewish scientists, who with few exceptions were denied professorships within Austrian universities with few exceptions. Though the SEE differed from the BVA in important respects, each station's scientific practice was undergirded by a particular mode of experimental biological research, and both were impacted by political entanglements and interests in relation to the state; these conditions provided common ground for their two very different directors.

The BVA's scientists were at the vanguard of new methods of quantification and probability in biological research (Coen 2006, 493–496). Their proprietary laboratory technology attracted international interest, as did their controversial findings. The institute's provocative research on the inheritance of acquired characteristics long engaged Davenport's interest over the course of his correspondence with Przibram. Several of the institute's scientists, most notably the biologist Paul Kammerer, saw a direct positive correlation between eugenics and the inheritance of acquired characteristics. In his letters to Przibram and others, Davenport commended the institute's scientists for their groundbreaking research on the influence of environmental conditions (particularly temperature modification) on pigmentation, color adaptation, and rejuvenation. Historians remember Davenport primarily as a eugenicist and adamant promoter of Weismann and Mendel, however, an examination of his relationship with the BVA complicates this characterization by revealing his avid, if qualified, interest in the institute's evidence of the influence of environment on the germ plasm.

In addition to augmenting our historical memory of the science of heredity in the United States, Davenport, Przibram, and their experimental stations highlight the presence of the state and politics in experimental biological research in the years leading up to the Great War and in the interwar period. The research and findings of these scientists was inherently and, at times, explicitly political. By considering the process of scientific knowledge production through a close examination of the relationship of two institutional leaders working in two different countries, this chapter examines the important role played by the transnational exchange of people, ideas, and funds associated with international experimental stations in the production and legitimization of scientific knowledge in national contexts.

Experimental Biological Stations

Biographies of Davenport identify him as one of the primary leaders of the eugenics movement in the United States, and an abundance of literature supports this characterization (Witkowski and Inglis 2008; Allen 1986; Kevles 1985; Riddle 1947; MacDowell 1946).[3] Though his leadership played a formative role in that movement, he was more multifaceted than is generally acknowledged. Born in 1866 in Stamford, Connecticut, Davenport was the eighth of eleven children. He received his AB and PhD from Harvard University in 1889 and 1892, respectively (Witkowski and Inglis 2008, 36–38). He served thereafter as an instructor of zoology at Harvard prior to assuming the directorship of Cold Spring Harbor. His *Experimental Morphology*, published in two volumes in 1897 and 1899, in the words of one of his biographers, "served to stimulate the movement [of experimental biology], already in progress in Europe and America, to apply experimental methods to zoological and embryological materials" (Riddle 1947, 78). A Darwinian, he was one of the first supporters of Mendel in the United States.

Davenport has been identified as one of a cohort of biologists who helped to effect a shift in the place and practice of biological research in the United States around the turn of the twentieth century (Kohler 2002, 30–32; Witkowski and Inglis 2008; 39). Believing strict laboratory research practices to be too limited or narrow, these scientists sought to integrate field research with the technology and resources of the laboratory. Their efforts resulted in the establishment of new kinds of institutions that encouraged an integration of field and laboratory research: biological and marine stations, as well as vivaria (Kohler 2002, 24–25). The institutional innovation of the experimental scientific research station served as a foundation for changes in the methodology of research in evolutionary biology at the turn of the twentieth century. These new institutions brought about changes in both the place and practice of experimental biological research, as specimens were transported from the field into the laboratory where they were maintained under controlled conditions, often intended to replicate or imitate their natural environments.

One of these experimental stations, Przibram's BVA, was responsible for leading an epistemological and technological redirection in experimental biology. Popularly known as the Vivarium for its location in the old vivarium of the Vienna Prater, the BVA boasted new quantitative, interdisciplinary approaches and unique laboratory conditions under which to study the mechanisms of heredity (Coen 2006, 493–495). Vivaria were generally more contained than marine stations, which were often situated on coastal locations close to the environment from which they took their samples. As such, vivaria required highly controlled laboratory conditions for preservation of their specimens (Kohler 2002, 48). The BVA possessed advanced climate-control technology and drew on the quantitative and statistical tools of physicists. With this sophisticated laboratory technology, scientists could examine organisms within the context of their environment and thereby observe the dynamic effects of environment on organisms' development and evolution; thus the innovative technology enabled the institute's research on the influence of environment in the inheritance of acquired characteristics.

In addition to the conscious effort to integrate laboratory and field, many experimental biological stations were explicitly international, intended and imagined to supersede the boundaries and limitations of the nation-state in the pursuit of scientific knowledge. Notably these international spaces were often located in major European and United States coastal cities. The stations drew scientists from all over the globe. In 1909, the United States Bureau of Education commissioned a survey of European experimental stations from the American zoologist Charles Atwood Kofoid of the University of California, Berkeley, who was at the time engaged in research at the Naples Zoological Station. He had previously assisted the biologist William Emerson Ritter in the selection and construction of a site in San Diego for a marine biological station on the California coast (Goldschmidt 1951, 122–127). Kofoid described the experimental biological station as "a unique agency … indispensable in the equipment of a nation for the upbuilding of leaders in biological teaching and in the development and expansion of the spirit of research." He related the value and potential of research at experimental stations directly to their extent of internationality, concluding: "Not the least valuable factor is the stimulus of contact with other researchers ... the larger the station, the more international clientele, the greater its value to the investigator" (Kofoid 1910, 6). These stations were thus simultaneously national and international in their conception. While necessarily the product of international cooperation and collaboration, they performed a valuable service for the nation.

The Naples Zoological Station was the model for most of the international experimental stations that followed it. It was founded in 1872, under the directorship of the German zoologist Anton Dohrn, as a marine biological experimental station intended to draw researchers from many nations. Founded independently of any nation-state, this pioneering model was a private institution that had "escaped the evils of bureaucratic control" and maintained its intellectual autonomy, attracting scientists from the four corners of the globe. Its internationalism was nearly fetishized. Kofoid asserted, almost tautologically, that it was "this feature of the Naples Station which has made it from the beginning an international institution and has drawn investigators to it from practically all civilized lands" (Kofoid 1910, 13). A central feature of the Naples station and many of its imitators was the "table system," a model for the allocation of experimental space that enabled the internationalism celebrated by Kofoid and others. Dorhn's station "rented" tables to scientists and intuitions at the rate of $500 per annum as the primary means of financing the station's expenses and broader research program (Kofoid 1910, 13). It was this system of renting tables to scientists, private foundations, and governments that funded the entire enterprise, making Naples "the Mecca of the biological world." It was this system that enabled the station to remain independent of the claims and prerogatives of any one of its patrons (Kofoid 1910, 9). In this manner, the table system theoretically isolated an institutional space from state interests, inviting and encouraging collaborative and often cross-disciplinary research, independent of the claims of the nation-state. However, the state was never truly absent from these spaces. Governmental institutions, such as the Smithsonian Institution of the United States, often funded tables, thereby directly involving national governments (and government funding) in the daily research practices of these allegedly independent stations. The BVA also operated on the table system, attracting scientists from diverse places and supporting research at the institute through the rental of tables. The only explicit condition guiding the research of visiting scientists was that their research must be experimental in its methods and approach, in keeping with the institute's mission to further experimental biological research.

Experimental biological stations had existed before the development of the international station as modeled in Naples. In the United States, agricultural stations dated from the nineteenth century and were generally affiliated with land-grant universities (Oleson and Voss 1979, 213). Marine biological stations similarly dated from the second half of the nineteenth century. Among the most notable of these was the Marine Biological Station at Woods Hole, Massachusetts, established in 1888 (Dexter 1988, 3–4). This station, modeled after Naples, inspired the construction of similar marine biological stations, including the Long Island Biological Station at Cold Spring Harbor in 1890, of which Davenport became director in 1898, after having spent many of his summers as a student at Woods Hole (Witkowski and Inglis 2008, 38–42). Davenport had also conducted research at Naples in 1902, before his Department of Genetics at the SEE was established in 1903 under the auspices of the Carnegie Institution of Washington (Witkowski and Inglis 2008, 43–44; Cook 1999, 429; Kofoid 1910, 17). Davenport established another notable institution at the SEE, the Eugenics Record Office, in 1910, and much of his research and writings on the subject of inheritance were informed by the many family histories collected by that office's eugenics field workers (Allen 1986, 226).

As a scientist, Davenport in many ways operated as a businessman, forging connections between individuals and institutions in Europe and the rest of the Americas. He was extremely well connected and often took it upon himself to intercede on behalf of these associates when requesting funds or favors from foundations, donors, or the state. He used his connections to influence policy indirectly by promoting favored individuals, such as Harry H. Laughlin, superintendent of the Eugenics Record Office, to serve on state advisory commissions or provide testimony on policy addressing social concerns, like efforts to restrict immigration (Allen 1986, 247). Davenport harbored great credulity toward his own celebrity and ability to provide patronage, and that belief drove his indefatigable efforts to establish ties with scientists at home and abroad. He was deliberate in his promotions and quick to seek out opportunities to expand his network of associates.

It was the controversial research on the inheritance of acquired characteristics conducted by scientists at the BVA that captured Davenport's particular interest. In his exchanges with Przibram, Davenport applauded the latter's station and the work it was producing.[4] He expressed particular interest in Przibram's seven-volume *Experimental Zoology* as well as the work on rejuvenescence (the science of restoring and rejuvenating organisms to reverse the effects of age, illness, or impairment) and the inheritance of acquired characteristics by Eugen Steinach and Paul Kammerer.[5] Both Steinach and Kammerer conducted research on the relation of environment to the development of sexuality and the onset of puberty in organisms (Coen 2006, 508; Logan 2001, 297–303). Kammerer's research on the role of temperature in the inheritance of coloration and pigmentation, particularly his research on the nuptial pads of midwife toads, earned him international recognition and Davenport's special attention. Kammerer's initial research focus diverged from this particular development with the midwife toads, although it was nevertheless that aspect of his research that gained him recognition, and eventually notoriety, in the United States (see chapter 8). Prior to the outbreak of the First World War, Davenport wrote to Przibram: "I note with interest Kammerer's … work of which I saw something at Vienna. It seems quite incredible and I am anxious to test his results. This is fine work that your laboratory is turning out."[6] From the early years of their acquaintance, their exchange of their colleagues' and institutions' scientific papers was focused on the matter of research in heredity, and they also discussed and commended the expansion and progress of the growing network of experimental stations in Europe and the United States.[7]

In 1909, Davenport first visited the BVA.[8] During his time in Vienna, he stayed with Przibram's family and was introduced to the institute, its scientists, its specimens, and the proprietary laboratory technology that enabled the controlled modification of environmental conditions in the institute's research on the influence of environment in the inheritance of acquired characteristics. He and Przibram continued their correspondence, operating within a broader network of intellectual exchange, up until the outbreak of the First World War. The war and the ensuing change in political climate irreversibly altered the political stakes of their research, as well as the futures of their experimental stations.

The Great War

The First World War halted all exchange between the experimental biological stations of warring nations and across the Atlantic, resulting in the interruption of communication between the BVA and the SEE. Davenport and Przibram's correspondence did not resume until two years after the cessation of hostilities. When they did renew their correspondence, their mutual and immediate priority was the exchange of scientific literature, which had been impossible during the time of war (figure 13.1). Davenport supplied the BVA with past and current subscriptions to *Genetics* and, through the zoologist Ross G. Harrison of Yale's Osborn Zoological Laboratory, the *Journal of Experimental Zoology*. Przibram forwarded volumes of his own *Experimental Zoology* and scientific papers from his institute.[9] As their correspondence revived, so too did Davenport's particular interest in Kammerer's research and its relevance for the inheritance of acquired traits.

Przibram was equally preoccupied with financial and other practical concerns. The conditions in interwar Vienna were dire. The lack of coal, cold winters, and the general scarcity of resources presented considerable challenges for his institute. He asked Davenport for his opinion on the probability of finding buyers in the United States for various books in the institute's library. He wrote: "We have had to combat with want of coal and economic difficulties … Perhaps I will also make up my mind to sell the old books (including some with copper-plates) and atlases in my library to get cash and room for new periodicals. If thou think that somebody of your acquaintance may have interest in these things I would send a list of the books."[10] In another of his letters, he confided to Davenport: "I am sorry to say that our Institution is in a very bad way. I don't know yet how we may be able to heat this winter. When the war had ended, we thought the worst was over and we would be able to continue as before. But affairs are getting worse and worse." Such challenges necessarily interfered with the basic research capabilities of the station. "You will already have heard, I suppose, of Steinach's achievements," he continued. "Last

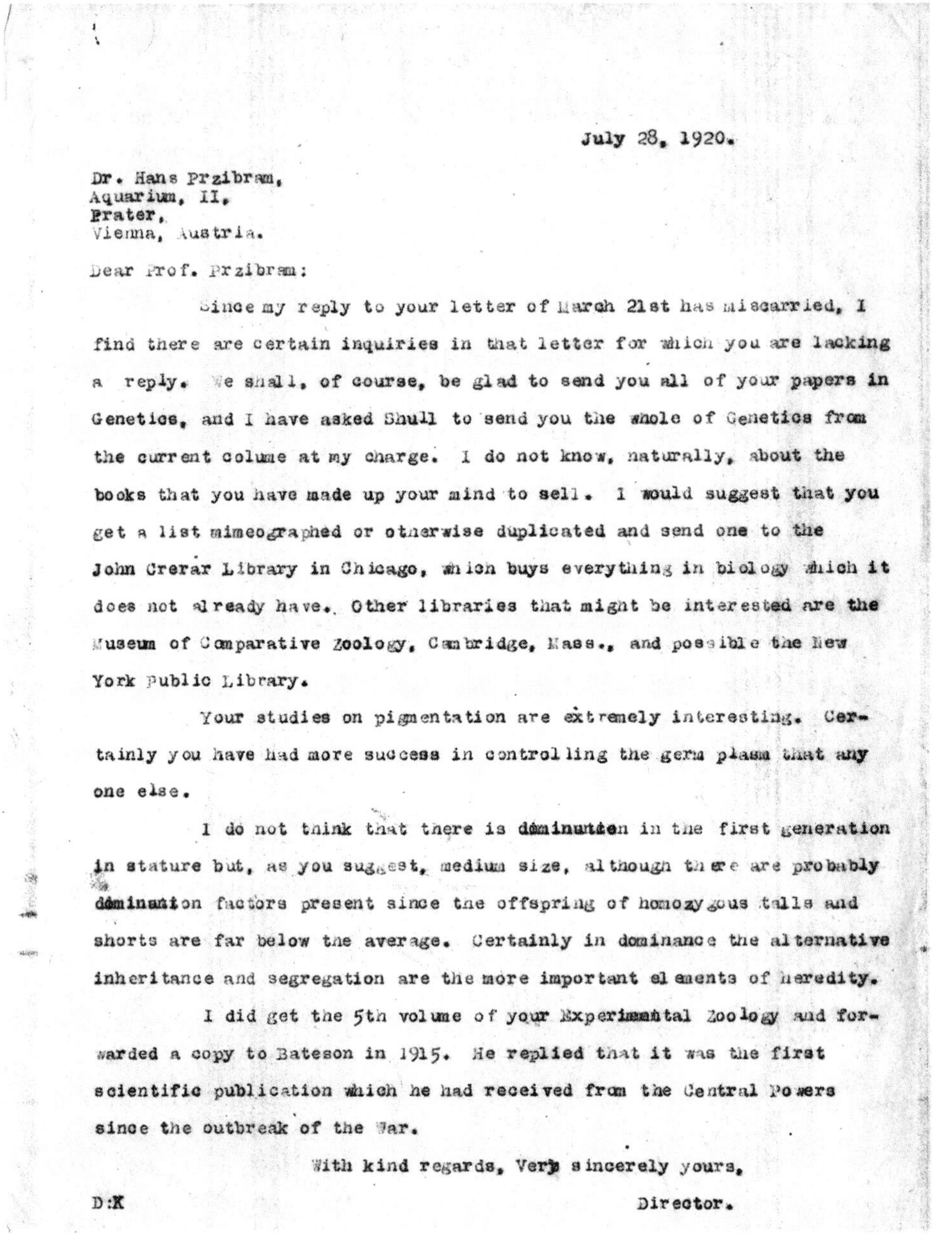

July 28, 1920.

Dr. Hans Przibram,
Aquarium, II,
Prater,
Vienna, Austria.

Dear Prof. Przibram:

Since my reply to your letter of March 21st has miscarried, I
find there are certain inquiries in that letter for which you are lacking
a reply. We shall, of course, be glad to send you all of your papers in
Genetics, and I have asked Shull to send you the whole of Genetics from
the current colume at my charge. I do not know, naturally, about the
books that you have made up your mind to sell. I would suggest that you
get a list mimeographed or otherwise duplicated and send one to the
John Crerar Library in Chicago, which buys everything in biology which it
does not already have. Other libraries that might be interested are the
Museum of Comparative Zoology, Cambridge, Mass., and possible the New
York Public Library.

Your studies on pigmentation are extremely interesting. Certainly you have had more success in controlling the germ plasm that any
one else.

I do not think that there is diminution in the first generation
in stature but, as you suggest, medium size, although there are probably
domination factors present since the offspring of homozygous talls and
shorts are far below the average. Certainly in dominance the alternative
inheritance and segregation are the more important elements of heredity.

I did get the 5th volume of your Experimental Zoology and forwarded a copy to Bateson in 1915. He replied that it was the first
scientific publication which he had received from the Central Powers
since the outbreak of the War.

With kind regards, Very sincerely yours,

D:K   Director.

Figure 13.1
Letter by Charles Davenport to Hans Przibram on 28 July 1920. Courtesy of the American Philosophical Society.

year he had entirely to leave off experimenting, as we could not procure the necessary animals (rats and guineapigs), food and fuel, nor sufficient assistance. We are trying now to induce the state to see to this affair, as his discovery of the rejuvenescence by experimental ways has also much practical interest especially now, where [there] are so many men prematurely invalidated by war and worry."[11] If resources had been scarce during and on the eve of war, the immediate postwar years proved very nearly fatal to the BVA. Indeed, the institute had been presented to the Austrian Academy of Sciences in May 1914 (Logan and Brauckmann 2015, 212), a move Przibram had "thought secured" the BVA. However, when he wrote to Davenport in 1922, Przibram confessed: "Now misfortune … is menacing to destroy our creation."[12]

The institute's proprietary laboratory technology made possible the experiments in temperature and climate control conducted at the BVA that studied the relationship between environment and the inheritance of acquired characteristics. The control of environment of was of prime importance for the quantitative methodology applied by the institute's scientists' in their endeavor to identify precisely the impact of environmental changes on the organism so as to detect patterns of evolutionary biological development (Coen 2006, 498–499). The availability of technology to control the conditions of experimentation within one to two degrees Celsius, for instance, significantly shaped the type of knowledge yielded by the work in this unique institutional setting. Indeed, much of the research produced by the BVA focused intently on the interplay between environment and development (Coen 2006, 498–500).

Davenport encouraged Przibram and assured the latter of his confidence in the valuable creative potential of the BVA even in its reduced circumstances. "I am glad to learn from your letter of February 25th that you are able to keep up with your researches," he replied, "despite the difficulties of carrying on research with limited funds and the difficulties of publication … I have no doubt that even with lack of apparatus, provided you have enough to live on, you can do important work. Some of the greatest discoveries have been made by intensive thinking over old material. I think, for example, of Weismann's germ plasm."[13] In an earlier letter, Davenport had similarly commented that "certainly you have had more success in controlling the germ plasm [than] any one else."[14] The station's research on the germ plasm was of particular relevance for research on the inheritance of acquired characteristics, and Davenport's interest in its progress, after years of silence during the war, suggests his continued, if noncommittal, investment in the possibility of the inheritance of acquired characteristics.

The theories of the German evolutionary biologist August Weismann—believed to be the first to refer to the internal or nuclear substance of cells believed to determine inheritance as "germ plasm" in 1883—had long been applied as disproof of the inheritance of acquired characteristics, in conjunction with the rediscovery of Mendel's research at the turn of the century by Hugo de Vries, Carl Correns, and Erich von Tschermak (Coleman 1965, 151). However, in his earliest writings, Weismann conceived of the possible influence of environmental factors on the germ plasm. These early writings allowed for the inheritance of acquired characteristics. It was around 1883, in response to strong criticism, that Weismann's interpretations shifted and he disavowed any influence of the environment on the germ plasm (Coleman 1965, 151–154). It is possible, however, that Davenport was familiar with Weismann's earlier beliefs. In a review of a book published in 1903, Davenport penned: "There is evidence coming from Standfuss, Fischer, et al that the modification in color pattern of the *Lepidoptera* young brought by heat are seen in the next generation, reared under normal conditions of life. Weismann admits the facts and maintains that the temperature acts on the germ cells *directly* and that germ cells are, consequently, modified *before* the modified coloration appears in the first modified adult generation."[15] Thus, Davenport's references to Weismann and control over the germ plasm in his letters to Przibram may well harken to Weismann's earlier views.

Though Davenport himself was eager to restore ties with former wartime enemies in the interest of advancing science and, possibly, repairing his network of personal and professional connections, this sentiment was not readily shared by many scientists in the United States. Many Allied scientists were reluctant to resume cooperation with those hailing from the Central Powers. On this matter, the United States National Research Council released a "Declaration in Regard to Enemy Nations" in 1919, condemning the wartime offenses of the Central Powers and stating: "In order to restore the confidence without which no scientific intercourse can be fruitful, the Central Powers must renounce the political methods which have led to the atrocities that have shocked the civilized world" (Stebbens 1919, 196). Some scientists in the United States remained highly critical of the "chauvinism" that animated German and Austrian science in the years leading up to and immediately following the First World War (Fangerau 2007). Many remained highly suspicious of the impetus that fueled the production of scientific knowledge in these former Central Power nations; these scientists viewed nationalism and militarism in connection with science as irresponsible and even dangerous. In 1922, Davenport wrote to Przibram, "I have no doubt that it will be some years before Americans flock to Europe for study as they did before the war."[16] This negative sentiment proved significant for the future relationship of the BVA to United States scientists and institutions.

Science and Politics during the Interwar Years

The end of the First World War marked the beginning of a change in the nature of the relationship between the BVA and the SEE. That change reflected the altered political environment of the interwar period, the financial scarcity engendered by postwar conditions, and the politicization of certain individuals' research at the BVA. The increasing notoriety of Kammerer in particular instigated alterations to United States views of the BVA. Kammerer endeavored to engage a United States audience in the postwar years even as his relationship with both Przibram and the BVA began to deteriorate (Logan and Brauckmann 2015, 218). Kammerer's publication in the United States of an English translation of his research went so far as to project a message of a benevolent and optimistic eugenics, which he couched in the rhetoric of individual moral uplift. He suggested that the inheritance of acquired traits might operate via Mendelian mechanisms in which the germ plasm was modified by the environmental conditions, thereby satisfying the disciples of Mendel and Weismann while preserving the optimistic possibilities he perceived in the inheritance of acquired traits (Coen 2007, 13–31; Coen 2006, 493–496; Gliboff 2006, 526–557). Kammerer's optimistic eugenics may have been of particular interest to American scientists who were invested in the inheritance of acquired traits at a time when their government faced sociopolitical anxiety over the policy implications of eugenics, including concerns about degeneracy and assimilation. However, Kammerer and Davenport did not meet until Kammerer visited Cold Spring Harbor during his tour of the United States in 1923. Kammerer's work had been of particular interest to Davenport over the years of the latter's acquaintance with the BVA, and Kammerer in turn had strongly encouraged Davenport to test his experiments on rejuvenation and pigmentation with salamanders.[17]

Policies predicated on belief in the immutable heritability of defect, including lifelong institutional segregation or sterilization of the unfit, were seen as viable alternatives in a world where scientific belief in the reformability of so-called degenerates had been discredited by the findings of Weismann and the rediscovery of Mendel at the turn of the century. Kammerer's findings suggested that scientific belief in the reformability of society's degenerates might be renewed. He was not alone in his interpretations. Though not as wholly optimistic, French *puériculture* and Cuban *homicultura* were both predicated on a belief in the great importance of hygiene and welfare (directed primarily at children and mothers) in the development and uplift of individuals and populations. For instance, many Latin American scientists perceived an important link between eugenics and *homicultura*, seeing them as part of a joint approach to the racial uplift and formation of states. Scientists in the United States, specifically Davenport and Laughlin, were in conversation with Latin American proponents of these beliefs during the interwar years, particularly as they applied to immigration (Gobierno de la República de Cuba 1928; Schneider 1990).

In the United States publication of his book *The Inheritance of Acquired Characteristics*, Kammerer warned of the dangers of thinking in "absolutes," which "makes it so much easier to consider the course of inheritance as something utterly unchangeable, independent from foreign influences." Such thinking was dangerous because "the Great War ... almost the world over, but especially in Central Europe, has resulted in an intensification of nationalistic and racial consciousness. This nationalistic self-confidence, however, is opposed to the theory of the inheritance of acquired characteristics, because it teaches that inheritance—the passing on of proudly referred-to race and caste characteristics inherited from forebears—is not everything" (Kammerer 1924, 16–17). Thus, Kammerer recognized in the inheritance of acquired characteristics "a message of salvation" (Kammerer 1924, 362). Nations could elevate the lowest of their populations through reform and education. His optimistic interpretation of the mechanisms of heredity, in his own view, opened the door to a positive and uplifting eugenics that could improve the lives of every race in every nation in the world.

Kammerer's book propounded a highly politicized, and even moral, mandate for scientists. He implored that "we face the necessity of developing new ethics; a new conscience has become necessary—a conscience of the generation and of the race, but never in that antiquated form of a *single* race striving to forge aggressively ahead at the expense of other races." He saw this worldview as severely threatened by the practice of science in much of Europe, particularly among the Central Powers. He warned that "we still may observe in the universities of Europe, especially in Germany and France and even Austria, a most unfortunate outburst of nationalistic movement, a relapse to race-hatred and revenge-propaganda" (Kammerer 1924, 359). In one respect, Kammerer's theories ran counter to nationalism, yet they simultaneously promised the uplift of the races of nations and therefore occupied a tenuous idealistic space between the national and international.

However, Kammerer's positive interpretation of his theory of the inheritance of acquired characteristics was by no means universally accepted, nor was belief in the inheritance of acquired characteristics a guarantor of a particular political ideology. This is evidenced by the conservative politics of the British biologist and eugenicist E. W. MacBride. One of Kammerer's greatest public supporters, MacBride's radically conservative interpretation of Kammerer's findings could not have been more politically inimical to Kammerer's optimistic proposals of uplift. Rather, MacBride perceived a deterministic message in Kammerer's research on heredity that indicated—in his interpretation—the solidification of racial differences and threatened the degeneration of nations, thus demonstrating how politically divergent interpretations of the science of heredity could be (Bowler 1984, 245, 254).

Because the BVA was funded by the table system, the remnant hostilities between warring nations were cause enough for many scientists to abstain from ready reintegration into international institutions that required cooperation with former political enemies. Indeed, while some prominent scientists, like the English geneticist William Bateson, called on science to serve as a promoter of "international amity," many others, such as the American biologist Thomas Hunt Morgan, protested the attendance of scientists from formerly hostile nations at the proposed international congresses on eugenics and genetics in 1921.[18] These divisions, along with wariness over the political activism of certain members of the BVA, and the economic scarcity caused by the war, proved financially debilitating to the BVA and its scientists.

Financial Crises in the "Roaring" Twenties

The financial uncertainty of the BVA's future dominated the thoughts Przibram shared with Davenport in the early twenties. Przibram made repeated appeals to Davenport for financial assistance. He also enlisted Davenport in the circulation of an advertisement for the rental of tables at the institute, in the hopes of recruiting either individual scientists or institutions to fund research in the way it had been supported before the war. He confided to Davenport:

During the hardships of war and the still greater disaster of peace, we have continued work, and, as you know, were lucky enough to enter upon some new pathways hopeful for scientific progress, if we could only procure means to follow out our work. Also as before we would desire others to share our institutions and methods of work. As it would not be possible to let foreign scientists partake of these *gratis*, and relatively small contributions in money of high value would be able to support the existence of the "Biologische Versuchsanstalt" the salvation of this institution could perhaps be secured.[19]

Davenport eagerly intervened to play the role of patron; later that year Przibram issued requests to the Rockefeller Foundation and the Smithsonian Institution through Davenport, rather than sending them directly. There was already a strong precedent for United States governmental institutions and private foundations to fund tables at experimental biological research stations in other countries. The Smithsonian Institution had funded a table at the Naples Station since 1893, and the Carnegie Foundation had likewise funded a table there since 1903 (Kofoid 1910, 4).

In addition to forwarding Przibram's requests, Davenport personally appealed to his Smithsonian and Rockefeller contacts on Przibram's behalf, tailoring his appeals to the particular interests and prerogatives of each organization. These letters focused on the utility and medical relevance of the work at the station. To Dr. George E. Vincent, president of the Rockefeller Foundation, Davenport acclaimed that

the work of Steinach on rejuvenation in mammals and man has been carried on there. Dr. Przibram is now conducting remarkable experiments on replantation of extirpated eyes in mammals and these apparently become functional. This experiment has naturally important biological bearings for man. In fact nearly all of the experiments going on there are of prime importance for medicine and surgery which rest largely upon the foundations of such biological experiments.[20]

In this manner Davenport foregrounded the medical applications of the BVA's research to the human devastation wrought by war. He highlighted the work on replantation and rejuvenescence and its potential importance for the medical treatment of veterans, many of whom had suffered loss of limb or disabling injuries in combat.

To Dr. Charles D. Walcott, Secretary to the Smithsonian Institution, with whom Davenport enjoyed an enduring personal and professional relationship, he expressed his qualified interest in the BVA's findings on the inheritance of acquired traits, remarking too on the station's advances in its broader scientific research. He wrote: "There is a lot of remarkable work on modification of color in insects by external conditions and some work on inheritance of acquired characters which is so remarkable that it must be done elsewhere with similar results before it can be generally accepted. There is, however, no doubt about the extraordinarily high quality of the work done at the Institute."[21] In this way Davenport appealed to the research interests of the Smithsonian-funded tables at similar institutions in the past. These declarations indicate the perseverance of Davenport's great respect for the work produced at the station, as well as his willingness to call upon his connections on behalf of Przibram and the institute.

In spite of Davenport's appeals, the Smithsonian Institution and Rockefeller Foundation declined Przibram's requests. Both organizations cited the lack of funds as the primary reason for their denial.[22] These claims may reasonably be accepted, with some skepticism. The overall character of Allied sentiment toward German and Austrian scientists was deeply influenced by the recent memory of war. Suspicion of the Central Powers' militaristic nationalism, and its presumed corrupting influence on the production of scientific knowledge, should not be underestimated. Indeed, Morgan protested an integrated second International Eugenics Congress:

I suppose it will only be decent to invite the Germans and the Austrians to the Eugenics Conference, but I do not find myself approving the matter with any enthusiasm. As far as scientific intercourse between these countries and ours is concerned, there should of course be no barriers, but the Congress is largely a social matter, and I foresee that if you invite the Germans and they come the rest of the civilized world will stay away or else there will be rows.[23]

Disagreeing with Davenport's insistence that the congress should be "international," for that would require the invitation of the Austrians and Germans, Morgan instead proposed the convening of an "interallied" or "neutral" congress.[24] It is also possible that anti-Semitism and the controversial character of Kammerer's work were factors in the decision to deny financial assistance. Walcott's apologetic letter to Davenport said simply, "[I] do not find a very sympathetic feeling toward the work being carried on by Professor Przibram."[25]

Przibram's ties with scientists in the United States did, however, ultimately serve his institute. In 1923 he received a degree of monetary support from the US anthropologist Franz Boas's Emergency Society for German and Austrian Science and Art, which provided relief specifically to Austrian and German scientists after the war and later assisted with evacuations to the United States in the years preceding World War II as the Nazis rose to power.[26] Przibram also applied to the German-born United States biologist Jacques Loeb, who had ties to the Rockefeller Foundation and Woods Hole, and to leaders of various biological and zoological stations in the United States in an effort to preserve the financial viability of his institution.[27]

Activist Science

Experimental scientific research in evolutionary biology during the interwar years was almost always in some sense political. It was nearly impossible to truly divorce science from the state, even in those institutions that claimed such independence. Institutions, like the BVA, that consciously tried to separate themselves from the state did so for political reasons. Although both Davenport and Przibram conceived both political and practical applications for their scientific research, their views diverged with respect to the proper relation of science to policy and the state. Davenport believed that the role of scientists in society and in relation to the state was one of suggestion. He conceived of the scientist as expert and advisor vis-à-vis the state. From the content of their letters and publications it is evident that Przibram, and others at his institute, perceived a more direct role for science in politics. In a way this is unsurprising. The United States enjoyed prosperity for much of the interwar period, as well as increasing international recognition and respect for its practice of science. Przibram and a majority of the scientists at the BVA, on the other hand, were ostracized by the racially anti-Semitic Viennese academy, while they continued to shoulder anti-German and anti-Austrian sentiments following the First World War. These conflicting views on the relation of politics and activism to science may have been a factor in the eventual dissolution of Przibram and Davenport's professional relationship and friendship, which is reflected in the sharp decline in their correspondence in the years following the first meeting between Kammerer and Davenport in 1923.

Several scientists—and Morgan in particular—may also have mobilized professional opinion against the BVA based on their disapproval of Kammerer's activities during his tour of the United States. Many scientists derided his display as debasing to science or as a cheap ploy to profit off of the public. Morgan's general criticism of Kammerer's public antics, if not his demeanor of disdain, was shared even by more temperate skeptics in the United States. One such skeptic was the zoologist, geneticist, and eugenicist Herbert Spencer Jennings. Jennings concurred in part with Morgan's reproaches, writing to Morgan:

You are quite justified in your criticisms of his activities,—or perhaps those of his manager—since [he] has come to America. The only question is as to how seriously that is to be considered; how heavily it is to be weighed against his scientific investigations … [I] feel that those poor devils over there are to be forgiven much. What Kammerer has done is to launch out an attempt to support himself by lecturing and writing books, and as a necessary step toward this, to put himself in the hands of a manager. Thereupon, he is driven steadily toward advertising and sensationalism, and his ultimate scientific ruin is doubtless assured.[28]

Thus Jennings, though more sympathetic to the difficulty of Kammerer's situation, similarly regretted his fanaticism. The Czech biologist Hugo Iltis remembered Kammerer to his son as one who "had many enemies, both for personal and political reasons. He was [an] outstanding and conspicuous personality. Outspokenly handsome, he was a ladies man. He paid great attention to his appearance and was almost vain."[29] Kammerer's "sensationalism," as it is here described, consisted of publicizing both his incredible scientific results and the prescriptive applications of his theories on the heredity of acquired characteristics while on tour in the United States. This practice of science for profit attracted scorn and disapproval on many fronts, in part for what was perceived as the stark politicization of science and in part for cheapening scientific research with the gaudy trappings of publicity.

Kammerer's credibility was all but extinguished when he was exposed as having allegedly committed scientific fraud. The scandal arose from Kammerer's experiments with midwife toads (see chapter 8). His experiments had demonstrated that midwife toads, which reproduced on land in their natural state, developed nuptial pads on their forelegs after one generation had been bred in water. These nuptial pads appeared in subsequent generations that were bred on land (Gliboff 2006, 525–526). To some, Kammerer's midwife toads represented convincing proof of the inheritance of an acquired characteristic (but see chapter 8); it was believed that the environment had altered the toads' germ cells, thus substantiating the belief that acquired traits could be passed on to future generations. The discovery, in 1926, that India ink had been injected into the only remaining specimen proved Kammerer's undoing. To this day it is uncertain whether Kammerer himself injected the ink, and whether the development of the nuptial pads was indeed a fabrication or an artificial enhancement. Some of his contemporaries, including Iltis, defended him, though they were in the minority. In apparent despair of his disgrace, Kammerer committed suicide soon after the scandal broke, an act many of his contemporaries took as proof of his guilt. The scandal discredited the inheritance of acquired traits irreversibly (Gliboff 2006, 525–528). This in turn opened to door to more general criticism of the scientific practices and political activism of the BVA, in spite of Kammerer's official disassociation from the institution in the years following the First World War (Logan and Brauckmann 2015, 218).

It is plausible that any remaining interest of Davenport's in the inheritance of acquired characteristics might have withered with the ostensible disproof of Kammerer's findings and that the scandal influenced Davenport's opinion of the other work completed at the BVA. Through the 1920s, Przibram and Davenport continued to facilitate visits and introductions between their institutions. Davenport wrote at least two letters of introduction addressed to Przibram for two of his field-workers from Cold Spring Harbor's Department of Genetics, Miss Grace Allen and Miss Mary Chantler, in 1926, and for his son-in-law, the director of the Biological Laboratory at Cold Spring Harbor, Reginald G. Harris, in 1929.[30] The last surviving letter exchanged was sent in 1930. Both Davenport's and Przibram's lives ended in 1944, though under dramatically different circumstances. Davenport expired from illness in the same year that Przibram died in the Nazi concentration camp Theresienstadt (Witkowski and Inglis 2008, 54; Coen 2006, 496).

Although little of their correspondence explicitly reveals Davenport's or Przibram's political beliefs, in a letter to the Rockefeller Foundation's Dr. George E. Vincent, Davenport lamented the rise to power of what he designated that "inexperienced group" in the Austrian government and expressed his wish that they attain, in his words, "more experience and higher ideals."[31] Davenport wrote, in response to a claim proffered by Przibram, "If, in accordance with your suggestion, biologists were to be entrusted with politics, I fear that biology would suffer, even if politics were improved. However, I believe it is very desirable for biologists to insist directly and indirectly, thru [sic] their work on the value of biological conceptions to society."[32] In addition to the controversial reception their research encountered, the public and political activism—or even "sensationalism"—of certain scientists of the BVA attracted scorn or apprehension from certain scientists in the United States, undoubtedly discouraging financial investment in the institute on the part of United States scientists and organizations.[33] Their divergent perspectives on the political applications of scientific knowledge distinguished these two scientists, who operated in very different political and social contexts. Their relationship and exchange nevertheless underscores the complex interplay between international and local concerns in the production of scientific knowledge that the BVA encouraged and enabled through its unique practices, its laboratory setting, and its situation in interwar Vienna.

Theories of inheritance were deeply entangled with politics during the interwar years, as both the work of the SEE and the BVA and the relationship between their directors reveal. Changes in the practice and setting of biological research established epistemological commonalities that led to the creation of internationally directed networks of individuals and institutions. These networks functioned to facilitate the movement of both intellectual and fiscal capital, as well as people. Though Davenport never expressly changed his beliefs on the mechanisms of heredity, and never wholly accepted the BVA scientists' findings, his marked interest and cautious acceptance of their research unsettles characterizations of Davenport as an unwavering proponent of Weismann and Mendel and highlights the critical importance of laboratory place and practice in the spread and assimilation, as well as the production, of scientific knowledge both nationally and transnationally.

Acknowledgments

I would like to thank especially my advisors, Angus Burgin, Ronald G. Walters, and Nathaniel Comfort, for their support and feedback on numerous drafts. I also owe thanks to the Johns Hopkins University Twentieth Century Seminar, Richard Nash, Jenna Tonn, Charles Greifenstein and the reference staff at the American Philosophical Society, and the editors and fellow contributors to the volume.

Notes

1. The research for this chapter is based primarily on the surviving correspondence exchanged between Hans Przibram and Charles B. Davenport, dated from 1907 to 1930. All correspondence was written in English and is presently in the keeping of the American Philosophical Society.

2. Hans Przibram to Charles B. Davenport, September 25, 1907, Charles B. Davenport Papers, American Philosophical Society, Philadelphia, PA (hereafter cited as APS).

3. Much of the recent historiography on the eugenics movement in the United States has progressed beyond its more narrow characterization as an East Coast phenomenon that began to fade by the late interwar years. This literature has expanded the timeline and the geographic scope of the eugenics movement by examining different regions, actors, and periods (Bashford and Levine 2010; Stern 2012; Stern 2005; Comfort 2012; Larson 1995).

4. Charles B. Davenport to Hans Przibram, February 27, 1907, Charles B. Davenport Papers, APS.

5. Charles B. Davenport to Hans Przibram, November 5, 1910, Charles B. Davenport Papers, APS.

6. Ibid.

7. Charles B. Davenport to Hans Przibram, April 1, 1910, Charles B. Davenport Papers, APS.

8. Charles B. Davenport to Hans Przibram, April 6, 1909, Charles B. Davenport Papers, APS.

9. Hans Przibram to Charles B. Davenport, March 21, 1920; Hans Przibram to Charles B. Davenport, June 23, 1920; Charles B. Davenport to Hans Przibram, July 20, 1920; Charles B. Davenport to Hans Przibram, July 28, 1920. Charles B. Davenport Papers, APS.

10. Hans Przibram to Charles B. Davenport, March 21, 1920, Charles B. Davenport Papers, APS.

11. Hans Przibram to Charles B. Davenport, August 15, 1920, Charles B. Davenport Papers, APS.

12. Enclosure, Hans Przibram to Charles B. Davenport, April 8, 1922, Charles B. Davenport Papers, APS.

13. Charles B. Davenport to Hans Przibram, May 14, 1921, Charles B. Davenport Papers, APS.

14. Charles B. Davenport to Hans Przibram, July 28, 1920, Charles B. Davenport Papers, APS.

15. Charles B. Davenport, "The Inheritance of Acquired Dynamical Qualities," 1903, Charles B. Davenport Papers, APS.

16. Charles B. Davenport to Hans Przibram, April 20, 1922, Charles B. Davenport Papers, APS.

17. Paul Kammerer to Charles B. Davenport, December 20, 1923, Charles B. Davenport Papers, APS.

18. Tomas Hunt Morgan to Herbert Spencer Jennings, April 16, 1920; Thomas Hunt Morgan to William Bateson, April 17, 1920; William Bateson to Thomas Hunt Morgan, May 19, 1920. Herbert Spencer Jennings Papers, APS.

19. Enclosure, Hans Przibram to Charles B. Davenport, April 8, 1922, Charles B. Davenport Papers, APS.

20. Charles B. Davenport to George E Vincent, November 3, 1922, Charles B. Davenport Papers, APS.

21. Charles B. Davenport to Charles D. Walcott, November 3, 1922, Charles B. Davenport Papers, APS.

22. Charles D. Walcott to Charles B. Davenport, November 11, 1922; George E. Vincent to Charles B. Davenport, December 19, 1922. Charles B. Davenport Papers, APS.

23. Thomas Hunt Morgan to Charles B. Davenport, October 13, 1919, Charles B. Davenport Papers, APS.

24. Thomas Hunt Morgan to Herbert Spencer Jennings, April 16, 1920, Herbert Spencer Jennings Papers, APS.

25. Charles D. Walcott to Charles B. Davenport, November 11, 1922, Charles B. Davenport Papers, APS.

26. Franz Boas to Hans Przibram, January 29, 1923, Franz Boas Papers, APS.

27. Hans Przibram to Charles B. Davenport, December 19, 1920, Charles B. Davenport Papers, APS.

28. Herbert Spencer Jennings to Thomas Hunt Morgan, February 2, 1924, Herbert Spencer Jennings Papers, APS.

29. Hugo H. Iltis to Arthur Koestler, August 30, 1972, Paul Kammerer Papers, APS.

30. Charles B. Davenport to Hans Przibram, June 14, 1926, and September 11, 1929, Charles B. Davenport Papers, APS.

31. Charles B. Davenport to George E. Vincent, December 20, 1922, Charles B. Davenport Papers, APS.

32. The letter to which Davenport writes in response was regrettably not among Davenport's papers at the American Philosophical Society. Charles B. Davenport to Hans Przibram, May 14, 1921, Charles B. Davenport Papers, APS.

33. Herbert Spencer Jennings to Thomas Hunt Morgan, February 2, 1924, Charles B. Davenport Papers, APS.

References

Allen, G. E. 1986. The Eugenics Record Office at Cold Spring Harbor, 1910–1940: An essay in institutional history. *Osiris. Second Series* 2:225–264.

Bashford, A., and P. Levine. 2010. *The Oxford Handbook of the History of Eugenics*. New York: Oxford University Press.

Bowler, P. J. 1983. *The Eclipse of Darwinism: Anti-Darwinian Evolution Theories in the Decades around 1900*. Baltimore: Johns Hopkins University Press.

Bowler, P. J. 1984. E.W. MacBride's Lamarckian eugenics and its implications for the social construction of scientific knowledge. *Annals of Science* 41 (3): 245–260.

Coen, D. R. 2006. Living precisely in fin-de-siècle Vienna. *Journal of the History of Biology* 39 (3): 493–523.

Coen, D. R. 2007. *Vienna in the Age of Uncertainty: Science, Liberalism, and Private Life*. Chicago: University of Chicago Press.

Coleman, W. 1965. Cell, nucleus, and inheritance: An historical study. *Proceedings of the American Philosophical Society* 109 (3): 124–158.

Comfort, N. 2012. *The Science of Human Perfection: How Genes Became the Heart of American Medicine*. New Haven: Yale University Press.

Cook, G. M. 1999. Neo-Lamarckian experimentalism in America: Origins and consequences. *Quarterly Review of Biology* 74 (4): 417–437.

Dexter, R. W. 1988. History of American marine biology and marine biology institutions. Introduction: Origins of American marine biology. *American Zoologist* 28 (1): 3–6.

Fangerau, H. 2007. Biology and war: American biology and international science. *History and Philosophy of the Life Sciences* 29:395–428.

Gliboff, S. 2006. The case of Paul Kammerer: Evolution and experimentation in the early twentieth century. *Journal of the History of Biology* 39 (3): 525–563.

Gobierno de la República de Cuba. 1928. *Actas de la primera Conferencia Panamericana de Eugenesia y Homicultura de las Repúblicas Americanas*. Havana: Gobierno de la República de Cuba.

Goldschmidt, R. B. 1951. Charles Atwood Kofoid, 1865–1947. In *National Academy of Sciences* 26:119–151. Washington, DC: National Academy of Sciences.

Kammerer, P. 1924. *The Inheritance of Acquired Characteristics*. New York: Boni and Liveright Publishers.

Kevles, D. J. 1985. *In the Name of Eugenics: Genetics and the Uses of Human Heredity*. Cambridge: Harvard University Press.

Kofoid, C. A. 1910. *The Biological Stations of Europe*. Washington, DC: Government Printing Office.

Kohler, R. E. 2002. *Landscapes and Labscapes: Exploring the Lab-Field Border in Biology*. Chicago, London: The University of Chicago Press.

Larson, E. J. 1995. *Sex, Race, and Science: Eugenics in the Deep South*. Baltimore: Johns Hopkins University Press.

Lawrence, C., and G. Weisz. 1998. *Greater than the Parts: Holism in Biomedicine, 1920–1950*. New York: Oxford University Press.

Logan, C. A. 2001. "[A]re Norway rats … things?": Diversity versus generality in the use of albino rats in experiments on development and sexuality. *Journal of the History of Biology* 34:287–314.

Logan, C. A., and S. Brauckmann. 2015. Controlling and culturing diversity: Experimental zoology before World War II and Vienna's *Biologische Versuchsanstalt*. *Journal of Experimental Zoology* 323A:211–226.

MacDowell, E. C. 1946. Charles Benedict Davenport, 1866–1944: A study of conflicting influences. *Bios* 17 (1): 2–50.

Maienschein, J. 1991. *Transforming Traditions in American Biology, 1880–1915*. Baltimore, London: The Johns Hopkins University Press.

Oleson, A., and J. Voss. 1979. *The Organization of Knowledge in Modern America, 1860–1920*. Baltimore: Johns Hopkins University Press.

Riddle, O. 1947. *Biographical Memoir of Charles Benedict Davenport, 1866-1944*. vol. 25. Washington, DC: National Academy of Sciences of the United States of America Biographical Memoirs.

Schneider, W. H. 1990. *Quality and Quantity: The Quest for Biological Regeneration in Twentieth-Century France*. New York: Cambridge University Press.

Stebbens, J. 1919. *Minutes of Joint Meeting of the Executive Board and the Council of the National Academy of Sciences*. Washington, DC: National Research Council.

Stern, A. M. 2005. *Eugenic Nation: The Faults and Frontiers of Better Breeding*. Berkeley: University of California Press.

Stern, A. M. 2012. *Telling Genes: The Story of Genetic Counselling in America*. Baltimore: Johns Hopkins University Press.

Witkowski, J. A., and J. R. Inglis. 2008. *Davenport's Dream: 21st Century Reflections on Heredity and Eugenics*. Cold Spring Harbor, NY: Cold Spring Harbor Laboratory Press.

Contributors

Heiner Fangerau
Department of the History, Philosophy, and Ethics of Medicine, Heinrich-Heine-University, Düsseldorf, Germany

Johannes Feichtinger
Austrian Academy of Sciences, Wien, Austria

Georg Gaugusch
Wilhelm Jungmann & Neffe, Wien, Austria

Manfred D. Laubichler
School of Life Sciences and Global Biosocial Complexity Initiative, Arizona State University, Tempe, Santa Fe Institute, Santa Fe, and KLI Klosterneuburg, Austria

Cheryl A. Logan
Departments of Psychology and History, University of North Carolina, Greensboro

Gerd B. Müller
Department of Theoretical Biology, University of Vienna, and KLI Klosterneuburg, Austria

Tania Munz
Linda Hall Library of Science, Engineering, and Technology, Kansas City

Kärin Nickelsen
History of Science, Ludwig Maximilian University Munich, Germany

Christian Reiß
Professur für Wissenschaftsgeschichte, Universität Regensburg, Germany

Kate E. Sohasky
Department of History, Johns Hopkins University, Baltimore

Heiko Stoff
Institut für Geschichte, Ethik und Philosophie der Medizin, Medizinische Hochschule, Hannover, Germany

Klaus Taschwer
Der Standard, Wien, Austria

Index